Tensor Analysis for Engineers and Physicists - With Application to Continuum Mechanics, Turbulence, and Einstein's Special and General Theory of Relativity

Meinhard T. Schobeiri

Tensor Analysis for Engineers and Physicists - With Application to Continuum Mechanics, Turbulence, and Einstein's Special and General Theory of Relativity

Springer

Meinhard T. Schobeiri
Department of Mechanical Engineering
Texas A&M University
College Station, TX, USA

ISBN 978-3-030-35738-2 ISBN 978-3-030-35736-8 (eBook)
https://doi.org/10.1007/978-3-030-35736-8

This Springer imprint is published by the registered company Springer Nature Switzerland AG
The registered company address is: Gewerbestrasse 11, 6330 Cham, Switzerland

Preface

Tensor analysis presented in the current book contains understanding, operation and application of tensors in a broad spectrum of disciplines including engineering and physics. It is an indivisible integral part of continuum mechanics that includes fluid mechanics, solid mechanics and turbulence. It also plays an important role in electrical engineering dealing with electromagnetic field. Special and general theory of relativity exhibit another field of application of tensors, without which the theory of relativity could never be developed.

In the past three decades working at Texas A&M University, I have been teaching Fluid Mechanics Graduate Core Course (MEEN-621), Advanced Fluid Mechanics,(MEEN-622), and Turbomachinery Course (MEEN-646). Major chapters of these courses require knowledge of tensor, its application and operation. In an effort to provide the students with the basics of tensor analysis to follow the above courses for each course I tailored a tensor chapter to the specific needs of the students. Encouraged by the comments of my students, I decided to write a book that not only unifies the above tensor chapters in a systematic way but it also reflects an enhancement that goes far beyond those special chapters mentioned above.

Tensor analysis as such is an integral part of the mathematics curriculum. The graduate math courses offered for engineering students treat tensor analysis only peripherally and within the scope of general graduate math courses. In this regard, there have been countless textbooks published. The majority of these textbooks are written by math instructors with their inherent mathematical approach that is fundamentally different from that of engineers. The current book is introduced by an engineer but it addresses a wide range of applications relevant to engineers and physicists. It presents the subject matter in a clear, concise and easily understandable manner. Moreover, it enables the reader to obtain a solid knowledge about the tensor analysis and its application without spending too much time on mathematical formalism and proofs. The latter is presented whenever a detailed understanding of the subject requires it. While there are numerous texts on the market for engineering, physics students and instructors to choose from, there are only limited texts that comprehensively address the particular needs of graduate courses. To complement the lecture materials, the instructors more often recommend several texts, each of which treats special topics. This circumstance and the need to have a textbook that covers the materials needed in the engineering and physics courses gave the impetus to provide the community with a coherent textbook that comprehensively addresses their needs. Although this textbook is primarily aimed at engineering and physics students, it is equally suitable for self-study provided that the reader has sufficient knowledge of calculus and differential equations.

The book is structured in a systematic way. Chapters 1, 2 and 3 deal with tensors, their transformation and their differential operation. Chapters 4 and 5 treat the tensor operations in continuum mechanics. Chapter 6 treats the tensor operation in the curvilinear coordinate system. Chapter 7 deals with the tensor application in continuum mechanics, where the different cases are studied. Chapter 8 gives in a compact manner an overview of the differential geometry specially the curves, surfaces and geodesics. The derivation of the differential equation for geodesics in this chapter enables the reader to get a better understanding of the Einstein Field Equation, treated in Chap. 11. It shows that Newton's gravitation which is based on absolute time and absolute space assumption can be derived from Einstein's General Theory of Relativity as a special case.

An important topic, namely turbulence and its modeling, occupies Chap. 9. In this chapter, the quintessence of turbulence required for a graduate level course is presented. The use of a coordinate invariant form is particularly essential in understanding the underlying physics of the turbulence and its implementation into equations of motion and energy. Furthermore, it presents the necessary mathematical manipulations to arrive at different correlations. The resulting correlations are the basis for the following turbulence modeling. It is worth noting that in standard textbooks of turbulence, index notations are used throughout with almost no explanation of how they were brought about. This circumstance adds to the difficulty in understanding the nature of turbulence by readers who are freshly exposed to the problematics of turbulence. Introducing the coordinate invariant approach makes it easier for the reader to follow step-by-step the mathematical manipulations and arrive at the index notation and the component decomposition. A detailed derivation of the exact two-equation models $k - g$, $k - T$ and their approximate models are presented and critically discussed. The combination of these two models along with a few examples concludes this chapter.

One of the more complex tensor equations is the Einstein Field Equation with the spacetime geometry tensor on the left-hand side and the mass–momentum–energy tensor on the right. This tensor equation has to be an integral part of any advanced tensor book. It is the basis of the General Theory of Relativity (GTR), which follows the Special Theory of Relativity (STR). Starting with the STR, Chap. 10 prepares the groundwork for GTR that is treated in Chap. 11. While in STR the laws of physics are the same in all inertial frames, and the speed of light is the same for all observers, the GTR deals with the gravitation in a four-dimensional spacetime frame. The development of the GTR took Einstein ten years after completion of the STR in 1905. Explaining the GTR in Chap. 11, first a few operators specific to the GTR are explained. These include, among other things, commutator and parallel transport that are used to obtain the Riemann tensor. Contracting the Riemann tensor and adding the curvature scalar make up spacetime geometry, which is the left-hand side of the

Einstein Field Equation. How Einstein arrived at the right-hand side with the mass, momentum and energy tensor is explained in Chap. 11.

In order to keep the number of pages within an acceptable limit, at the end of a few chapters, there is a section that entails problems and projects. Instead of assigning problems that have homework format, the readers are advised to derive equations detailed in the book.

In typing several thousand equations, errors may occur. I tried hard to eliminate typing, spelling and other errors, but I have no doubt that some remain to be found by readers. In this case, I sincerely appreciate the reader notifying me of any mistakes found; the electronic address is on the title page. I also welcome any comments or suggestions regarding the improvement of future editions of the book.

College Station, USA
August 2019

Meinhard T. Schobeiri

Contents

Nomenclature

Symbols

$\mathbf{A}$	Acceleration vector
$\mathbf{A} = e_i A_i$	Vector $\mathbf{A}$ its index notation
$\mathbf{A} = e'_j A'_j$	Vector $\mathbf{A}$ transformed
c	Speed of light
c	Speed of sound
c_w	Specific heat capacities of a hot wire
C_D	Drag coefficient
C_f	Friction coefficient
C_p	Pressure coefficient
C_i	Constants
e	Specific total energy
E, F, G, I	Coefficients for first fundamental form, Chapter 8
E	Total energy
$E(k)$	Energy spectrum
$E(t, \mathbf{X})$	Event $E = f(t, \mathbf{X})$
f_S	Sampling frequency
F_S	Surface force vector
$\mathbf{F}$	Force vector
g	Gravitational constant
$\mathbf{g}_i, \mathbf{g}^i$	Co-, contravariant base vectors
$\mathbf{g}_i^*, \mathbf{g}^{*i}$	Co-, contravariant physical base vectors
g_{ij}, g^{ij}	Co-, contravariant metric coefficients
$g_{\mu,\nu}$	Metric tensor in spacetime coordinate
g_i^j	Mixed metric coefficient
$g_{i,j}$	Derivative of covariant base vector with respect to j-coordinate
G_i	Temperature function
$\mathbf{G}_i$	Transformation vector
h	Specific static enthalpy
H	Specific total enthalpy $H = h + 1/2V^2$

$I(\mathbf{x}, t)$	Intermittency function
$\mathbf{I}$	Unit tensor $\mathbf{I} = e_j e_j \delta_{ij}$
I_1, I_2, I_3	Principle invariants of deformation tensor
J	Jacobian transformation function
k	Thermal conductivity
k	Specific kinetic energy
$\mathbf{k}$	Wave number vector
K	Kinetic energy
L, M, N, II	Coefficients of the second fundamental form, Chapter 9
L_0	Rest length
$L_{ij}(\mathbf{x}, t)$	Turbulence length scale
m	Mass
m_0	Rest mass
m_U	Mass of moving particles
$\dot{m}$	Mass flow
$\mathbf{M}$	Vector of moment of momentum
$\mathbf{M}_a$	Axial vector of moment of momentum
$\mathbf{n}$	Normal unit vector
N	Navier-Stokes operator
Nu	Nusselt number
p	Static pressure
$\mathbf{P}$	Four momentum vector
P_μ	Components of four momentum vector
p'	Random pressure fluctuation
P, p_0	Total (stagnation) pressure, $P = p + DV^2/2$
Pr	Prandtl number $Pr = \mu c_p/\kappa$
Pr_e	Effective Prandtl number
Pr_t	Turbulent Prandtl number
$\mathbf{q}$	Specific thermal energy
$\dot{\mathbf{q}}$	Heat flux vector
Q	Thermal energy
$Q, \bar{Q}, Q'$	An arbitrary quantity, its time averaged, its fluctuation
Q_{ij}	Transformation matrix
$\mathbf{Q}, \mathbf{Q}^\mathbf{T}$	Transformation matrix and its transposed
$\dot{\mathbf{Q}}$	Time change of transformation matrix
R	Radius of curvature
Re	Reynolds number
$\mathbf{R}$	Position vector in absolute frame
$R_{ij}(\mathbf{x}, t, \mathbf{r}\tau)$	Correlation tensor
$\mathbf{R}(\mathbf{x}, t, \mathbf{r}, \tau)$	Correlation second order tensor
$\mathbf{R}'$	Position vector in relative frame
$\dot{\mathbf{R}}$	Time change of absolute position vector
$\dot{\mathbf{R}}'$	Time change of position vector in relative frame
$\ddot{\mathbf{R}}$	Acceleration in relative frame
s	Specific entropy

$\mathbf{S}$	Stress vector
$S = S(x, y)$	A planar curve
$S, S(t)$	Fixed, time dependent surface
t	Time
$\mathbf{t}$	Tangential unit vector
T	Static temperature
T_W	Wall temperature
Tr	Trace of second order tensor
T_u	Turbulence intensity
$\mathbf{T}$	Tangent vector
$\mathbf{T}$	Friction stress tensor (also called extra stress tensor $\mathbf{S}$
$\mathbf{T} = T^{kl}\mathbf{g}_k\mathbf{g}_l$	Second order tensor with contravariant component
$\mathbf{T} = T^{*kl}\mathbf{g}_k^*\mathbf{g}_l^*$	Second order tensor with contravariant physical component
$\mathbf{T}' = \rho\overline{V'V'}$	Reynolds stress tensor
u	Specific internal energy
u	Velocity
U	Undisturbed potential velocity
$\mathbf{U}$	Velocity vector
$\mathbf{U_M}$	Velocity of moving inertial frame
v	Specific volume
V	Volume
V_0	Fixed volume
$V(t)$	Time dependent volume
$\mathbf{V}$	Absolute velocity vector, eigenvector
$\mathbf{V}_T$	Velocity vector, turbulent solution
$\tilde{\mathbf{V}}$	Deterministic velocity fluctuation vector
$\overline{V}$	Mean velocity vector
$\mathbf{V}'$	Random velocity fluctuating vector
$\overline{\mathbf{V}'\mathbf{V}'}$	Reynolds stress tensor
V_i, V^j	Co- and contravariant component of a velocity vector
$< \mathbf{V} >$	Ensemble averaged velocity vector
$\mathbf{W}$	Relative velocity vectors
$\mathbf{X}$	Absolute position vector
$\dot{\mathbf{X}}$	Time change of position vector in absolute frame
$\ddot{\mathbf{X}}$	Acceleration of position vector in absolute frame

Greek Symbols, Operators

α	Heat transfer coefficient
γ	Lorentz correction transformation factor
Γ_{ijk}	Christoffel symbol of first kind
Γ^i_{jk}	Christoffel symbol of second kind

δ^{jk}	Kronecker delta
δ_{ij}	Matrix of Kronecker delta in cartesian coordinate system
δ_i^j	Mixed Kronecker delta
ε	Third order permutation tensor
ε	Dissipation
ε_c	Complete dissipation
ε^{ijk}, ε_{ijk}	Permutation symbol in curvilinear coordinate system
ε_{ijk}	Cartesian permutation symbol
ζ	Total pressure loss coefficient
η	Kolmogorov's length scale
θ	Transformation angle
$\Theta_{ij}(k_1, t)$	One-dimensional spectral function
κ	Curvature
κ	Thermal conductivity
κ	Ratio of specific heats
κ_1, κ_2	Principal curvatures
λ	Wave length
λ	Eigenvalue
λ	Taylor micro length scale
λ	Tangent unit vector
μ	Absolute viscosity
$\overline{\mu}$	Balk viscosity in Chapter 9
ν	Kinematic viscosity
ξ	Curvilinear coordinate
π	Pressure ratio
Π	Stress tensor, $\Pi = e_i e_j \pi_{ij}$
$\Pi_{,k}$	Differentiation of the stress tensor with respect to ξ_k
ρ	Density
$\rho_{ij}(\mathbf{x}, t, \mathbf{r}, \tau)$	Correlation coefficient
τ	Kolmogorov's time scale
τ	Proper time (eigenzeit)
τ	Torsion in Chapter 8
τ_1, τ_2, τ_3	Principal shear stress
τ_0, τ_W	Wall shear stress
υ	Kolmogorov's velocity scale
$\mathbf{\Phi}$	A second order tensor in Chapter 9
$\mathbf{\Phi}(k, t)$	Spectran tensor in Chapter 9
$\boldsymbol{\omega}$	Vorticity vector $\omega = \nabla \times \mathbf{V}$
$\mathbf{\Omega}$	Rotation tensor
∇	Angular velocity
$\boldsymbol{\nabla}$	Vorticity vector $= \boldsymbol{\nabla} \times \mathbf{V}$

Subscripts, Superscripts

∞	FREESTREAM
a, t	Axial, tangential
ex	Exit
in	Inlet
M	Moving frame
max	Maximum
min	Minimum
s	Stationary frame
s	Isentropic
t	Turbulent
w	Wall
$-$	Time averaged
$/$	Random fluctuation
$*$	Dimensionless

Operators

d	Convective differential operator
D	Total differential operator in absolute frame of reference
D_R	Total differential operator in relative frame of reference
N	Navier–Stokes operator
∇	Nabla, vector differential operator vector
Δ	Laplace scalar differential operator $= \nabla \cdot \nabla$

Chapter 1
Vectors and Tensors

1.1 Introduction

Tensor analysis is a powerful tool that enables the students of engineering and physics to study and to understand more effectively the fundamentals of computational fluid mechanics, continuum mechanics, solid mechanics, bio-engineering, civil engineering, electrical engineering, and the necessary operation required in general theory of relativity, to name just a few areas of tensor applications. The instructors teaching the above subjects have to spend a substantial amount of time to teach the students the very basics of tensor analysis necessary for following and understanding the actual subject matter. The graduate math courses offered for engineering students treat tensor analysis only peripherally and within the scope of general graduate math courses. Moreover, the majority of the existing tensor textbooks are written by math instructors with their inherent mathematical approach that is fundamentally different from that of engineers. The current book is introduced by an engineer and for engineers, physicists and presents the subject matter without too much mathematical formalism and ballasts. It is tailored for engineering understanding. Having understood the course content, for the physics students it will be easy to follow the course of the Special and General Theory of Relativity taught by Physics Instructors.

1.1.1 Space, Euclidean Space, Curved Space

Tensors are defined within certain categories of spaces that must be defined first. Generally, a space is the region of collection of objects. From a physical point of view a space may be one, two or three dimensional. In a one dimensional space the position of objects relative to a fixed origin or to each other is given by only one dimension of a three-dimensional coordinate system, for example x-coordinate. To determine the position of an object in a two dimensional space, two dimensions are

M. T. Schobeiri, *Tensor Analysis for Engineers and Physicists - With Application to Continuum Mechanics, Turbulence, and Einstein's Special and General Theory of Relativity*, https://doi.org/10.1007/978-3-030-35736-8_1

required for example x- and y-coordinate. Finally in a three dimensional space all thee dimensions are required to identify the position of an object. These different spaces are categorized in flat and curved spaces. While the flat space is defined by the Euclidean geometry, the Riemann geometry defines the curved space. Starting with the Euclidean geometry, Euclid (Efkleídis), an ancient Greek mathematician (300 BCE) introduced the geometry of flat spaces. The element of Euclidean geometry can be drawn on flat piece of paper. His geometry includes a few postulates that we know from our high school time. The relevant ones are:

1. The shortest distance between two points is one unique straight line.
2. The sum of the angles in any triangle equals 180 degrees.
3. Given a line L and a point P on a surface, among the infinite numbers of lines through the point P, only one line can be drawn that is parallel to the line L.

The concepts in Euclid geometry are valid for flat spaces they are no longer applicable to curved spaces. The groundbreaking original work by the German mathematician Johann Carl Friedrich Gauss (30 April 1777–23 February 1855) introduced a new geometry, the non-Euclidean geometry. This new geometry was further enhanced by one of the talented students of Gauss, Georg Friedrich Bernhard Riemann (17 September 1826–20 July 1866), who made lasting and revolutionary contributions to analysis, number theory, and differential geometry. He developed the Riemann geometry, which is the basis of all theories that deal with curved space such as the Field Equation developed by Albert Einstein (14 March 1879–18 April 1955).

Einstein was a German theoretical physicist, who developed the special and general theory of relativity, one of the two pillars of modern physics. Another mathematician, who contributed to the non-Euclidean geometry is Nikolai Ivanovich Lobachevsky (1792–1856), a Russian mathematician, who developed the elliptic geometry. The non-Euclidean space includes, among others, the elliptic and spherical geometry. The Riemann concept of curved space contradicts the concept of Euclidean flat space. Here are a few examples:

1. In a curved space, there is no unique straight line that connects two points on a curved surface.
2. In a curved space, for example a sphere Fig. 1.3, the shortest distance between two points is a segment of a great circle called a geodesic. There are infinite number of geodesics between the north and south poles of any sphere. They intersect at the north and south poles.
3. On a surface with positive curvature such as a sphere, Fig. 1.4, the sum of the angles in any triangle is $\pi \leq \alpha + \beta + \gamma \leq 2\pi$.
4. On a surface with negative curvature such as a hyperbolic paraboloid sphere, Fig. 1.5 the sum of the angles in any triangle may be $\alpha + \beta + \gamma =\leq \pi$.

Fig. 1.1 Johann Carl Friedrich Gauss (1777–1855)

1.1.2 Cartesian Tensors in Three-Dimensional Euclidean Space

In this section, we introduce tensors, their significance to engineering problems and their applications. The tensor analysis enables the reader to study and to understand more effectively the fundamentals of engineering mentioned above. Once the basics of tensor analysis are understood, the reader will be able to derive all equations relative to engineering laws without memorizing. In this section, we focus on the tensor analytical application rather than mathematical details and proofs that are not primarily relevant to engineering and science students. To directly connect to the knowledge that the students have from vector analysis, we present the definition of tensors from a unified point of view and use first the three–dimensional Euclidean space, also called flat space with $N = 3$ as the number of dimensions.

Following this chapter, we will treat tensors in curvilinear coordinate system. But before getting into the tensor analysis, we need to understand what the meaning of the Euclidean space is and what other spaces are there and which ones are relevant for engineering applications.

Fig. 1.2 Georg Friedrich Bernhard Riemann (1826–1866)

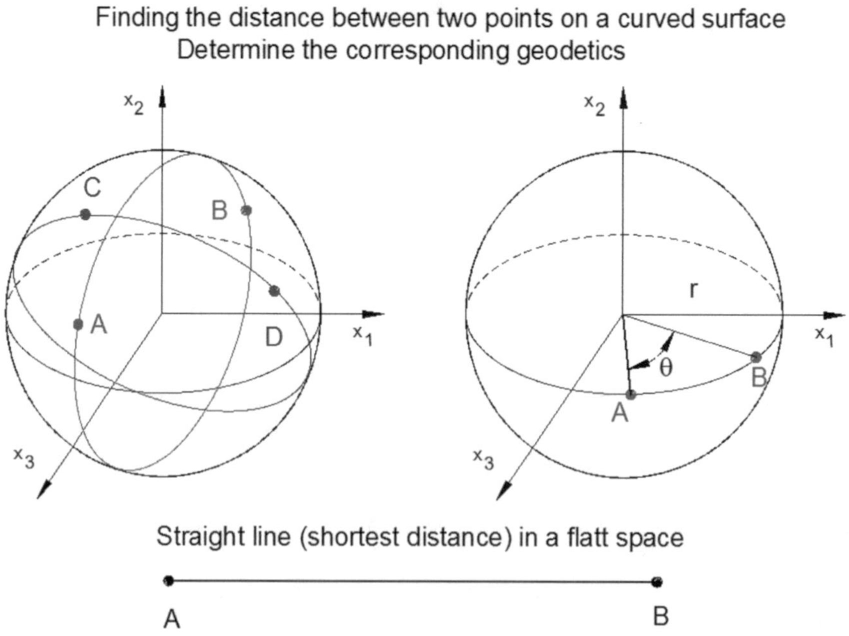

Fig. 1.3 Finding the shortest distance on a curved suface versus flat surface

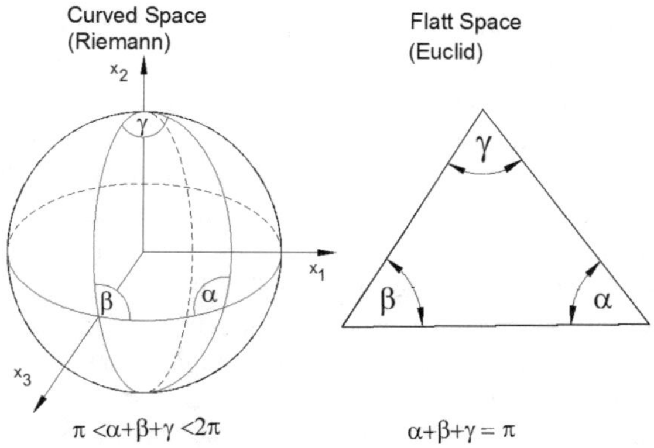

Fig. 1.4 Curved surface, positive curvature

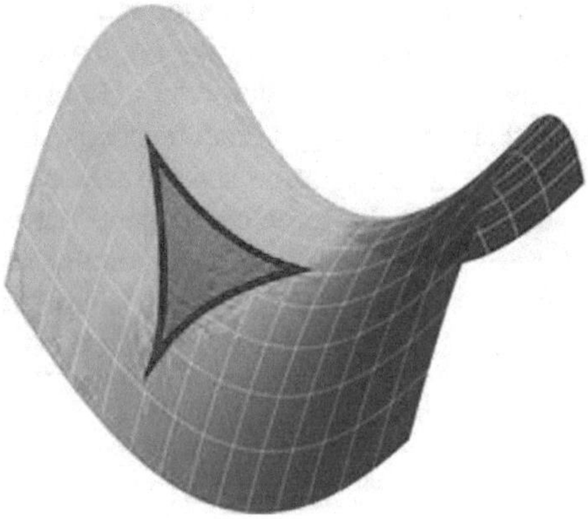

Fig. 1.5 Curved surface, negative curvature

1.1.3 Physical Quantities, Order of Tensors

The quantities encountered in engineering are tensors. A physical quantity which has a *definite magnitude* but not a *definite direction* exhibits a *zeroth-order tensor*, which is a special category of tensors. In a N-dimensional Euclidean space with $N = 3$, a zeroth-order tensor has $N^0 = 3^0 = 1$ component, which is basically its magnitude. In physical sciences, this category of tensors is well known as a *scalar* quantity, which has a definite magnitude but not a definite direction. Examples are quantities such as mass m, volume v, thermal energy Q (heat), mechanical energy W (work) and the entire thermo-fluid dynamic properties such as density ρ, temperature T, enthalpy h, entropy s, speed of sound, speed of light. etc. In electrical engineering, we encounter current, voltage, electrical potential as zeroth order tensors or scalars. All scalar quantities are written in small Latin and Greek letters such as h, s, ρ etc.

In contrast to the zeroth-order tensor, a *first-order tensor* encompasses physical quantities with a *definite magnitude* with $N^1 (N^1 = 3^1 = 3)$ components and a *definite direction* that can be decomposed in $N^1 = 3$ directions. This special category of tensors is known as *vectors*. All vector quantities are written in bold capital Latin letters. Distance $\boldsymbol{X}$, velocity $\boldsymbol{V}$, acceleration $\boldsymbol{A}$, force $\boldsymbol{F}$ and moment of momentum $\boldsymbol{M}$ are few examples. A vector quantity is *invariant* with respect to a given category of coordinate systems. Changing the coordinate system by applying certain transformation rules, the vector components undergo certain changes resulting in a new set of components that are related, in a definite way, to the old ones. As we will see later, the order of the above tensors can be reduced if they are multiplied with each other in a scalar manner. The mechanical energy $W = \boldsymbol{F}.\boldsymbol{X}$ is a representative example, that shows how a tensor order can be reduced. The reduction of order of tensors is called *contraction*. All vectors are written in bold roman capital letter such as $\boldsymbol{U}$, $\boldsymbol{V}$, $\boldsymbol{F}$ etc. All vector components are written in small roman letters such as $\boldsymbol{u}_i$, $\boldsymbol{v}_i$, $\boldsymbol{f}_i$, etc.

A *second-order tensor* is a quantity, which has N^2 definite components and N^2 definite directions. In three-dimensional Euclidean space a second order tensor has $N^2 = 9$ components. General stress tensor $\boldsymbol{\Pi}$, normal stress tensor $\boldsymbol{\Sigma}$, shear stress tensor $\boldsymbol{T}$, deformation tensor $\boldsymbol{D}$ and rotation tensor $\boldsymbol{\Omega}$ are a few examples. Unlike the zeroth and first order tensors (scalars and vectors), the second and higher order tensors cannot be directly geometrically interpreted. However, they can easily be interpreted by looking at their pertinent first order tensor. An example should clarify this statement: the stress distribution on the six surfaces of a differential volume is the result of the distribution of the force components acting on the differential volume. Dividing the force components by the corresponding surfaces results in nine components of the total stress tensor as seen later in Chap. 3. All tensor quantities are written in bold capital Greek or Latin letters such as $\boldsymbol{\Pi}$, $\boldsymbol{\Phi}$, $\boldsymbol{\Omega}$, $\boldsymbol{D}$, etc.

1.1.4 Index Notation

In a three-dimensional Euclidean space, any arbitrary first order tensor or vector can be decomposed into three components. In a Cartesian coordinate system shown in Fig. 1.6, the base vectors in x_1, x_2, x_3 directions $\boldsymbol{e}_1, \boldsymbol{e}_2, \boldsymbol{e}_3$ are perpendicular to each other and have the magnitude of unity, therefore, they are called *orthonormal unit vectors*. Furthermore, these base vectors are not dependent upon the coordinates, therefore, their derivatives with respect to any coordinates are identically zero. In contrast, in a general curvilinear coordinate system the base vectors do not have the magnitude of unity. They depend on the curvilinear coordinates, thus, their derivatives with respect to the coordinates do not vanish.

Starting with vector $\boldsymbol{A}$ in a Cartesian coordinate system, its components A_1, A_2 and A_3 are shown in Fig. 1.6 and written as:

$$\boldsymbol{A} = e_1 A_1 + e_2 A_2 + e_3 A_3 = \sum_{i=1}^{N=3} e_i A_i. \tag{1.1}$$

According to Einstein's summation convention, Eq. (1.1), it can be written as:

$$\boldsymbol{A} = e_i A_i. \tag{1.2}$$

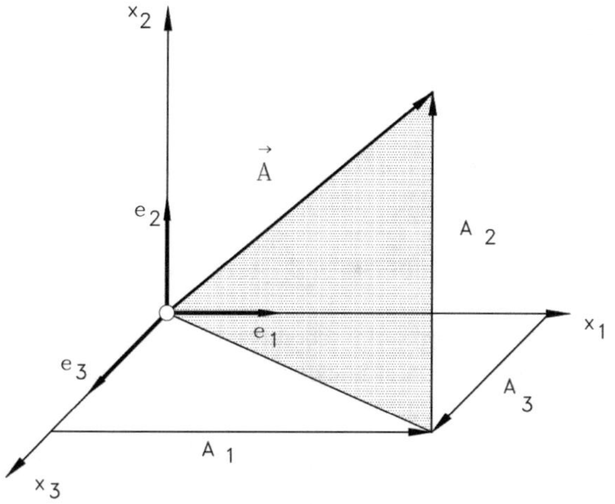

Fig. 1.6 Base vectors and the components in Cartesian coordinate system

The above form is called the *index notation*. Whenever the same index (in the above equation i) appears twice, the summation is carried out from 1 to $N(N = 3$ for Euclidean space). In an expression, where the same index is repeated twice, the index is called *dummy index* and can be replaced by any other letter that does not appear in the same expression. As an example, die indices in Eq. (1.2) can be replaced by $e_i A_i = e_j A_j = e_k A_k$. In case the index appears alone, it calls *free index* and cannot be renamed.

1.2 Vector Operations

Operation with vectors involves scalar, vector and tensor products that we discuss in the following sections.

1.2.1 Scalar Product

Throughout this book, the scalar product of two vectors is separated by a dot $(\cdot)$. Scalar or dot product of two vectors results in a scalar quantity $\boldsymbol{A} \cdot \boldsymbol{B} = C$. We apply the Einstein's summation convention defined in Eq. (1.2) to the above vectors:

$$(e_i A_i) \cdot (e_j B_j) = C. \tag{1.3}$$

In Eq. (1.3) we have used index i for the vector $\boldsymbol{A}$ and j for the vector $\boldsymbol{B}$ because it is not allowed to use the same index for both vectors. This would result in the same expression with four identical indices which violates the summation convention.

Now we rearrange the unit vectors and the components separately:

$$(e_i \cdot e_j)A_i B_j = C. \tag{1.4}$$

In Cartesian coordinate system, the scalar product of two unit vectors is called *Kronecker delta*, which is defined as:

$$\delta_{ij} = e_i \cdot e_j = 1 \text{ for } i = j,\ \delta_{ij} = e_i \cdot e_j = 0 \text{ for } i \neq j. \tag{1.5}$$

The symbol delta δ_{ij} is the matrix of a second order tensor with $3^2 = 9$ components from which only 3 are non-zero. It is called the identity tensor or unit tensor with $\boldsymbol{I} = e_i e_j \delta_{ij}$. Its matrix is given below:

$$(\delta_{ij}) = \begin{bmatrix} 1 & 0 & 0 \\ 0 & 1 & 0 \\ 0 & 0 & 1 \end{bmatrix}. \tag{1.6}$$

Using the Kronecker delta, we get:

$$(e_i \cdot e_j)A_i B_j = \delta_{ij} A_i B_j. \tag{1.7}$$

The non-zero components are found only for $i = j$, or $\delta_{ij} = 1$, which means in the above equation the index j must be replaced by i (or i by j) resulting in:

$$\boldsymbol{A} \cdot \boldsymbol{B} = A_i B_i = A_1 B_1 + A_2 B_2 + A_3 B_3 = C \tag{1.8}$$

with scalar C as the result of a scalar multiplication. Equation (1.7) can be used to find the magnitude of a vector. Given the vector $\boldsymbol{A} = e_i A_i$, the magnitude is calculated

$$\boldsymbol{A} \cdot \boldsymbol{A} = (e_i \cdot e_j)A_i A_j = \delta_{ij} A_i A_j = A_i A_i \tag{1.9}$$

and the magnitude is

$$\sqrt{\boldsymbol{A} \cdot \boldsymbol{A}} = \sqrt{A_i A_i} = \sqrt{A_1^2 + A_2^2 + A_3^2}. \tag{1.10}$$

Finally the angle between two vectors $\phi\sphericalangle(\boldsymbol{A}, \boldsymbol{B})$ is calculated from:

$$\boldsymbol{A} \cdot \boldsymbol{B} = |A|\,|B| \cos \boldsymbol{\Phi} \tag{1.11}$$

or

$$\cos \phi = \frac{A_i B_i}{\sqrt{A_i A_i}\,\sqrt{B_i B_i}}. \tag{1.12}$$

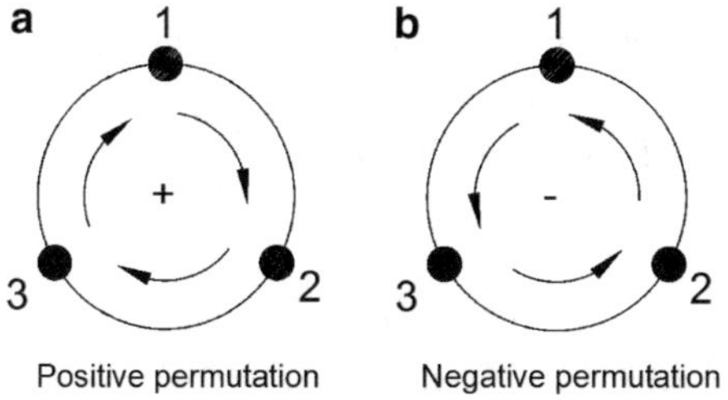

Fig. 1.7 Permutation symbol, **a** positive, **b** negative permutation

1.2.2 Vector or Cross Product

The vector product of two vectors is a vector that is perpendicular to the plane described by those two vectors. Example:

$$\boldsymbol{F} \times \boldsymbol{R} = \boldsymbol{M} \text{ or } \boldsymbol{A} \times \boldsymbol{B} = \boldsymbol{C} \tag{1.13}$$

with $\boldsymbol{C}$ as the resulting vector. We apply the index notation to Eq. (1.13):

$$\boldsymbol{A} \times \boldsymbol{B} = (e_i A_i) \times (e_j B_j) = \epsilon_{ijk} e_k A_i B_j \tag{1.14}$$

with ϵ_{ijk} as the permutation symbol introduced by Levi Civita, an Italian mathematician (1873–1941). The following definition is illustrated in Fig. 1.7

$\epsilon_{ijk} = 0$ for $i = j,\ j = k$ or $i = k$: (e.g. 122)
$\epsilon_{ijk} = 1$ for cyclic permutation: (e.g. 123)
$\epsilon_{ijk} = -1$ for anticyclic permutation: (e.g. 132)

The Levi Civita permutation symbols ϵ_{ijk} is actually a third order tensor with $3^3 = 27$ components. Of these 27 components, 21 components are zero. Of the remaining 6 components 3 components are equal to $+1$ and 3 components are equal to -1. Using the above definition, the vector product is given by:

$$\boldsymbol{C} = e_k C_k = \epsilon_{ijk} e_k A_i B_j. \tag{1.15}$$

Expanding Eq. (1.14) gives:

$$\begin{aligned} \boldsymbol{C} = e_{kC_k} =& \epsilon_{123} e_3 A_1 B_2 + \epsilon_{231} e_1 A_2 B_3 + \epsilon_{312} e_2 A_3 B_1 + \\ & \epsilon_{132} e_2 A_1 B_3 + \epsilon_{321} e_1 A_3 B_2 + \epsilon_{213} e_3 A_2 B_1. \end{aligned} \tag{1.16}$$

Inserting the corresponding values for ϵ_{ijk} we obtain:

$$\boldsymbol{C} = e_3(A_1 B_2 - A_2 B_1) + e_2(A_3 B_1 - A_1 B_3) + e_1(A_2 B_3 - A_3 B_2). \tag{1.17}$$

It is noted that while the commutative law for scalar product of two vectors is valid, it fails for cross product since $\boldsymbol{A} \times \boldsymbol{B} = -\boldsymbol{B} \times \boldsymbol{A}$.

There is a relationship between the permutation symbol ϵ_{ij} and the Kronecker delta δ_{ij} as given below:

$$\epsilon_{ijk}\epsilon_{imn} = \delta_{jm}\delta_{kn} - \delta_{jn}\delta_{kn}. \tag{1.18}$$

1.2.3 Tensor Product

The tensor product is a product of two or more vectors, where the unit vectors are not subject to scalar or vector operation. Consider the following *tensor operation*:

$$\boldsymbol{\Phi} = \boldsymbol{AB} = (e_i A_i)(e_j B_j) = e_i e_j A_i B_j = e_i e_j \Phi_{ij}. \tag{1.19}$$

The tensor product of two or more vectors do not have any dot $(\cdot)$ or cross $(\times)$ in between. In continuum mechanics they are separated by the symbols $\otimes$. For our applications, this notation is impractical, since we will be dealing with tensors of higher order.

In a three dimensional Euclidean space, the result of this purely mathematical operation is a second order tensor with nine components:

$$\begin{aligned}\boldsymbol{\Phi} =& e_1(e_1 A_1 B_1 + e_2 A_1 B_2 + e_3 A_1 B_3)+ \\ & e_2(e_1 A_2 B_1 + e_2 A_2 B_2 + e_3 A_2 B_3)+ \\ & e_3(e_1 A_3 B_1 + e_2 A_3 B_2 + e_3 A_3 B_3).\end{aligned} \tag{1.20}$$

In Eq. (1.20), the matrix $(A_i B_j)$ can be set $(A_i B_j) = (\Phi_{ij})$ with:

$$(A_i B_j) = \begin{bmatrix} A_1 B_1 & A_1 B_2 & A_1 B_3 \\ A_2 B_1 & A_2 B_2 & A_2 B_3 \\ A_3 B_1 & A_3 B_2 & A_3 B_3 \end{bmatrix} = \Phi_{ij} = \begin{bmatrix} \Phi_{11} & \Phi_{12} & \Phi_{13} \\ \Phi_{21} & \Phi_{22} & Q_{23} \\ \Phi_{31} & \Phi_{32} & \Phi_{33} \end{bmatrix}. \tag{1.21}$$

As seen in Eq. (1.19), the second order tensor can be thought of as an association of two vectors between which there is no dot $(\cdot)$ and no cross $(\times)$. Higher order tensors can also be thought of as an association of more than two vectors. As an example, we construct a third order tensor as shown below:

$$\boldsymbol{T} = \boldsymbol{ABC} = e_i e_j e_k A_i B_j C_k = e_i e_j e_k T_{ijk}. \tag{1.22}$$

The operation with any tensor such as the above second or higher order ones acquires a physical meaning if it is multiplied with a vector (or another tensor) in scalar manner. Consider the scalar product of the vector $\boldsymbol{C}$ and the second order tensor $\boldsymbol{\Phi}$. The result of this operation is a *first order tensor* or a *vector*. The following example should clarify this:

$$\boldsymbol{D} = \boldsymbol{C} \cdot \boldsymbol{\Phi} = \boldsymbol{C} \cdot (\boldsymbol{AB}) = e_k C_k \cdot (e_i e_j) A_i B_j. \tag{1.23}$$

Rearranging the unit vectors and the components separately:

$$\boldsymbol{D} = \boldsymbol{C} \cdot \boldsymbol{\Phi} = \boldsymbol{C} \cdot (\boldsymbol{AB}) = e_k \cdot (e_i e_j) C_k A_i B_j. \tag{1.24}$$

It should be pointed out that in the above equation, the unit vector $\boldsymbol{e}_k$ must be multiplied with the **closest unit vector** namely $\boldsymbol{e}_i$

$$\boldsymbol{D} = \boldsymbol{C} \cdot \boldsymbol{\Phi} = \boldsymbol{C} \cdot (\boldsymbol{AB}) = \delta_{ki} e_j C_k A_i B_j = e_j C_i A_i B_j. \tag{1.25}$$

The result of this tensor operation is a vector with the same direction as vector $\boldsymbol{B}$. Different results are obtained if the positions of the terms in a dot product of a vector with a tensor are reversed as shown in the following operation:

$$\boldsymbol{E} = \boldsymbol{\Phi} \cdot C = (\boldsymbol{AB}) \cdot \boldsymbol{C} = e_i A_i B_j \delta_{jk} C_k = e_i A_i B_j C_j. \tag{1.26}$$

The result of this operation is a vector in direction of $\boldsymbol{A}$. Thus, the products $\boldsymbol{E} = \boldsymbol{\Phi} \cdot \boldsymbol{C}$ is different from $\boldsymbol{D} = \boldsymbol{C} \cdot \boldsymbol{\Phi}$. This is another example, where the commutative law of multiplication fails.

Given two second (or higher) order tensors that are multiplied with each other as:

$$\boldsymbol{\Xi} = \boldsymbol{\Phi} \cdot \boldsymbol{\Psi} = e_i e_j \Phi_{ij} \cdot e_k e_n \boldsymbol{\Psi}_{kn} = e_i e_n \delta_{jk} \Phi_{ij} \Psi_{kn} = e_i e_n \Phi_{ij} \Psi_{jn}. \tag{1.27}$$

The result of Eq. (1.27) is a second order tensor. If the position of the above two tensors are switched, the result is a different tensor as shown below:

$$\boldsymbol{\Lambda} = \boldsymbol{\Psi} \cdot \boldsymbol{\Phi} = e_i e_j \boldsymbol{\Psi}_{ij} \cdot e_k e_n \boldsymbol{\Phi}_{kn} = e_i e_n \delta_{jk} \Psi_{ij} \Phi_{kn} = e_i e_n \Psi_{ij} \Phi_{jn}. \tag{1.28}$$

As can be examined easily, the two second order tensors $\boldsymbol{\Lambda}$, $\boldsymbol{\Xi}$ are different. This means that the commutative law of multiplication fails in case of tensor multiplication.

1.2.4 Symmetric, Antisymmetric Behavior of Tensors

Considering the second order tensor with the matrix $\boldsymbol{\Phi}_{ij}$ and its transpose $(\boldsymbol{\Phi}_{ij})^T$:

$$(\Phi_{ij}) = \begin{bmatrix} \Phi_{11} & \Phi_{12} & \Phi_{13} \\ \Phi_{21} & \Phi_{22} & \Phi_{23} \\ \Phi_{31} & \Phi_{32} & \Phi_{33} \end{bmatrix}, (\Phi_{ij})^T = \begin{bmatrix} \Phi_{11} & \Phi_{21} & \Phi_{31} \\ \Phi_{12} & \Phi_{22} & \Phi_{32} \\ \Phi_{13} & \Phi_{23} & \Phi_{33} \end{bmatrix}. \tag{1.29}$$

If the above matrices are equal: $(\Phi_{ij}) = (\Phi_{ij})^T$, then the tensor is called symmetric with $\Phi_{ij} = \Phi_{ji}$ and the tensor is written as $\boldsymbol{\Phi} = \boldsymbol{\Phi}^T$. On the other hand if the matrix of a tensor is $\Phi_{ij} = -\Phi_{ji}$ the tensor is antisymmetric and it is written as $\boldsymbol{\Phi} = -\boldsymbol{\Phi}^T$.

1.3 Contraction of Tensors

Contraction of a tensor is the reduction of its order. In engineering and physics we encounter cases, where the order of a tensor is reduced to construct another tensor. In Fluid Mechanics and Continuum Mechanics, contraction is applied to several tenors when energy balance is treated. In general Theory of Relativity, the second order Ricci tensor used in Einstein tensor is the contraction of the fourth order Riemann's tensor. In the following, we discuss different method of contracting tensors.

1.3.1 Contraction of a Second Order Tensor

Consider two first order tensors $\boldsymbol{A} = e_i A_i$ and $\boldsymbol{B} = e_j B_j$ and their tensor product $\boldsymbol{AB} = e_i e_j A_i B_j$ which is a second order tensor. The scalar multiplication of these tensors results in scalar quantity, which is a zeroth order tensor.

$$\boldsymbol{A} \cdot \boldsymbol{B} = \delta_{ij} A_i B_j = A_i B_i = A_1 B_1 + a_2 B_2 + A_3 B_3 = C. \tag{1.30}$$

Constructing a second order tensor that includes a position vector $\boldsymbol{X} = e_i X_i$ and a force vector $\boldsymbol{F} = e_j F_j$. Their tensor product $\boldsymbol{XF} = e_i e_j X_i F_j$ is a second order tensor. The contraction of this tensor results in mechanical energy which is a scalar quantity, which is a zeroth order tensor:

$$\boldsymbol{X} \cdot \boldsymbol{F} = e_i \cdot e_j X_i F_j = \delta_{ij} X_i F_j = X_i F_i = W. \tag{1.31}$$

In both cases, the order of the second tensor has reduced from one to zero. Also, as shown previously, the scalar product of a second order tensor with a first order one is a first order tensor or a vector.

1.3.2 Trace of a Second Order Tensor

Taking the *trace of a second order tensor* is another way of contracting the order of a tensor. The result of this contraction is a scalar quantity.

$$\mathrm{Tr}(\boldsymbol{\Phi}) = e_i \cdot e_j \Phi_{ij} = \delta_{ij} \Phi_{ij} = \delta_{ij} \Phi_{ij} = \Phi_{11} + \Phi_{22} + \Phi_{33}. \tag{1.32}$$

As seen easily, the trace of a second order tensor is the sum of the diagonal element of the *matrix* Φ_{ij}.

If the tensor $\boldsymbol{\Phi}$ itself is the result of a contraction of two second order tensors $\boldsymbol{\Pi}$ and $\boldsymbol{D}$:

$$\boldsymbol{\Phi} = \boldsymbol{\Pi} \cdot \boldsymbol{D} = (e_i e_j \Pi_{ij}) \cdot (e_k e_l D_{kl}) = e_i e_l \delta_{jk} \Pi_{ij} D_{kl} = e_i e_l \Pi_{ik} D_{kl} \tag{1.33}$$

then the *Trace* $\boldsymbol{\Phi}$ is:

$$\mathrm{Tr}(\boldsymbol{\Phi}) = e_i \cdot e_l \Pi_{ik} D_{kl} = \delta_{il} \Pi_{ik} D_{kl} = \Pi_{lk} D_{kl}. \tag{1.34}$$

1.3.3 Product of Two Second Order Tensors

As seen in Eq. (1.33) a single dot $(\cdot)$ caused the juxtaposed unit vectors to multiply scalarly leading to a second order tensor. Taking a *double dot* $(:)$ further reduces the order of the tensor resulting in scalar.

$$\boldsymbol{\Pi} : \boldsymbol{D} = e_i e_j \Pi_{ij} : e_k e_l D_{ki} = \delta_{il} \delta_{jk} \Pi_{ij} D_{kl} = \Pi_{ik} D_{ki}. \tag{1.35}$$

Note: In some literature terms such as *inner* and *outer product* of tensors are used. Outer product is the result of the tensor product of two tensors. For example a tensor of order m multiplied with another tensor of order n results in a tensor with the order $m + n$. Considering the same two tensors, their inner product is formed by contracting their outer product resulting in a tensor of order $m + n - 2$. In this and the following sections we avoid using these terms.

1.3.4 Contraction of Higher Order Tensors

Given a third order tensor:

$$\boldsymbol{T} = e_i e_j e_k T_{ijk} \tag{1.36}$$

it can be contracted in different ways as shown below

$$\begin{aligned} \boldsymbol{T} &= e_i \cdot e_j e_k T_{ijk} = \delta_{ij} e_k T_{ijk} = e_k T_{iik} \\ \boldsymbol{T} &= e_i e_j \cdot e_k T_{ijk} = \delta_{jk} e_k T_{ijk} = e_k T_{ikk}. \end{aligned} \tag{1.37}$$

In both cases the summation is performed differently as shown below:

$$\begin{aligned} T_{iik} &= T_{11k} + T_{22k} + T_{33k} \\ T_{ikk} &= T_{i11} + T_{i22} + T_{i33} \end{aligned} \tag{1.38}$$

and the order of tensor has been reduced from 3 to 1. For an arbitrary tensor of the order m, the contraction results in a new order of $m - 2$.

Note: In all contraction cases presented above, we always carried along the Kronecker symbol. This was to ensure that the reader easily understands how to contract tensors. In contracting tensors in non-curved space, the use of Kronecker symbol can be omitted altogether. Take for example Eq. (1.37): we arrive at the same results by setting in T_{ijk} either $i = j$ resulting in T_{iik} or $j = k$ resulting in T_{ikk} the without using the Kronecker delta.

1.4 Decomposition of Second Order Tensors

Any non-symmetric second order tensor can be decomposed into a symmetric and an antisymmetric tensor:

$$\boldsymbol{T} = \tfrac{1}{2}(\boldsymbol{T} + \boldsymbol{T}^T) + \tfrac{1}{2}(\boldsymbol{T} - \boldsymbol{T}^T) = \boldsymbol{D} + \boldsymbol{\Omega} \tag{1.39}$$

with $\boldsymbol{D}$ as the deformation tensor and $\boldsymbol{\Omega}$ rotation tensor as they are called in continuum mechanics, we have:

$$T_{ij} = \tfrac{1}{2}(T_{ij} + T_{ji}) + \tfrac{1}{2}(T_{ij} - T_{ji}). \tag{1.40}$$

The index notations of these tensors are:

$$D_{ij} = \tfrac{1}{2}(T_{ij} + T_{ji}) \text{ and } \boldsymbol{\Omega}_{ij} = \tfrac{1}{2}(T_{ij} - T_{ji}). \tag{1.41}$$

The product of a symmetric and an antisymmetric tensor such as the above ones results in:

$$\boldsymbol{D} : \boldsymbol{\Omega} = 0. \tag{1.42}$$

1.5 Inverse of a Tensor

Given a second order tensor $\boldsymbol{T}$ with a non-zero determinant det $\boldsymbol{T} \neq 0$, so there exist an inverse $\boldsymbol{T}$ defined as $\boldsymbol{T}^{-1}$ such that

$$\boldsymbol{T}\boldsymbol{T}^{-1} = \boldsymbol{I} \tag{1.43}$$

with I as the unitary tensor $I = e_i e_j \delta_{ij}$.

Problems

1: The vectors $\boldsymbol{A}$ and $\boldsymbol{B}$ are given by:

$$\begin{aligned} \boldsymbol{A} &= e_i A_i = e_1 x_1 x_2 + e_2 x_2^2 + 2e_3 \\ \boldsymbol{B} &= e_j B_j = e_1 x_1^2 - e_2 x_1 x_2 + 2e_3 x_3. \end{aligned} \tag{1.44}$$

2: (1) determine the scalar product $\boldsymbol{A}.\ \boldsymbol{B}$, (2) the vector product $\boldsymbol{A} \times \boldsymbol{B}$ and (3) the tensor product $\boldsymbol{\Phi} = \boldsymbol{AB}$. First use index notation, Kronecker delta δ_{ij}, and permutation symbol ϵ_{ijk}. After you did all the rearrangements then plug the components into the equations and get the results in equation form. (4) For $x_1 = 1,\ x_2 = 2,\ x_3 = 3$, calculate: $A \cdot \boldsymbol{B},\ \boldsymbol{A} \times \boldsymbol{B},\ \boldsymbol{AB}$.

3: Given the vector:

$$\boldsymbol{C} = e_k C_k = e_1 x_1^2 x_2 + e_3 x_3^2. \tag{1.45}$$

Find the volume V by using the above vectors $\boldsymbol{A}$ and $\boldsymbol{B}$:

$$V = \boldsymbol{C} \cdot (\boldsymbol{A} \times \boldsymbol{B}). \tag{1.46}$$

4: Given the vector:

$$\boldsymbol{C} = e_k C_k = e_1 x_1^2 x_2 + e_3 x_3^2 \tag{1.47}$$

and from Problem 1 the vectors

$$\begin{aligned} \boldsymbol{A} &= e_i A_i = e_1 x_1 x_2 + e_2 x_2^2 + 2e_3 \\ \boldsymbol{B} &= e_j B_j = e_1 x_1^2 - e_2 x_1 x_2 + 2e_3 x_3 \end{aligned} \tag{1.48}$$

(a) Find their second order tensor product

$$\boldsymbol{\Phi} = \boldsymbol{AB}. \tag{1.49}$$

(b) Find the vector products $\boldsymbol{D} = \boldsymbol{C} \cdot \boldsymbol{\Phi}$ and $\boldsymbol{E} = \boldsymbol{\Phi} \cdot \boldsymbol{C}$ and show that $\boldsymbol{D} \neq \boldsymbol{E}$. For $x_1 = 1,\ x_2 = 2,\ x + 3 = 3$, calculate: $\boldsymbol{D},\ \boldsymbol{E}$.

5: Show that the product of a symmetric and an antisymmetric tensor $\boldsymbol{D} : \boldsymbol{\Omega} = 0$.

References

1. Aris R (1962) Vector, tensors and the basic equations of fluid mechanics. Prentice-Hall. Inc, Englewood Cliffs, NJ
2. Brand L (1947) Vector and tensor analysis. Wiley, New York
3. Klingbeil E (1966) Tensorrechnung für Ingenieure. Bibliographisches Institut, Mannheim
4. Lagally M (1944) Vorlesung über Vektorrechnung, dritte edn. Akademische Verlagsgeselschaft, Leipzig
5. Vavra MH (1960) Aero-thermodynamics and flow in turbomachines. Wiley, Hoboken

Chapter 2
Transformation of Tensors

In the previous chapter we discussed several aspects of operating with tensors without discussing the properties that a mathematical quantity must have in order to qualify as a tensor. Using an inertial frame of reference that moves with a velocity that is negligible compared to the speed of light, $V << C$, we intuitively assumed that scalar quantities such as mass, volume, density, and the thermodynamic properties are independent of any coordinate system. The characteristic of a tensor, however is its invariance with respect to any coordinate system that may experience translation, rotation or both. In this chapter we present criteria that a tensor of any order has to fulfill to be classified as a tensor.

2.1 Transformation of a First Order Tensor

A first order tensor or a vector quantity is *invariant* with respect to a given category of coordinate systems. Changing the coordinate system by applying certain transformation rules, the vector components undergo certain changes resulting in a new set of *components* that are related, in a definite way, to the old ones. In the following, we investigate two coordinate system transformations. In the first case these two coordinate systems are embedded in a stationary frame of reference, where one coordinate system is rotated around a fixed axis at a given fixed angle. In the second case one coordinate system is embedded in a stationary frame, where the second one rotates around an arbitrary axis with a constant angular velocity ω.

M. T. Schobeiri, *Tensor Analysis for Engineers and Physicists - With Application to Continuum Mechanics, Turbulence, and Einstein's Special and General Theory of Relativity*, https://doi.org/10.1007/978-3-030-35736-8_2

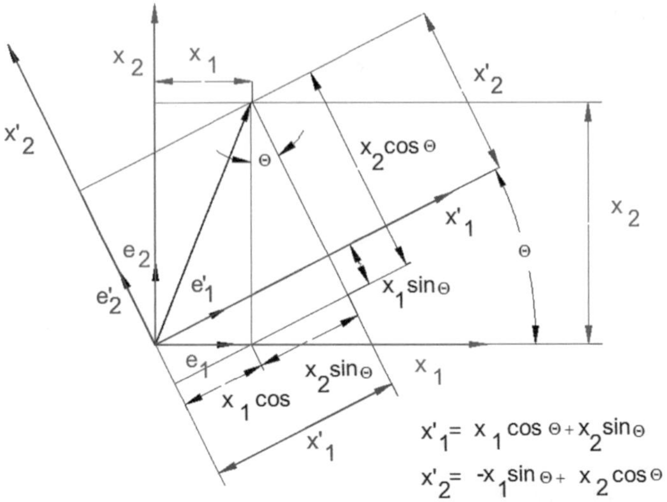

Fig. 2.1 Transformation of vector

2.1.1 *Coordinate Transformation in an Absolute Frame of Reference*

We start with a Cartesian coordinate system defined as $x_i = (x_1, x_2, x_3)$ with a set of unit vectors $e_i = (e_1, e_2, e_3)$. The new Cartesian coordinate system $x'_j = (x'_1, x'_2, x'_3)$ with a set of new unit vectors $e'_j = (e'_1, e'_2, e'_3)$ is rotated relative to x_i as shown in Fig. 2.1. The components of the vector A in the x_i-coordinate system, namely A_i will change, if the same vector is decomposed in its new components in the e'_i-coordinate system. Without loss of generality, we may assume that both coordinate systems have the same origin but have a rotation angle θ about the x_3-axis. The vector A can be written in x_i-coordinate system as:

$$\boldsymbol{A} = e_i A_i = e_1 A_1 + e_2 A_2 + e_3 A_3 \tag{2.1}$$

with the components A_i. The same vector decomposed in x'_i system is:

$$\boldsymbol{A} = e'_j A'_j = e'_1 A'_1 = e'_2 A'_2 + e'_3 A'_3 \tag{2.2}$$

with the components A'_i. In Eqs. (2.1) and (2.2) the unit vectors e_i and e'_j are associated with the coordinates x_i and x'_i. Both unit vectors are orthonormal and have the magnitude $|e_i| = |e'_j| = 1$. These two unit vectors are *linearly coupled* by the following relation:

$$e_i = Q_{ij} e'_j, \qquad e'_i = Q_{ji} e_j \tag{2.3}$$

with Q_{ij} as *transformation coefficients*. These coefficients are thought of as the elements of a transformation matrix $[\boldsymbol{Q}]$. Equation (2.3) expanded is written as;

$$\begin{aligned} e_i &= Q_{ij}e'_j \\ e_1 &= Q_{11}e'_1 + Q_{12}e'_2 + Q_{13}e'_3 \\ e_2 &= Q_{21}e'_1 + Q_{22}e'_2 + Q_{23}e'_3 \\ e_3 &= Q_{31}e'_1 + Q_{32}e'_2 + Q_{33}e'_3. \end{aligned} \tag{2.4}$$

Similarly the unit vectors e'_i can be expressed in terms of the unit vectors e_j of the original coordinate system.

$$\begin{aligned} e'_i &= Q_{ij}e_j \\ e'_1 &= Q_{11}e_1 + Q_{12}e_2 + Q_{13}e_3 \\ e'_2 &= Q_{21}e_1 + Q_{22}e_2 + Q_{23}e_3 \\ e'_3 &= Q_{31}e_1 + Q_{32}e_2 + Q_{33}e_3. \end{aligned} \tag{2.5}$$

The elements Q_{ij} of the transformation matrix $[\boldsymbol{Q}]$ are related to the angles between the old and the new coordinate system as shown below:

$$Q_{ij} = e_i \cdot e'_j = \cos(e_i, e'_j). \tag{2.6}$$

The coefficients in Eq. (2.6) are the direction cosines of the coordinate transformation, they are written as:

$$Q_{ij} = \begin{bmatrix} \cos(e_1, e'_1) & \cos(e_1, e'_2) & \cos(e_1, e'_3) \\ \cos(e_2, e'_1) & \cos(e_2, e'_2) & \cos(e_2, e'_3) \\ \cos(e_3, e'_1) & \cos(e_3, e'_2) & \cos(e_3, e'_3) \end{bmatrix}. \tag{2.7}$$

As an example, how to apply Eq. (2.6), we consider the two dimensional coordinate system shown in Fig. 2.1 with the rotation angle θ:

$$\begin{aligned} Q_{11} &= e_1 \cdot e'_1 \equiv \cos(e_1, e'_1) = \cos\theta \\ Q_{12} &= e_1 \cdot e'_2 \equiv \cos(e_1, e'_2) = \cos(\tfrac{\pi}{2} + \theta) = -\sin\theta \\ Q_{21} &= e_2 \cdot e'_1 \equiv \cos(e_2, e'_1) = \cos(\tfrac{\pi}{2} - \theta) = +\sin\theta \\ Q_{22} &= e_2 \cdot e'_2 \equiv \cos(e_2, e'_2) = \cos\theta. \end{aligned} \tag{2.8}$$

Using the above coefficient, we now calculate the components V'_1, V'_2, V'_3 in x'_i coordinate system. Given is the vector $\boldsymbol{V} = e_i V_i$ with its components V_1, V_2, V_3 in coordinate system x_i. The same vector has in x'_i-coordinates the components V'_1, V'_2, V'_3. Without loss of generality, we assume that the two coordinate systems have the same origin, but have a rotation angle θ about the x_3 as shown in Fig. 2.1. The components are related to each other via the following transformation equations:

$$
\begin{aligned}
V_1 &= \cos\theta V_1' - \sin\theta V'2 \\
V_2 &= \sin\theta V_1' + \cos\theta V'2 \\
V_1' &= \cos\theta V_1 + \sin\theta V_2 \\
V_2' &= -\sin\theta V_1 + \cos\theta V_2.
\end{aligned}
\tag{2.9}
$$

In matrix form, Eq. (2.9) can be written as:

$$
\begin{bmatrix} V_1' \\ V_2' \end{bmatrix} = \begin{bmatrix} \cos\theta & \sin\theta \\ -\sin\theta & \cos\theta \end{bmatrix} \begin{bmatrix} V_1 \\ V_2 \end{bmatrix} \mapsto \begin{bmatrix} V_1 \\ V_2 \end{bmatrix} = \begin{bmatrix} \cos\theta & -\sin\theta \\ \sin\theta & \cos\theta \end{bmatrix} \begin{bmatrix} V_1' \\ V_2' \end{bmatrix}. \tag{2.10}
$$

The same calculation procedure is applied to any three-dimensional vector such as $\boldsymbol{A} = e_i A_i$. In matrix form, the transformation is:

$$
\boldsymbol{A} = \begin{bmatrix} A_1 \\ A_2 \\ A_3 \end{bmatrix} = \begin{bmatrix} Q_{11} & Q_{12} & Q_{13} \\ Q_{21} & Q_{22} & Q_{23} \\ Q_{31} & Q_{32} & Q_{33} \end{bmatrix} \begin{bmatrix} A_1' \\ A_2' \\ A_3' \end{bmatrix}. \tag{2.11}
$$

Likewise we obtain the component of the Vector A in the new coordinate x_i' is expressed as:

$$
\boldsymbol{A}' = \begin{bmatrix} A_1' \\ A_2' \\ A_3' \end{bmatrix} = \begin{bmatrix} Q_{11} & Q_{12} & Q_{13} \\ Q_{12} & Q_{22} & Q_{32} \\ Q_{31} & Q_{32} & Q_{33} \end{bmatrix}^{-1} \begin{bmatrix} A_1 \\ A_2 \\ A_3 \end{bmatrix}. \tag{2.12}
$$

Equations (2.11) and (2.12) are written in symbolic form:

$$
\begin{aligned}
[A] &= [\boldsymbol{Q}][A'] \\
[A'] &= [\boldsymbol{Q}^T][A] \\
A_i &= \boldsymbol{Q}_{ij} A_j' \\
A_i' &= \boldsymbol{Q}_{jl} A_j.
\end{aligned}
\tag{2.13}
$$

with $[\boldsymbol{Q}^T]$ as the transpose of $[Q]$. From Eq. (2.13) follows that:

$$
[\boldsymbol{Q}][\boldsymbol{Q}^T] = [\boldsymbol{Q}^T][\boldsymbol{Q}] = [I] \tag{2.14}
$$

with

$$
[\boldsymbol{I}] = Q_{ij} Q_{kj} = \delta_{ik} \tag{2.15}
$$

as the unitary matrix. Equation (2.15) reveals an important property of the transformation matrix, namely its orthogonality. It should be noted that, one can generate any arbitrary matrix and its inversion and multiply the matrix with its inversion to arrive at a unitary matrix.

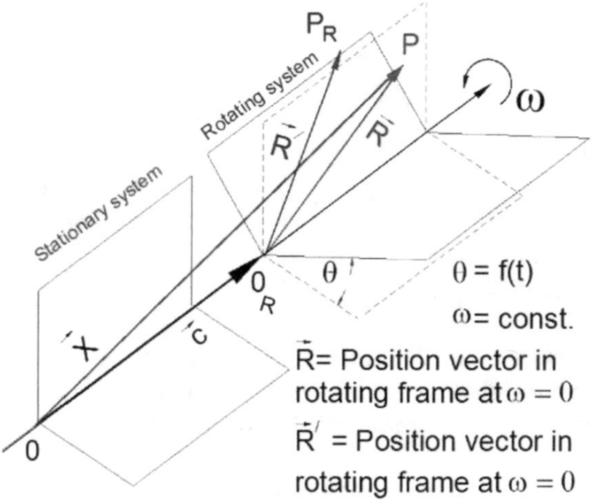

Fig. 2.2 Stationary and rotating frames

In the above equations we purposefully enclosed the transformation matrix Q within brackets to emphasize its matrix character. In general, when a matrix is associated with any vector, for example vector A in Eq. (2.13), it will change the vector components but not the vector itself. We may call this operation a passive transformation. In contrast, if a matrix is a part of a tensor such as ϕ_{ij} in the second order tensor $\boldsymbol{\Phi} = e_i e_j \phi_{ij}$ or the tensor itself is associated with another tensor, the result is a completely different tensor. Equations (1.25) to (1.28) showed this fact: the scalar multiplication of a second order tensor Φ with a first order tensor (vector) $\boldsymbol{C} = e_k c_k$ resulted in $\boldsymbol{\Phi} \cdot \boldsymbol{C} = \boldsymbol{D}$ which is another vector. In this case the tensor acts as an operator. In the following, for the sake of simplicity, we omit the brackets.

2.1.2 Transformation from an Absolute Frame into a Relative Frame

In Sect. 2.1.1 we assumed that the transformation angle θ is constant. In this section, however, we assume that entire coordinate system rotates with a time dependent rotation angle $\theta(t)$, Fig. 2.2.

Also, without loss of generality, we assume that the third axis x_3 is chosen as the axis of rotation around which the coordinate rotates with an angular velocity ω that may be constant or a function of time. This transformation has a wide range of engineering applications. We chose the transformation of conservation laws of motion from a stationary frame to a rotating one. As we will see, this transformation creates forces such as Coriolis force and centrifugal force that do not exist in a stationary frame. Starting with the transformation process, Fig. 2.2, we first modify the transformation matrix $\boldsymbol{Q}$, Eq. (2.8), and replace θ with $\theta(t) = \omega t$ as follows:

$$\boldsymbol{Q} = \begin{bmatrix} \cos\omega t & -\sin\omega t & 0 \\ \sin\omega t & \cos\omega t & 0 \\ 0 & 0 & 1 \end{bmatrix}. \tag{2.16}$$

Referring to Fig. 2.2, we consider a particle situated at the point $P = P(t = 0)$ with $\boldsymbol{Q}\boldsymbol{Q}^T = \boldsymbol{Q}^T\boldsymbol{Q}$, det $\boldsymbol{Q} = 1$ thus $\boldsymbol{Q}^{-1} = \boldsymbol{Q}^T$. An observer located in the stationary frame of reference at time $t = 0$ (frozen rotor) has the position vector $\boldsymbol{X}$ extending from the origin O to the point P. The corresponding position vector in a frozen rotating frame registered by the same observer is $\boldsymbol{R}$. At the time $t \neq 0$, because of rotation $\theta = \omega t$, the same particle moves to another location $P_R = P_R(t)$ described by the relative position vector $\boldsymbol{R}'$. Similar to Eq. (2.13) the components of this vector are related to the components of $\boldsymbol{R}$ by the following transformation matrix. For the reason explained above the matrix brackets are omitted:

$$\begin{aligned} \boldsymbol{R}' &= \boldsymbol{Q}\boldsymbol{R} \\ \boldsymbol{R} &= \boldsymbol{Q}^T\boldsymbol{R}'. \end{aligned} \tag{2.17}$$

In the course of transforming the conservation law of motion from a stationary to a rotating frame we need the first and second derivatives of $\boldsymbol{R}$ with respect to time:

$$\begin{aligned} \frac{d\boldsymbol{R}}{dt} &= \dot{\boldsymbol{R}} = \dot{\boldsymbol{Q}}^T\boldsymbol{R}' + \boldsymbol{Q}^T\dot{\boldsymbol{R}}' \\ \frac{d_2\boldsymbol{R}}{dt^2} &= \ddot{\boldsymbol{R}} = \ddot{\boldsymbol{Q}}^T\boldsymbol{R}' + 2\dot{\boldsymbol{Q}}^T\dot{\boldsymbol{R}}' + Q^T\ddot{\boldsymbol{R}}'. \end{aligned} \tag{2.18}$$

To perform the differentiation procedure in Eq. (2.18), we consider the first and the second derivative of Q with respect to time:

$$\begin{aligned} \dot{\boldsymbol{Q}} &= \frac{d\boldsymbol{Q}}{dt} = \omega \begin{bmatrix} -\sin\omega t & -\cos\omega t & 0 \\ \cos\omega t & -\sin\omega t & 0 \\ 0 & 0 & 0 \end{bmatrix} \\ \ddot{\boldsymbol{Q}} &= \frac{d\dot{\boldsymbol{Q}}}{dt} = \omega^2 \begin{bmatrix} -\cos\omega t & \sin\omega t & 0 \\ -\sin\omega t & -\cos\omega t & 0 \\ 0 & 0 & 0 \end{bmatrix}. \end{aligned} \tag{2.19}$$

Furthermore we differentiate the unitary matrix $\boldsymbol{Q}^T\boldsymbol{Q} = I$ which gives:

$$\dot{\boldsymbol{Q}}^T\boldsymbol{Q} + \boldsymbol{Q}^T\dot{\boldsymbol{Q}} = 0 \tag{2.20}$$

leading to:

$$\dot{\boldsymbol{Q}}^T\boldsymbol{Q} = -\boldsymbol{Q}^T\dot{\boldsymbol{Q}}. \tag{2.21}$$

With Eqs. (2.17), (2.19) and (2.21) we arrive at:

$$\boldsymbol{Q}^T\dot{\boldsymbol{Q}} = \omega \begin{bmatrix} \cos\omega t & \sin\omega t & 0 \\ -\sin\omega t & \cos\omega t & 0 \\ 0 & 0 & 1 \end{bmatrix} \begin{bmatrix} -\sin\omega t & -\cos\omega t & 0 \\ \cos\omega t & -\sin\omega t & 0 \\ 0 & 0 & 0 \end{bmatrix}$$
$$= \omega \begin{bmatrix} 0 & -1 & 0 \\ 1 & 0 & 0 \\ 0 & 0 & 0 \end{bmatrix}. \tag{2.22}$$

As seen from Eq. (2.22), the product $\boldsymbol{Q}^T\dot{\boldsymbol{Q}}$ is an antysymmetric matrix that can be thought of as the coefficient matrix of an antysymmetric second order tensor. Any second order tensor can be decomposed in a symmetric and an antysymmetric part. Consider an arbitrary second order tensor Ψ:

$$\begin{aligned} \boldsymbol{\Psi} &= \tfrac{1}{2}(\boldsymbol{\Psi} + \boldsymbol{\Psi}^T) + \tfrac{1}{2}(\boldsymbol{\Psi} - \boldsymbol{\Psi}^T) \\ \boldsymbol{\Psi} &= \boldsymbol{\Xi} + \boldsymbol{\Phi} \end{aligned} \tag{2.23}$$

with the first expression in the parentheses as the symmetric tensor renamed Ξ and the second expression as the antysymmetric one renamed Φ. The matrix of the antysymmetric tensor has generally three independent elements as seen below:

$$\Phi_{ij} = \begin{bmatrix} 0 & -\varphi_{12} & \varphi_{13} \\ \varphi_{21} & 0 & \varphi_{23} \\ -\varphi_{13} & -\varphi_{23} & 0 \end{bmatrix}. \tag{2.24}$$

In the case presented here with the rotation around only one axis, the matrix has only one independent element given below:

$$\Omega_{mn} = \boldsymbol{Q}^T\dot{\boldsymbol{Q}} = \begin{bmatrix} 0 & -\omega & 0 \\ \omega & 0 & 0 \\ 0 & 0 & 0 \end{bmatrix} = \omega \begin{bmatrix} 0 & -1 & 0 \\ 1 & 0 & 0 \\ 0 & 0 & 0 \end{bmatrix} \tag{2.25}$$

with ω as the element of the vector $\boldsymbol{\omega} = e_i\omega_i$ that is directly related to the rotation tensor $\boldsymbol{\Omega}$. Considering $\boldsymbol{\Omega} = e_m e_n \omega_{mn}$ as a second order antysymmetric tensor, we extract the axial vector (see also Chap. 4) using the product of the third order permutation tensor $\boldsymbol{\epsilon} = e_i e_j e_k \epsilon_{ijk}$ with the second order rotation tensor. Performing the contraction by using the Kroner delta and replacing indices we arrive at:

$$\boldsymbol{\epsilon} : \boldsymbol{\Omega} = \epsilon_{ijk} e_i e_J e_k : e_m e_n \Omega_{mn} = \epsilon_{ijk} e_i \Omega_{kj}. \tag{2.26}$$

Using Eq. (2.24) under consideration of Eq. (2.25) we find the rotation vector:

$$\boldsymbol{\epsilon} : \boldsymbol{\Omega} = 2e_3\omega. \tag{2.27}$$

The resulting vector Eq. (2.27) is a vector that is twice the axial vector. Thus the axial vector is:

$$\boldsymbol{\omega} = e_3 \omega. \tag{2.28}$$

The direction of the axial vector Eq. (2.28) is identical with the direction of a vector that results from a cross product of two vectors. As a result, whenever this vector is associated with another vector it acts as a cross product. As a result, any second order antisymmetric tensor with ω as its axial follows the relation: $\boldsymbol{\Omega} \cdot x = \boldsymbol{\omega} \times \boldsymbol{x}$.

With the above preparation, in the following we transform the second law of motion from a stationary frame into a rotating one. Without loss of generality, we assume that the origin of the absolute and the rotating frame O and O_R lie on the same axis, as shown in Fig. 2.2. The position vector of the particle in the absolute frame is:

$$\boldsymbol{X} = \boldsymbol{C} + \boldsymbol{R} \tag{2.29}$$

where the vector C is the distance vector from the origin O to the origin O_R. It may be a constant or function of time. In latter case the relative frame not only rotates, but it also undergoes a translation. The velocity and the accleration of a particle that has moved from point P to P_R in Fig. 2.2 are the first and second derivative of Eq. (2.29). Under consideration of Eqs. (2.17), (2.18) and (2.19) we have:

$$\begin{aligned}
\frac{d\boldsymbol{X}}{dt} &= \dot{\boldsymbol{C}} + \dot{\boldsymbol{Q}}^T \boldsymbol{R}' + \boldsymbol{Q}^T \dot{\boldsymbol{R}}' \\
\frac{d_2\boldsymbol{X}}{dt^2} &= \ddot{\boldsymbol{C}} + \ddot{\boldsymbol{Q}}^T \boldsymbol{R}' + 2\dot{\boldsymbol{Q}}^T \dot{\boldsymbol{R}}' + \boldsymbol{Q}^T \ddot{\boldsymbol{R}}'.
\end{aligned} \tag{2.30}$$

Inserting Eqs. (2.17) and (2.25) into Eq. (2.30), we arrive at:

$$\begin{aligned}
\frac{d\boldsymbol{X}}{dt} &= \dot{\boldsymbol{C}} + \Omega \boldsymbol{R} + \boldsymbol{Q}^T \dot{\boldsymbol{R}}' \\
\frac{d_2\boldsymbol{X}}{dt^2} &= \ddot{\boldsymbol{C}} + \dot{\Omega} \boldsymbol{R} + \Omega \dot{\boldsymbol{R}} + \dot{\boldsymbol{Q}}^T \dot{\boldsymbol{R}}' + \boldsymbol{Q}^T \ddot{\boldsymbol{R}}'.
\end{aligned} \tag{2.31}$$

Utilizing the identity $\boldsymbol{Q}^T \boldsymbol{Q} = \boldsymbol{Q}\boldsymbol{Q}^T = I$ and extending the term $\dot{\boldsymbol{Q}}^T \dot{\boldsymbol{R}}'$ in Eq. (2.31) as $\dot{\boldsymbol{Q}}^T \boldsymbol{Q}\boldsymbol{Q}^T \dot{\boldsymbol{R}}'$ we introduce the velocity vector $\boldsymbol{W} = \boldsymbol{Q}^T \dot{\boldsymbol{R}}'$ and obtain:

$$\begin{aligned}
\dot{\boldsymbol{Q}}^T \dot{\boldsymbol{R}}' &= \dot{\boldsymbol{Q}}^T \boldsymbol{Q}\boldsymbol{Q}^T \dot{\boldsymbol{R}}' \\
\dot{\boldsymbol{Q}}^T \dot{\boldsymbol{R}}' &= \boldsymbol{\Omega} W.
\end{aligned} \tag{2.32}$$

Inserting Eq. (2.32) into Eq. (2.31) and expressing $\boldsymbol{\Omega}\dot{\boldsymbol{R}}$ in terms of $\dot{\boldsymbol{R}}'$ and differentiating $\boldsymbol{R} = \boldsymbol{Q}^T \boldsymbol{R}'$ with respect to time, $\dot{\boldsymbol{R}} = \dot{\boldsymbol{Q}}^T \boldsymbol{R}' + \boldsymbol{Q}^T \dot{\boldsymbol{R}}'$, we obtain the relationship for the velocity and acceleration of a particle within a rotating frame of reference:

$$\frac{dX}{dt} = \dot{X} = \dot{C} + \boldsymbol{\Omega} \boldsymbol{R} + \boldsymbol{W}$$
$$\frac{d_2X}{dt^2} = \ddot{X} = \ddot{C} + \dot{\boldsymbol{\Omega}} \boldsymbol{R} + \boldsymbol{\Omega}(\boldsymbol{\Omega} \boldsymbol{R}) + 2(\boldsymbol{\Omega} \boldsymbol{W}) + \boldsymbol{Q}^T \ddot{\boldsymbol{R}}'. \tag{2.33}$$

Considering the relation $\boldsymbol{\Omega} \cdot x = \boldsymbol{\omega} \times \boldsymbol{x}$ showed previously, we insert now the rotation vector ω from Eq. (2.28) in Eq. (2.33) and consider the fact that, in context of this derivation, where ever ω appears, it acts as a cross product, thus we find:

$$\frac{dX}{dt} = \dot{X} = \dot{C} + \boldsymbol{\omega} \times \boldsymbol{R} + \boldsymbol{W}$$
$$\frac{d_2X}{dt^2} = \ddot{X} = \ddot{C} + \dot{\boldsymbol{\omega}} \times \boldsymbol{R} + \boldsymbol{\omega} \times (\boldsymbol{\omega} \times \boldsymbol{R}) + 2(\boldsymbol{\omega} \times \boldsymbol{W}) + \boldsymbol{A_R} \tag{2.34}$$

with $\boldsymbol{A_R} = \boldsymbol{Q}^T \ddot{\boldsymbol{R}}'$ as the acceleration. Equations in (2.34) exhibit the transformation of velocity and the acceleration vectors from a stationary frame into a rotating one. The C-terms describe the movement of the relative frame with respect to stationary one. The term $\boldsymbol{\omega} \times \boldsymbol{R}$ is the rotational velocity of the rotating frame; the term $\dot{\boldsymbol{\omega}} \times \boldsymbol{R}$ representing the acceleration of the rotating frame accounts for the changes of $\boldsymbol{\omega}$ with time. In the acceleration equation, two new terms appear, the centrifugal acceleration $\boldsymbol{\omega} \times \boldsymbol{\omega} \times \boldsymbol{R}$ and the Coriolis acceleration $2\boldsymbol{\omega} \times \boldsymbol{W}$. These two terms do not exist in a stationary frame and are the result of the transformation.

2.2 Transformation of a Second and Higher Order Tensor

A second order tensor can be thought of as a tensor product of two first order tensors:

$$\boldsymbol{\Psi} = \boldsymbol{AB} = e_i A_i e_j B_j = e_i e_j A_i B_j. \tag{2.35}$$

The components of the second order tensor in Eq. (2.35) are expressed in coordinate system $x_i = (x_1, x_2, x_3)$ with the unit vectors $e_i = (e_1, e_2, e_3)$. The matrix of the tensor Eq. (2.35) namely $A_i B_j$ can be renamed as ψ_{ij} such that we can re-write Eq. (2.35) as

$$\boldsymbol{\Psi} = \boldsymbol{AB} = e_i A_i e_j B_j = e_i e_j A_i B_j = e_i e_j \psi_{ij}. \tag{2.36}$$

Moving into a new coordinate system $x'_j = (x'_1, x'_2, x'_3)$ with a set of unit vectors $e'_i = (e'_1, e'_2, e'_3)$, Eq. (2.36) is written as:

$$\boldsymbol{\Psi} = e_i e_j \psi_{ij} = e'_p e'_q \psi'_{pq}. \tag{2.37}$$

Inserting Eqs. (2.4) and (2.5) into Eq. (2.37), we have:

$$\boldsymbol{\Psi} = e_i e_j \psi_{ij} = e_m e_n Q_{mp} Q_{nq} \psi'_{pq} \tag{2.38}$$

with $e_i e_j$, $e_m e_n$ as tensor products of two unit vectors. Since in Eq. (2.38) m and n are dummy indices they can be renamed to any letters with exception of p and q, therefore we may set $m = i$ and $n = j$. Thus Eq. (2.38) is written as:

$$\boldsymbol{\Psi} = e_i e_j \psi_{ij} = e_i e_j Q_{ip} Q_{jq} \psi'_{pq} \tag{2.39}$$

thus the components of the second order tensor ψ_{ij} in $x_i = x_1, x_2, x_3$-coordinate is expressed in $x'_j = x'_1, x'_2, x'_3$-coordinate by:

$$\psi_{ij} = Q_{ip} Q_{jq} \psi'_{pq}. \tag{2.40}$$

Accordingly, the components of the transformed second order tensor ψ'_{ij} can be expressed as

$$\psi'_{ij} = Q_{pi} Q_{qj} \psi_{pq}. \tag{2.41}$$

Equations (2.40) and (2.41) can be written in symbolic form:

$$[\boldsymbol{\Psi}] = [\boldsymbol{Q}][\boldsymbol{\Psi}'][\boldsymbol{Q}]^T$$
$$[\boldsymbol{\Psi}'] = [\boldsymbol{Q}]^T[\boldsymbol{\Psi}][\boldsymbol{Q}]. \tag{2.42}$$

The same transformation rules can be applied to a third or higher order tensor. Applying the transformation rule to a general third order tensor $\boldsymbol{T}$ with components in $x_i = x_1, x_2, x_3$-coordinates, its components in $x'_j = x'_1, x'_2, x'_3$ are $T'_{ijk} = Q_{mi} Q_{nj} Q_{lk} T_{mnl}$. In this context, it is of interest to transform the permutation tensor ϵ from one coordinate to another, which is rotated. Its transformation is $\epsilon'_{ijk} Q_{mi} Q_{nj} Q_{lk} \epsilon mnl$ with $Q_{mi} Q_{nj} Q_{lk} = \det(Q) = 1$ that leads to $\epsilon'_{ijk} = \epsilon_{ijk}$. This means that the components of the permutation tensor remain the same in all rotated coordinates. These type of tensors are called isotropic tensors. The Kronecker delta, which is a second order tensor is also isotropic, i.e. $\delta'_{ij} = \delta_{ij}$.

2.3 Eigenvalue and Eigenvector of a Second Order Tensor

Any second order tensor, for example stress tensor Π can be multiplied with any arbitrary first order tensor in a scalar manner. The result will be a first order tensor (or vector) which will have an arbitrary direction. However, there exists a particular vector $\boldsymbol{V}$ such that its scalar multiplication with Π results in a vector, which is parallel to $\boldsymbol{V}$ but has different magnitude:

$$\boldsymbol{\Pi} \cdot \boldsymbol{V} = \lambda \boldsymbol{V} \tag{2.43}$$

with $\boldsymbol{V}$ as the eigenvector and λ the eigenvalue of the second order tensor $\boldsymbol{\Pi}$. Since any vector can be expressed as a scalar product of the unit tensor and the vector itself

$\boldsymbol{I} \cdot \boldsymbol{V} = \boldsymbol{V}$, we may write:

$$\boldsymbol{\Pi} \cdot \boldsymbol{V} = \lambda \boldsymbol{V} = \lambda(\boldsymbol{I} \cdot \boldsymbol{V}) \tag{2.44}$$

that can be rearranged as:

$$(\boldsymbol{\Pi} - \lambda \boldsymbol{I}) \cdot \boldsymbol{V} = 0 \tag{2.45}$$

The index notation gives:

$$e_i(\pi_{ik} V_k - \lambda V_i) = 0 \tag{2.46}$$

or

$$\pi_{ij} V_j = \lambda V_i. \tag{2.47}$$

Expanding Eq. (2.47) gives a system of linear equations,

$$\begin{aligned} \pi_{11} V_1 + \pi_{12} V_2 + \pi_{13} V_3 &= \lambda V_1 \\ \pi_{21} V_1 + \pi_{22} V_2 + \pi_{23} V_3 &= \lambda V_2 \\ \pi_{31} V_1 + \pi_{32} V_2 + \pi_{33} V_3 &= \lambda V_3. \end{aligned} \tag{2.48}$$

A nontrivial solution of Eq. (2.48) is possible if and only if the following determinant vanishes:

$$\det(\boldsymbol{\Pi} - \lambda \boldsymbol{I}) = 0 \tag{2.49}$$

or in index notation:

$$\det(\pi_{ij} - \delta_{ij}\lambda) = 0. \tag{2.50}$$

Expanding Eq. (2.50) results in

$$\det \begin{pmatrix} \pi_{11} - \lambda & \pi_{12} & \pi_{13} \\ \pi_{21} & \pi_{22} - \lambda & \pi_{23} \\ \pi_{31} & \pi_{32} & \pi_{33} - \lambda \end{pmatrix} = 0. \tag{2.51}$$

After expanding the above determinant, we obtain an algebraic equation in λ in the following form:

$$\lambda^3 - I_{1D}\lambda^2 + I_{2D}\lambda - I_{3D} = 0 \tag{2.52}$$

where I_{1D}, I_{2D} and I_{3D} are invariants of the tensor $\boldsymbol{\Pi}$ defined as:

$$\begin{aligned} I_{1D} &= \operatorname{Tr} \boldsymbol{\Pi} = \pi_{ii} = \pi_{11} + \pi_{22} + \pi_{33} \\ I_{2D} &= \tfrac{1}{2}(I_{1D}^2 - \boldsymbol{\Pi} : \boldsymbol{\Pi}) = \tfrac{1}{2}(\pi_{ii}\pi_{ij} - \pi_{ij}\pi_{ij}) \\ I_{3D} &= \det(\pi_{ij}). \end{aligned} \tag{2.53}$$

Problems 1: The angles between the axes in two coordinate systems are given in the table below.

	x_1	x_2	x_3
x'_1	135°	60°	120°
x'_2	90°	45°	45°
x'_3	45°	60°	120°

Construct the corresponding transformation matrix $[Q]$, find the inverse and verify that it is orthogonal.

Problem 2: Given the transformation matrix

$$Q_{ij} = \begin{bmatrix} 0.866 & 0.5 & -0.5 \\ 0.0 & -0.707 & 0.707 \\ 0.5 & 0.5 & -0.5 \end{bmatrix}. \tag{2.54}$$

Find the angles and the inverse of the matrix.

Problem 3: Given the following tensor invariants, expand the invariant I_{1D}, I_{2D}, I_{3D} and

$$\begin{aligned} I_{1D} &= \text{Tr}\,\boldsymbol{\Pi} \\ I_{2D} &= \tfrac{1}{2}(I_{1D}^2 - \boldsymbol{\Pi} : \boldsymbol{\Pi}) \\ I_{3D} &= \det(\pi_{ij}). \end{aligned} \tag{2.55}$$

Chapter 3
Differential Operators in Continuum Mechanics

In continuum mechanics which is embedded in Newtonian physics, the particles of the working medium in general undergo a time dependent or unsteady motion. The flow quantities such as the velocity $\boldsymbol{V}$ and the thermodynamic properties of the working medium such as pressure p, temperature T, density ρ or any arbitrary flow quantity Q are generally functions of space and time:

$$\boldsymbol{V} = \boldsymbol{V}(\boldsymbol{x}, t), \; p = p(\boldsymbol{x}, t), \; T = T(\boldsymbol{x}, t), \; \rho = \rho(\boldsymbol{x}, t).$$

During the flow process, these quantities may change with respect to time and space. In this context and at this juncture we do not consider the time and space as one unit as Einstein formulated in his relativistic mechanics (four dimensional time space) but as two separate entities as Newton assumed, namely absolute space and absolute time. This assumption is acceptable as long as the velocity compared to the speed of light is negligibly small.

In engineering sciences there are a number of operators, spatial, temporal, and other differential operators are a few examples. The function of an operator is to act on the argument that follows the operator. The argument may be any physical quantity or any arbitrary function. The result of this operation is a function which has a different character as the original argument. Two simple example should clarify this statement: Consider a zeroth order tensor, a scalar quantity, such as density, specific volume, entropy etc. Taking the *gradient* of such scalar fields results in first order tensor or vector fields. This operation raised the order of the tensor. Another example: Given a second order tensor such as the total stress tensor. If we take the *divergence* of this stress field, we get the force components and the resulting force vector acting on a differential volume. Gradient, divergence, curls and others will be treated in this chapter. In the following, we start with operators that account for the *substantial, spatial,* and *temporal* changes of the quantities of a continuum.

M. T. Schobeiri, *Tensor Analysis for Engineers and Physicists - With Application to Continuum Mechanics, Turbulence, and Einstein's Special and General Theory of Relativity*, https://doi.org/10.1007/978-3-030-35736-8_3

3.1 Substantial Derivatives

The *temporal* and *spatial change* of the quantities mentioned above is described most appropriately by the *substantial* or *material derivative*. Generally, the substantial derivative of a flow quantity Q, which may be a scalar, a vector or generally a tensor valued function, is given by:

$$DQ = \frac{\partial Q}{\partial t} dt + dQ. \tag{3.1}$$

The operator D represents the *substantial* or *material* change of the quantity Q, the first term on the right hand side of Eq. (3.1) represents the *local* or *temporal change* of the quantity Q with respect to a fixed position vector $\boldsymbol{x}$. The operator d symbolizes the *spatial* or *convective change* of the same quantity with respect to a fixed instant of time. The convective change of Q may be expressed as:

$$dQ = \frac{\partial Q}{\partial x_1} dx_1 + \frac{\partial Q}{\partial x_2} dx_2 + \frac{\partial Q}{\partial x_3} dx_3. \tag{3.2}$$

A simple rearrangement of the above equation results in:

$$dQ = (e_1 dx_1 + e_2 dx_2 + e_3 dx_3) \cdot \left(e_1 \frac{\partial}{\partial x_1} + e_2 \frac{\partial}{\partial x_2} + e_3 \frac{\partial}{\partial x_3} \right) Q. \tag{3.3}$$

Scalar multiplication of the expressions in the two parentheses of Eq. (3.3) results in Eq. (3.2).

3.1.1 *Differential Operator* ∇

The expression in the second parenthesis of Eq. (3.3) is the *spatial differential operator* ∇ *(Nabla)* which has a vector character. In Cartesian coordinate system, the operator ∇ is defined as:

$$\nabla = \left(e_1 \frac{\partial}{\partial x_1} + e_2 \frac{\partial}{\partial x_2} + e_3 \frac{\partial}{\partial x_3} \right) = e_i \frac{\partial}{\partial x_i}. \tag{3.4}$$

The above operator acts on the argument that can be a scalar, a vector, a tensor or generally a quantity Q of any nature. Using the above differential operator, the change of the quantity Q is written as:

$$dQ = d\boldsymbol{x} \cdot \nabla Q. \tag{3.5}$$

The increment dQ of Eq. (3.5) is obtained either by applying the product $d\boldsymbol{x} \cdot \nabla$, or by taking the dot product of the vector $d\boldsymbol{x}$ and ∇Q. If Q is a scalar quantity,

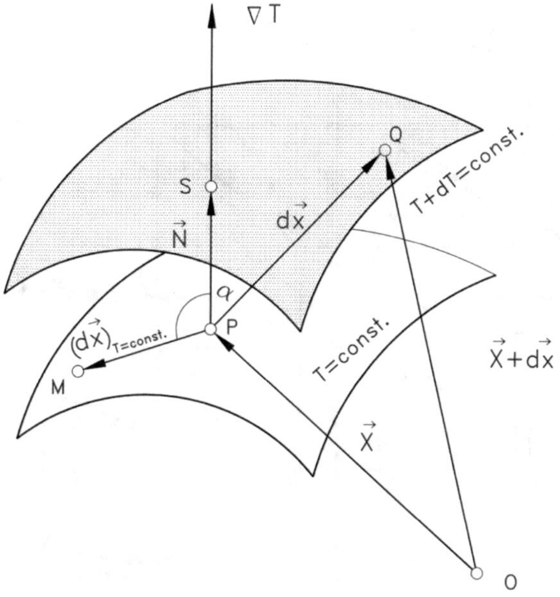

Fig. 3.1 Physical explanation of the gradient of scalar field

then ∇Q is a vector or a *first order tensor* with definite components. In this case, ∇Q is called the *gradient* of the scalar field. Equation (3.5) indicates that the spatial change of the quantity Q assumes a maximum if the vector ∇Q (*gradient* of Q) is parallel to the vector dx. If the vector ∇Q is perpendicular to the vector dx, their product will be zero. This is only possible, if the spatial change dx occurs on a surface with $Q =$*const.* Consequently, the quantity Q does not experience any changes. The physical interpretation of this statement is found in Fig. 3.1. The scalar field is represented by the point function temperature that changes from the surfaces T to the surface $T + dT$. In Fig. 3.1, the gradient of the temperature field is shown as ∇T, which is perpendicular to the surface $T =$*const.* at point P. The temperature probe located at P moves on the surface $T =$*const.* to the point M, thus measuring no changes in temperature ($\alpha = \pi/2,\ \cos\alpha = 0$). However, the same probe experiences a certain change in temperature by moving to the point Q, which is characterized by a higher temperature $T + dT (0 < \alpha < \pi/2)$. The change dT can immediately be measured, if the probe is moved parallel to the vector ∇T. In this case, the displacement dx (see Fig. 3.2) is the shortest ($\alpha = 0,\ \cos\alpha = 1$).

Performing the similar operation for a vector quantity as seen in Eq. (3.2) yields:

$$d\boldsymbol{V} = \frac{\partial \boldsymbol{V}}{\partial x_1}dx_1 + \frac{\partial \boldsymbol{V}}{\partial x_2}dx_2 + \frac{\partial \boldsymbol{V}}{\partial x_3}dx_3. \tag{3.6}$$

The right-hand side of Eq. (3.6) is identical with:

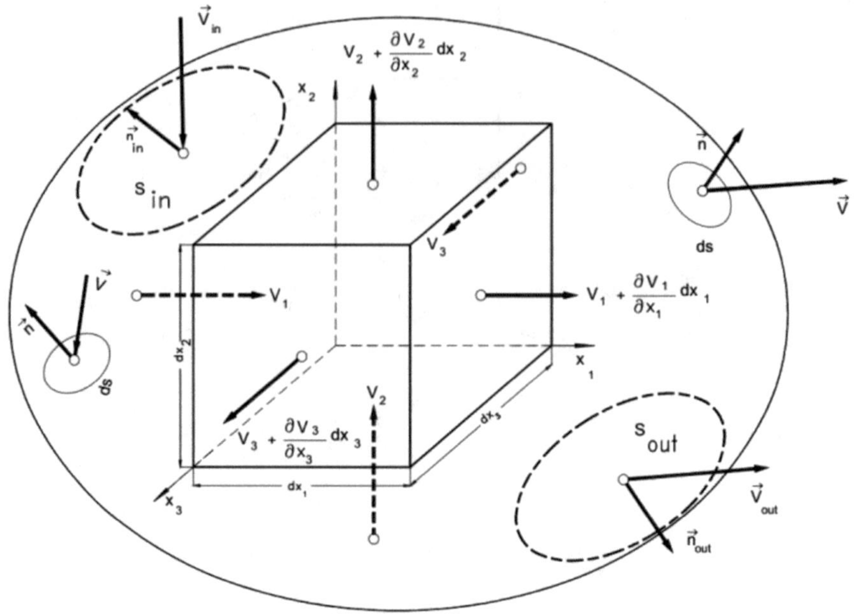

Fig. 3.2 Physical interpretation of $\nabla \cdot \boldsymbol{V}$

$$d\boldsymbol{V} = (d\boldsymbol{x} \cdot \nabla)\boldsymbol{V}. \tag{3.7}$$

In Eq. (3.7) the product $d\boldsymbol{x} \cdot \nabla$ can be considered as an operator that is applied to the vector $\boldsymbol{V}$ resulting in an increment of the velocity vector. Performing the scalar multiplication between $d\boldsymbol{x}$ and ∇ gives:

$$d\boldsymbol{V} = (d\boldsymbol{x} \cdot \nabla)\boldsymbol{V} = dx_i \frac{\partial \boldsymbol{V}}{\partial x_i} = e_j dx_i \frac{\partial V_j}{\partial x_i} = d\boldsymbol{x} \cdot \nabla\boldsymbol{V} \tag{3.8}$$

with $\nabla\boldsymbol{V}$ as the gradient of the vector field which is a second order tensor. To perform the differential operation, first the ∇ operator is applied to the vector $\boldsymbol{V}$, resulting in a second order tensor. This tensor is then multiplied with the vector dx in a scalar manner that results in a first order tensor or a vector. From this operation, it follows that spatial change of the velocity component can be expressed as the scalar product of the vector $d\boldsymbol{x}$ and the second order tensor $\nabla\boldsymbol{V}$, which represents the spatial gradient of the velocity vector. Using the spatial derivative from Eq. (3.8), the substantial change of the velocity is obtained by:

$$D\boldsymbol{V} = \frac{\partial \boldsymbol{V}}{\partial t} dt = d\boldsymbol{V} \tag{3.9}$$

where the spatial change of the velocity is expressed as:

$$d\boldsymbol{V} = d\boldsymbol{x} \cdot \nabla \boldsymbol{V}. \tag{3.10}$$

Dividing Eq. (3.10) by dt yields the *convective* part of the acceleration vector:

$$\frac{d\boldsymbol{V}}{dt} = \left(\frac{d\boldsymbol{x}}{dt}\right) \cdot (\nabla \boldsymbol{V}) = \boldsymbol{V} \cdot \nabla \boldsymbol{V}. \tag{3.11}$$

The substantial acceleration is then:

$$\frac{d\boldsymbol{V}}{dt} = \frac{D\boldsymbol{V}}{Dt} = \frac{\partial \boldsymbol{V}}{\partial t} + \boldsymbol{V} \cdot \nabla \boldsymbol{V}. \tag{3.12}$$

The differential dt may symbolically be replaced by Dt indicating the material character of the derivatives. Applying the index notation to the velocity vector and the Nabla operator, performing the vector operation, and using the Kronecker delta, the index notation of the *material acceleration* A is:

$$\boldsymbol{A} = e_i A_i = e_i \frac{\partial V_i}{\partial t} + e_i V_j \frac{\partial V_i}{\partial x_j}. \tag{3.13}$$

Equation (3.13) is valid only for Cartesian coordinate system, where the unit vectors do not depend upon the coordinates and are constant. Thus, their derivatives with respect to the coordinates disappear identically. To arrive at Eq. (3.13) with a unified index i, we renamed the dummy indices. To decompose the above acceleration vector into three components, we cancel the unit vector from both side in Eq. (3.13) and get:

$$A_i = \frac{\partial V_i}{\partial t} + V_j \frac{\partial V_i}{\partial x_j}. \tag{3.14}$$

To find the components in x_i-direction, the index i assumes subsequently the values from 1 to 3, while the summation convention is applied to the dummy index j. As a result we obtain the three components:

$$\begin{aligned}
A_1 &= \frac{\partial V_1}{\partial t} + V_1 \frac{\partial V_1}{\partial x_1} + V_2 \frac{\partial V_1}{\partial x_2} + V_3 \frac{\partial V_1}{\partial x_3} \\
A_2 &= \frac{\partial V_2}{\partial t} + V_1 \frac{\partial V_2}{\partial x_1} + V_2 \frac{\partial V_2}{\partial x_2} + V_3 \frac{\partial V_2}{\partial x_3} \\
A_3 &= \frac{\partial V_3}{\partial t} + V_1 \frac{\partial V_3}{\partial x_1} + V_2 \frac{\partial V_3}{\partial x_2} + V_3 \frac{\partial V_3}{\partial x_3}.
\end{aligned} \tag{3.15}$$

3.1.2 Transformation of Nabla Operator

In the course of changing the coordinate systems, we encounter expressions, where ∇-operator acts on tensors of zero, first, second or higher order. In these cases, ∇ itself must also be transformed from one coordinate system into another. In the following, the step by step instruction is given, how to perform the transformation. To make easier to understand the steps, we use coordinate systems that are embedded in Euclidean flat space as described in Chap. 2. Taking the equation from Eq. (3.4) and redefine it for a primed coordinate, we have:

$$\begin{aligned}
&\nabla = e_i \frac{\partial}{\partial x_i} && \text{(a)}\\
&\nabla' = e'_i \frac{\partial}{\partial x'_i} = e'_i \frac{\partial}{\partial x_j}\frac{\partial x_j}{\partial x'_i} && \text{(b) used chain rule of differentiation}\\
&\frac{\partial x_j}{\partial x'_i} = Q_{kj}\frac{\partial x'_k}{\partial x'_i} = Q_{kj}\delta_{ki} = Q_{ij} && \text{(c) the last ratio in (b) expanded}\\
&\nabla' = \left(e'_i \frac{\partial}{\partial x_j}\right) Q_{ij} = \left(e'_i Q_{ij}\frac{\partial}{\partial x_j}\right) && \text{(d) the result in (c) inserted in (b).}
\end{aligned} \tag{3.16}$$

Now we want to apply the result of Eq. (3.16) to a vector $\boldsymbol{V}$ which results in a scalar called in this case the divergence of the vector $\boldsymbol{V}$. As seen in Chap. 2, the vector is transformed as $V'_i = Q_{mi}V_m = Q_{ji}V_j$.

$$\nabla' \cdot \boldsymbol{V}' = \left(e'_i Q_{ij}\frac{\partial}{\partial x_J}\right) \cdot e'_k V'_k \quad \text{(a)}$$

$$\nabla' \cdot \boldsymbol{V}' = \delta_{ik} Q_{ij}\frac{\partial V'_k}{\partial x_j} = Q_{ij}\frac{\partial V'_i}{\partial x_j} \quad \text{(b)}$$

$$\nabla' \cdot \boldsymbol{V}' = Q_{ij}\frac{\partial Q_{ik}V_k}{\partial x_j} = Q_{ij}Q_{ik}\frac{\partial V_k}{\partial x_j} \quad \text{(c)}$$

$$\nabla' \cdot \boldsymbol{V}' = \delta_{kj}\frac{\partial V_k}{\partial x_j} = \frac{\partial V_j}{\partial x_j} = \nabla \cdot \boldsymbol{V} \quad \text{(d).} \tag{3.17}$$

Equation (3.17) shows that with coordinate transformation, the scalar has not changed its character.

3.1.3 Transformation of Gradient of a Scalar Function

As is well known from Chap. 1, a scalar quantity φ has a definite magnitude but not a definite direction. Obtaining the spatial changes of φ, namely $\nabla\varphi$ modified a scalar quantity into a vector. Since a vector is invariant with respect to coordinate transformation, the gradient $\nabla\varphi$ must also be invariant with respect to the coordinate transformation. To proof this statement, we look at the scalar quantity φ in coordinates x_i, x'_i

$$\varphi(x_i) = \varphi(x_1, x_2, x_3) = \varphi'(x'_i) = \varphi'(x'_1, x'_1, x'_1). \tag{3.18}$$

Since a vector is coordinate invariant, the following spatial differentiation of φ must also be invariant with respect to coordinate transformation.

$$\nabla'\varphi' = \nabla\varphi, \text{ with } \nabla' = e'_i\frac{\partial}{\partial x'_i}, \text{ and } \nabla = e_i\frac{\partial}{\partial x_i}. \tag{3.19}$$

Disregarding the unit vectors and focusing on the differentials:

$$\frac{\partial}{\partial x'_i}, \text{ and } \frac{\partial}{\partial x_i}, \tag{3.20}$$

it follows that

$$\frac{\partial\varphi'}{\partial x'_i} = \frac{\partial\varphi}{\partial x_j}\frac{\partial x_j}{\partial x'_i}. \tag{3.21}$$

The relation of the two coordinates are $x_j = Q_{kj}x'_k$ and their derivatives are determined by:

$$\frac{\partial x_j}{\partial x'_i} = Q_{kj}\frac{\partial x'_k}{\partial x'_i} = Q_{kj}\delta_{ki} = Q_{ij}. \tag{3.22}$$

Inserting Eq. (3.22) in to Eq. (3.21) we arrive at:

$$\frac{\partial\varphi'}{\partial x'_i} = \frac{\partial\varphi}{\partial x_j}Q_{ij}. \tag{3.23}$$

Equation (3.23) presents the proof that the gradient of a scalar quantity follows the transformation rule of vectors.

3.1.4 Laplace Operator Δ

The Laplace operator is defined as the scalar product of two Nabla operators leading to a scalar operator:

$$\Delta = \nabla\cdot\nabla = \left(e_i\frac{\partial}{\partial x_i}\right)\cdot\left(e_j\frac{\partial}{\partial x_j}\right) = \frac{\partial^2}{\partial x_j\partial x_j}. \tag{3.24}$$

It is a second order differential operator in Euclidean space (flat space) that acts on the argument that immediately follows it. The argument might be any tensor valued function that is twice differentiable. As an example the shear stress term in Navier–Stokes equation of motion is given by:

$$\Delta \boldsymbol{V} = \frac{\partial^2}{\partial x_i \partial x_i}(e_j V_j) = e_j \frac{\partial^2 V_j}{\partial x_i \partial x_i}. \tag{3.25}$$

If the argument is the gradient of a zeroth order tensor, first or higher order tensor $\boldsymbol{T}$, it is more appropriate to use the product $\nabla \cdot \nabla$ rather that Δ as shown in $\nabla \cdot (\nabla \boldsymbol{T})$.

3.2 Operator ∇ Applied to Different Functions

This section summarizes the applications of Nabla operator to different functions. As mentioned previously, the spatial differential operator ∇ has a vector character. If it acts on a scalar function, such as temperature, pressure, enthalpy etc., the result is a vector and is called the *gradient* of the corresponding scalar field, such as gradient of temperature, pressure, etc. (see also previous discussion of the physical interpretation of ∇Q). If, on the other hand, ∇ acts on a vector, three different cases are distinguished.

3.2.1 Scalar Product of ∇ and a Vector

This operation is called the *divergence of the vector* $\boldsymbol{V}$. The result is a zerothn t-order tensor or a scalar quantity. Using the index notation, the divergence of $\boldsymbol{V}$ is written as:

$$\nabla \cdot \boldsymbol{V} = e_i \frac{\partial}{\partial x_i} \cdot (e_j V_j) = \delta_{ij} \frac{\partial}{\partial x_i} V_j = \frac{\partial V_i}{\partial x_i}. \tag{3.26}$$

The physical interpretation of this purely mathematical operation is shown in Fig. 3.2. The mass flow balance for a steady incompressible flow through an infinitesimal volume $dv = dx_1 dx_2 dx_3$ is shown in Fig. 3.2. We first establish the entering and exiting mass flows through the cube side areas perpendicular to x_1-direction given by $dA_1 = dx_2 dx_3$:

$$\dot{m}_{x_{1en}} = \rho(dx_2 dx_3) V_1 \tag{3.27}$$

$$\dot{m}_{x_{1ex}} = \rho(dx_2 dx_3)\left(V_1 + \frac{\partial V_1}{\partial x_1} dx_1\right). \tag{3.28}$$

Repeating the same procedure for the cube side areas perpendicular to x_2 and x_3 directions given by $dA_2 = dx_3 dx_1$ and $dA_3 = dx_1 dx_2$ and subtracting the entering mass flows from the exiting ones, we obtain the net mass flow balances through the infinitesimal differential volume as:

$$\rho(dx_1dx_2dx_3)\left(\frac{\partial V_1}{\partial x_1}+\frac{\partial V_2}{\partial x_2}+\frac{\partial V_3}{\partial x_3}\right)=\rho dv(\nabla\cdot\boldsymbol{V})=0. \tag{3.29}$$

The right hand side of Eq. (3.29) is a product of three terms, the density ρ, the differential volume dv and the divergence of the vector $\boldsymbol{V}$. Since the first two terms are not zero, the divergence of the vector must disappear. As result, we find:

$$\nabla\cdot\boldsymbol{V}=0. \tag{3.30}$$

Equation (3.30) expresses the continuity equation for an incompressible flow, as we will see in the following chapters.

3.2.2 Transformation of Divergence of a Vector Function

The result of the scalar operation in Eq. (3.30) is obviously a scalar quantity, which is invariant with respect to any coordinate transformation. In this context, it is of interest to find out whether the transformation would confirm the claim. Using Eq. (3.30), the transformation must deliver;

$$\nabla'\cdot\boldsymbol{V}'=\nabla\cdot\boldsymbol{V}. \tag{3.31}$$

Expanding Eq. (3.31),

$$\nabla'\boldsymbol{V}'=\left(\frac{\partial V_i}{\partial x_i}\right)'=Q_{ij}\frac{\partial}{\partial x_i}(Q_{ik}V_k)=Q_{ij}Q_{ik}\left(\frac{\partial V_k}{\partial x_j}\right)=\left(\frac{\partial V_k}{\partial x_k}\right)=\nabla\cdot\boldsymbol{V}. \tag{3.32}$$

Equation (3.32) is identical with (3.17) derived previously.

3.2.3 Vector Product ∇ x V

This operation is called the *rotation or curl* of the velocity vector $\boldsymbol{V}$. Its result is a first-order tensor or a vector quantity. Using the index notation, the curl of V is written as:

$$\nabla\times\boldsymbol{V}=\left(e_i\frac{\partial}{\partial x_i}\right)\times(e_jV_j)=\epsilon_{ijk}e_k\frac{\partial V_j}{\partial x_i}=\boldsymbol{\omega}. \tag{3.33}$$

The curl of the velocity vector is known as *vorticity*, $\boldsymbol{\omega}=\nabla\times\boldsymbol{V}$. As we will see later, the vorticity plays a crucial role in fluid mechanics. It is a characteristic of a *rotational* flow. For viscous flows encountered in engineering applications, the curl $\boldsymbol{\omega}=\nabla\times\boldsymbol{V}$ is always different from zero. To simplify the flow situation and to solve the equation of motion, as we will discuss later, the vorticity vector $\omega = \nabla \times \boldsymbol{V}$, can under certain conditions, be set equal to zero. This special case is called *the irrotational flow*.

3.2.4 Tensor Product of ∇ and V

This operation is called the *gradient* of the velocity vector $\boldsymbol{V}$. Its result is a second tensor. Using the index notation, the gradient of the vector $\boldsymbol{V}$ is written as:

$$\nabla \boldsymbol{V} = \left(e_i \frac{\partial}{\partial x_i}\right)(e_j V_j) = e_i e_j \frac{\partial V_j}{\partial x_i}. \tag{3.34}$$

Equation (3.34) is a second order tensor with nine components and describes the deformation and the rotation kinematics of the fluid particle. As we saw previously, the scalar multiplication of this tensor with the velocity vector, $\boldsymbol{V} \cdot \nabla \boldsymbol{V}$ resulted in the convective part of the acceleration vector, Eq. (3.13). In addition to the applications we discussed, ∇ can be applied to a product of two or more vectors by using the Leibnitz's chain rule of differentiation:

$$\nabla(\boldsymbol{U} \cdot \boldsymbol{V}) = \boldsymbol{U} \cdot \nabla \boldsymbol{V} + \boldsymbol{V} \cdot \nabla \boldsymbol{U} + \boldsymbol{U} \times (\nabla \times \boldsymbol{V}) + \boldsymbol{V} \times (\nabla \times \boldsymbol{U}). \tag{3.35}$$

For $\boldsymbol{U} = \boldsymbol{V}$, Eq. (3.35) becomes $\nabla(\boldsymbol{V} \cdot \boldsymbol{V}) = 2\boldsymbol{V} \cdot \nabla \boldsymbol{V} + 2\boldsymbol{V} \times (\nabla \times \boldsymbol{V})$ or

$$\boldsymbol{V} \cdot \nabla \boldsymbol{V} = \tfrac{1}{2}\nabla(\boldsymbol{V} \cdot \boldsymbol{V}) - \boldsymbol{V} \times (\nabla \times \boldsymbol{V}) = \tfrac{1}{2}\nabla(V^2) - \boldsymbol{V} \times (\nabla \times \boldsymbol{V}). \tag{3.36}$$

Equation (3.36) is used to express the convective part of the acceleration in terms of the gradient of kinetic energy of the flow.

3.2.5 Scalar Product of ∇ and a Second Order Tensor

Consider a fluid element with sides dx_1, dx_2, dx_3 parallel to the axis of a Cartesian coordinate system, Fig. 3.3. The fluid element is under a general three-dimensional stress condition. The force vectors acting on the surfaces, which are perpendicular to the coordinates x_1, x_2, and x_3 are denoted by F_1, F_2, and F_3. The opposite surfaces are subject to forces that have experienced infinitesimal changes $F_1 + dF_1$, $F_2 + dF_2$, and $F_3 + dF_3$. Each of these force vectors is decomposed into three components F_{ij} according to the coordinate system defined in Fig. 3.3.

The first index i refers to the axis, to which the fluid element surface is perpendicular, whereas the second index j indicates the direction of the force component. We divide the individual components of the above force vectors by their corresponding area of the fluid element side. The results of these divisions exhibit the components of a second order stress tensor represented by $\boldsymbol{\Pi}$ as shown in Fig. 3.4. As an example, we

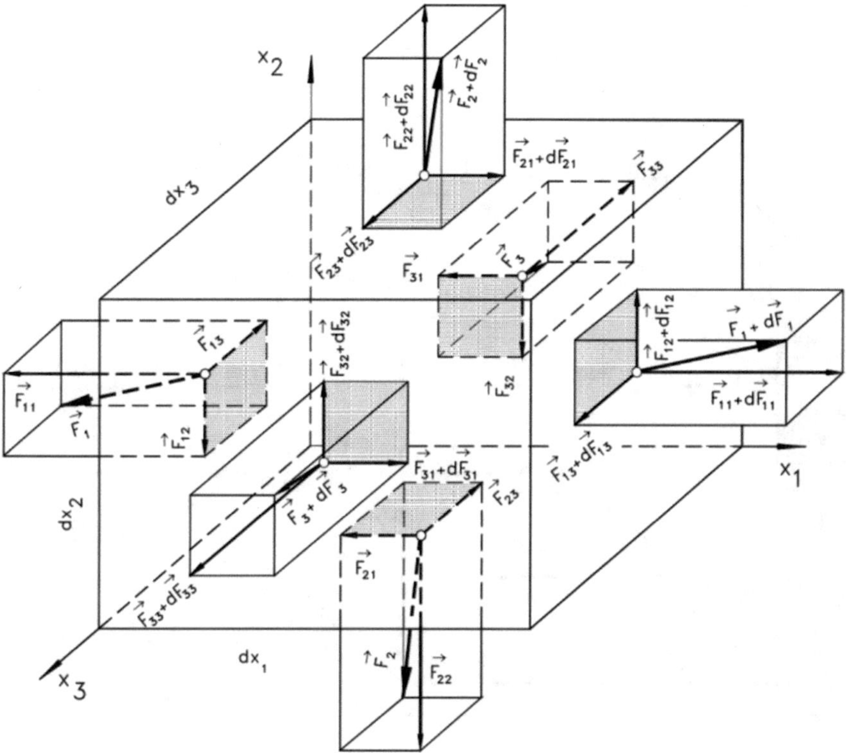

Fig. 3.3 Forces acing on fluid element under a general three-dimensional stress condition

take the force component F_{11} and divide it by the corresponding area dx_2dx_3 results in $\frac{F_{11}}{dx_2dx_3} = \pi_{11}$. Correspondingly, we divide the force component on the opposite surface $F_{11} + dF_{11}$ by the same area dx_2dx_3 and obtain $\frac{F_{11}+dF_{11}}{dx_2dx_3} = \pi_{11} + \frac{\partial \pi_{11}}{\partial x_1}dx_1$.

In a similar way we find the remaining stress components, which are shown in Fig. 3.4. The tensor $\mathbf{\Pi} = e_1 e_j \pi_{ij}$ has nine components π_{ij} as the result of forces that are acting on surfaces. Similar to the force components, the first index i refers to the axis, to which the fluid element surface is perpendicular, whereas the second index j indicates the direction of the stress component. Considering the stress situation in Fig. 3.4, we are now interested in finding the resultant force acting on the fluid particle that occupies the volume element $dv = dx_1dx_2dx_3$. For this purpose, we look at the two opposite surfaces that are perpendicular to the axis x_1 as shown in Fig. 3.5. As this figure shows, we are dealing with 3 stress components on each surface, from which one on each side is the *normal stress* component such as π_{11} and $\pi_{11} + \frac{\partial \pi_{11}}{\partial x_1}dx_1$. The remaining components are the shear stress components such as π_{12} and $\pi_{12} + \frac{\partial \pi_{12}}{\partial x_1}dx_1$. According to Fig. 3.5 the force balance in x_1-directions is:

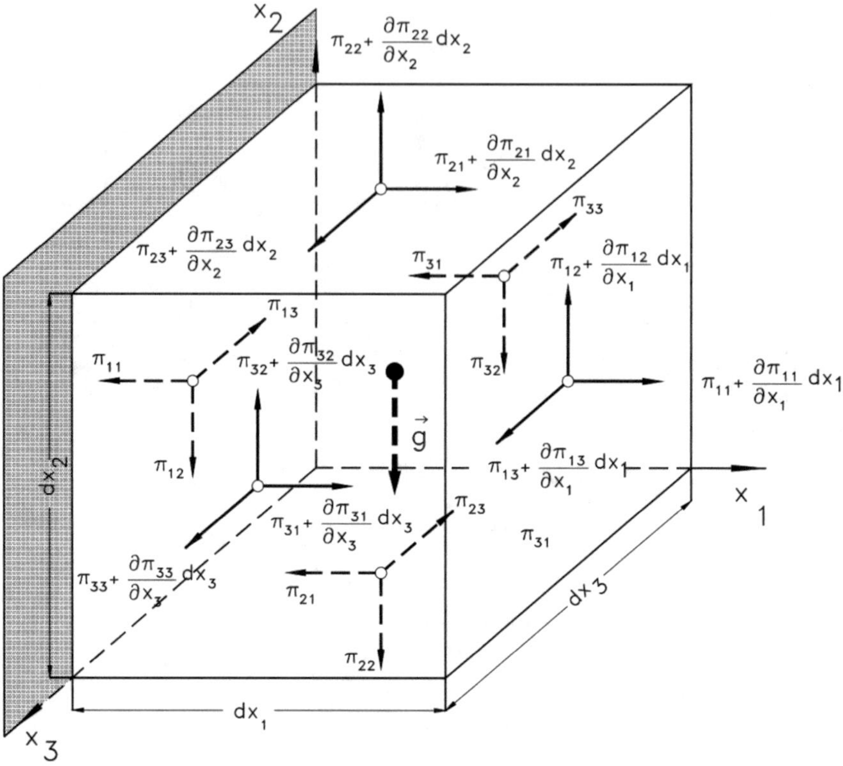

Fig. 3.4 Fluid element under a general three-dimensional stress condition

$$e_1\left(\pi_{11}+\frac{\partial \pi_{11}}{\partial x_1}dx_1-\pi_{11}\right)dx_2dx_3 = e_1\frac{\partial \pi_{11}}{\partial x_1}dx_1dx_2dx_3$$

and in x_2-direction, we find

$$e_2\left(\pi_{12}+\frac{\partial \pi_{12}}{\partial x_1}dx_1-\pi_{12}\right)dx_2dx_3 = e_2\frac{\partial \pi_{12}}{\partial x_1}dx_1dx_2dx_3.$$

Similarly, in x_3, we obtain

$$e_3\left(\pi_{13}+\frac{\partial \pi_{13}}{\partial x_1}dx_1-\pi_{13}\right)dx_2dx_3 = e_2\frac{\partial \pi_{13}}{\partial x_1}dx_1dx_2dx_3.$$

Thus, the resultant force acting on these two opposite surfaces is:

$$d\boldsymbol{F_1} = \left(e_1\frac{\partial \pi_{11}}{\partial x_1}+e_2\frac{\partial \pi_{12}}{\partial x_1}+e_3\frac{\partial \pi_{13}}{\partial x_1}\right)dx_1dx_2dx_3.$$

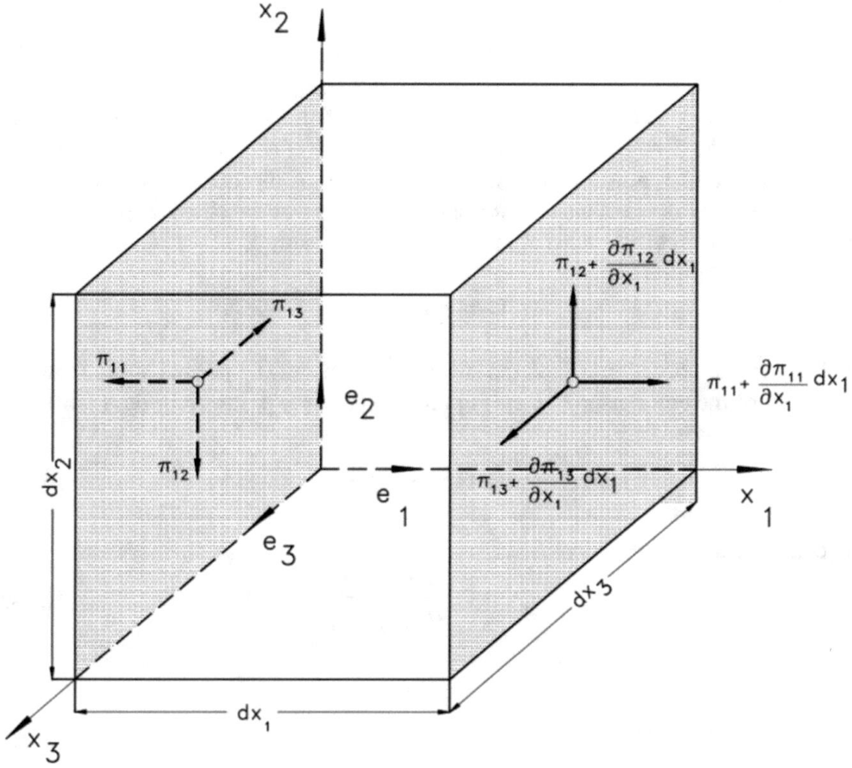

Fig. 3.5 Stress on opposite sides

In a similar way, we find the forces acting on the other four surfaces. The total resulting forces acting on the entire surface of the element are obtained by adding the nine components. Defining the volume element $dv = dx_1 dx_2 dx_3$, we divide the results by dv and obtain the resulting force vector that is acting on the volume element.

$$
\begin{aligned}
\frac{d\boldsymbol{F}}{dv} = e_1 &\left[\frac{\partial \pi_{11}}{\partial x_1} + \frac{\partial \pi_{12}}{\partial x_2} + \frac{\partial \pi_{13}}{\partial x_3}\right] + \\
+e_2 &\left[\frac{\partial \pi_{21}}{\partial x_1} + \frac{\partial \pi_{22}}{\partial x_2} + \frac{\partial \pi_{23}}{\partial x_3}\right] + \\
+e_3 &\left[\frac{\partial \pi_{31}}{\partial x_1} + \frac{\partial \pi_{32}}{\partial x_2} + \frac{\partial \pi_{33}}{\partial x_3}\right].
\end{aligned}
\tag{3.37}
$$

Since the stress tensor Π is written as:

$$
\boldsymbol{\Pi} = e_i e_j \pi_{ij} \tag{3.38}
$$

it can be easily shown that:

$$dF = \nabla \cdot \boldsymbol{\Pi} dv. \tag{3.39}$$

The expression $\nabla \cdot \boldsymbol{\Pi}$ is a scalar differentiation of the second order stress tensor and is called the divergence of the tensor field $\boldsymbol{\Pi}$. We conclude that the force acting on the surface of a fluid element is the divergence of its stress tensor. The stress tensor is usually divided into its normal stress $\boldsymbol{\Sigma}$ and shear stress $\boldsymbol{T}$ parts. For an incompressible fluid the normal stress is identical with $\boldsymbol{\Sigma} = -\boldsymbol{I}p$ and the stress tensor can be written as:

$$\boldsymbol{\Pi} = -\boldsymbol{I}p + \boldsymbol{T} \tag{3.40}$$

with $\boldsymbol{I}p$ as the normal and $\boldsymbol{T}$ as the shear stress tensor. The normal stress tensor is a product of the unit tensor $\boldsymbol{I} = e_i e_j \delta_{ij}$ and the pressure p. Inserting Eq. (3.40) into Eq. (3.39) leads to

$$\frac{d\boldsymbol{F}}{dv} = \nabla \cdot \boldsymbol{\Pi} = -\nabla p + \nabla \cdot \boldsymbol{T}. \tag{3.41}$$

Its components are

$$\frac{dF_i}{dv} = -\frac{\partial p}{\partial x_i} + \frac{\partial \tau_{ji}}{\partial x_j}. \tag{3.42}$$

3.2.6 Stress Vector

Stress Vector: To explain the relation between the stress tensor and stress vector, we cut the parallelepiped in two parts by a plane that does not coincide with any of the coordinate surfaces and its orientation is determined by the unit normal vector $\boldsymbol{n} = e_i n_i$ with n_i as the components of the unit vector $\boldsymbol{n}$ in x_1-direction. The stress vector $\boldsymbol{S}$ is determined by scalar multiplication of the second order stress tensor $\boldsymbol{\Pi}$ with the first order unit tensor $\boldsymbol{n}$:

$$\begin{aligned}
\boldsymbol{S} &= \boldsymbol{\Pi} \cdot \boldsymbol{n} = e_i e_j \pi_{ij} \cdot e_k n_k = e_i \delta_{jk} \pi_{ij} n_k = e_i \pi_{ij} n_j \\
S_1 &= \pi_{11} n_1 + \pi_{12} n_2 + \pi_{13} n_3, \\
S_2 &= \pi_{21} n_1 + \pi_{22} n_2 + \pi_{23} n_3, \\
S_3 &= \pi_{31} n_1 + \pi_{32} n_2 + \pi_{33} n_3.
\end{aligned} \tag{3.43}$$

Example: Given the stress state at a point with the stress tensor and its matrix (π_{ij}) and a section with a normal vector (n_j), where the values are given below:

$$(\pi_{ij}) = \begin{pmatrix} \pi_{11} & \pi_{12} & \pi_{13} \\ \pi_{21} & \pi_{22} & \pi_{23} \\ \pi_{31} & \pi_{32} & \pi_{33} \end{pmatrix} = \begin{pmatrix} 72 & -54 & 0 \\ -27 & -72 & 0 \\ 0 & 0 & 36 \end{pmatrix}, \ (n_j) = \begin{pmatrix} 2/3 \\ -2/3 \\ 1/3 \end{pmatrix}. \tag{3.44}$$

The resulting stress vector is:

$$S_i = \pi_{ij} n_j = \begin{pmatrix} 72 & -54 & 0 \\ -27 & -72 & 0 \\ 0 & 0 & 36 \end{pmatrix} \times \begin{pmatrix} 2/3 \\ -2/3 \\ 1/3 \end{pmatrix} = \begin{pmatrix} 84 \\ 30 \\ 12 \end{pmatrix} \tag{3.45}$$

with the components of the stress vector as

$$\begin{aligned} S_1 &= \pi_{11} n_1 + \pi_{12} n_2 + \pi_{13} n_3, = \tfrac{2}{3} \times 72 + \tfrac{2}{3} \times 54 = 84 \\ S_2 &= \pi_{21} n_1 + \pi_{22} n_2 + \pi_{23} n_3, = -\tfrac{2}{3} \times 27 + \tfrac{2}{3} \times 72 = 30 \\ S_3 &= \pi_{31} n_1 + \pi_{32} n_2 + \pi_{33} n_{23}, = \tfrac{1}{3} \times 36 = 12. \end{aligned} \tag{3.46}$$

The magnitude of the stress vector is calculated from

$$\begin{aligned} e_i S_i \cdot e_j S_j = e_i \cdot e_j S_i S_j = \delta_{ij} S_i S_j = S_i S_i = S^2 = S_1^2 + S_2^2 + S_3^2 \\ S = \sqrt{84^2 + 30^2 + 12^2} = 90.00 MPa. \end{aligned} \tag{3.47}$$

3.2.7 Mohr Circle

Given the total stress at an arbitrary point $x = x_1, x_2, x_3$ in a three-dimensional-coordinate system:

$$\mathbf{\Sigma} = e_i e_j \sigma_{ij} \tag{3.48}$$

and the stress matrix σ_{ij} expanded inserting the following given values:

$$\sigma_{ij} = \begin{pmatrix} \sigma_{11} & \sigma_{12} & \sigma_{13} \\ \sigma_{21} & \sigma_{22} & \sigma_{23} \\ \sigma_{31} & \sigma_{32} & \sigma_{33} \end{pmatrix} = \begin{pmatrix} 1 & 1 & 3 \\ 1 & 5 & 1 \\ 3 & 1 & 1 \end{pmatrix} \times 10^2 MP. \tag{3.49}$$

We find the eigenvalues and eigenvectors by applying the procedure outlined in Sect. 2.3 replacing the tensor D by the stress tensor π_{ij} or in index notation and Λ by σ

$$\begin{vmatrix} \sigma_{11} - \sigma & \sigma_{12} & \sigma_{13} \\ \sigma_{21} & \sigma_{22} - \sigma & \sigma_{23} \\ \sigma_{31} & \sigma_{32} & \sigma_{33} - \sigma \end{vmatrix} = 0. \tag{3.50}$$

After expanding the above determinant, we obtain an algebraic equation in σ in the following form

$$\sigma^3 - I_1 \sigma^2 + I_2 \sigma - I_3 = 0 \tag{3.51}$$

where I_1, I_2 and I_3 are *invariants* of the tensor $\Sigma = e_{ij}\sigma_{ij}$ defined as:

$$I_1 = \sigma_{ij} = \sigma_{11} + \sigma_{22} + \sigma_{33} \tag{3.52}$$

$$I_2 = \tfrac{1}{2}(I_1^2 - \Sigma : \Sigma) = \tfrac{1}{2}(\sigma_{ii}\sigma_{jj} - \sigma_{ij}\sigma_{ij}) \tag{3.53}$$

$$I_{3D} = \det(\sigma_{ij}). \tag{3.54}$$

The roots of Eq. (3.51) the principal stresses namely $\sigma_1, \sigma_2, \sigma_3$ are known as the eigenvalues of the tensor Σ. The subscripts 1, 2, 3 refer to the roots of Eq. (3.51)—not to be confused with the component of a vector. With the values of Eq. (3.49) the stresses are as follows:

$$\sigma_1 = 6.0 \times 10^2 MP,\ \sigma_2 = 3.0 \times 10^2 MP,\ \sigma_3 = -2.0 \times 10^2 MP. \tag{3.55}$$

With the above values the Mohr circle can be constructed. The radii of the three circles are calculated as follows:

$$\begin{aligned} R_1 &= \tfrac{1}{2}(\sigma_1 - \sigma_3) = 4.0 \times 10^2 MPa, \\ R_2 &= \tfrac{1}{2}(\sigma_1 - \sigma_2) = 1.5 \times 10^2 MPa, \\ R_3 &= \tfrac{1}{2}(\sigma_2 - \sigma_3) = 2.5 \times 10^2 MPa. \end{aligned} \tag{3.56}$$

The transformation into the principal axes system delivers:

$$\sigma = \begin{pmatrix} \sigma_1 & 0 & 0 \\ 0 & \sigma_2 & 0 \\ 0 & 0 & \sigma_3 \end{pmatrix} \tag{3.57}$$

and the invariants are expressed in terms of principal stresses

$$\begin{aligned} I_1 &= \sigma_1 + \sigma_2 + \sigma_3 \\ I_2 &= \sigma_1\sigma_2 + \sigma_2\sigma_3 + \sigma_3\sigma_1 \\ I_3 &= \sigma_1\sigma_2\sigma_3. \end{aligned} \tag{3.58}$$

The so called principal shear stress is obtained from:

$$\begin{aligned} \tau_1 &= \tfrac{1}{2}(\sigma_2 - \sigma_3) \\ \tau_2 &- \tfrac{1}{2}(\sigma_3 - \sigma_1) \\ \tau_3 &= \tfrac{1}{2}(\sigma_1 - \sigma_2). \end{aligned} \tag{3.59}$$

The maximum shear stress is:

$$\tau_{\max} = \tfrac{1}{2}(\sigma_1 - \sigma_3). \tag{3.60}$$

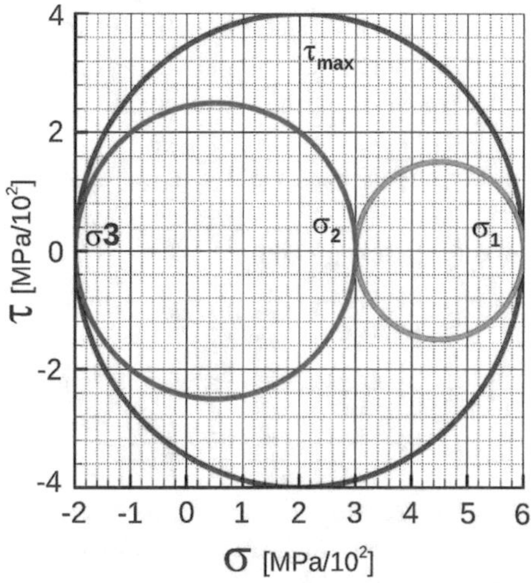

Fig. 3.6 The Mohr circle

With the above information, the Mohr circle is constructed and show in Fig. 3.6.

Problems

Problem 3.1: Show that $\nabla \times (\nabla\varphi) = 0$ with φ as a scalar function.
Problem 3.2: Show that $\nabla \times \nabla \times \boldsymbol{V}) = 0$ with $\boldsymbol{V}$ as a vector function.
Problem 3.3: Show that $\nabla \times (\varphi\boldsymbol{A}) = \varphi(\nabla \times \boldsymbol{A}) + (\nabla\varphi) \times \boldsymbol{A}$ with φ as a scalar and $\boldsymbol{A}$ a vector function.
Problem 3.4: Show that $\nabla \times (\nabla \times \boldsymbol{A}) = -\nabla^2\boldsymbol{A} + \nabla(\nabla \cdot \boldsymbol{A})$.
Problem 3.5: A scalar function is given as $r = \sqrt{x_1^2 + x_2^2 + x_3^2}$, find ∇r.
Problem 3.6: Show that $\nabla \times (\varphi\boldsymbol{A}) = \varphi(\nabla \times \boldsymbol{A}) + (\nabla\varphi) \times \boldsymbol{A}$ with φ and $\boldsymbol{A}$ as a scalar, vector function, respectively.
Problem 3.7: A scalar function is given as $f(x_1, x_2, x_3) = 2x_1^3x_2^2x_3^4$ find ∇f and $\nabla \cdot \nabla f$.
Problem 3.8: An incompressible flow field with water as the working fluid is given by the following vector function, where the coordinates are measured in meters.

$$\boldsymbol{V} = e_1(x_1 + 2x_2)e^{-t} - e_2 2x_2 e^{-t}$$

(a) Find the substantial acceleration of a fluid particle in vector form.
(b) Decompose the acceleration into the components, specify the nature of the flow.
(c) Using the Euler equation of motion:

$$\frac{D\boldsymbol{V}}{Dt} = -\tfrac{1}{\varrho}\nabla p + g, \text{ where } g = -e_3 g = -\nabla(gz)$$

(d) Find the pressure gradient at the $p(x_1, x_2) = (1, 2)$.

Problem 3.9: Starting from the above Euler equation of motion for inviscid incompressible flow obtain: (a) the energy equation by multiplying the equation of motion with a differential displacement using the vector identity

$$\boldsymbol{V} \cdot \boldsymbol{V} = \nabla(\boldsymbol{V} \cdot \boldsymbol{V})/2 - \boldsymbol{V} \times (\nabla \times \boldsymbol{V})$$

Problem 3.10: The velocity field is given by:

$$\begin{aligned} u_1 &= 0 \\ u_2 &= a(x_1 x_2 - x_3^2)e^{-B(t-t_0)} \\ u_3 &= a(x_2^2 - x_1 x_3)e^{-B(t-t_0)} \end{aligned}$$

(a) with B as a constant. Determine the components of the velocity gradient tensor. Start with the coordinate invariant form of the tensor, use index notation, write components and then plug functions in.
(b) Determine the components of the deformation tensor, use index notation, write components and then plug functions in.
(c) Determine the components of the rotation tensor. Start with the coordinate invariant form of the tensor, use index notation, decompose into components and then plug the values in.

Problem 3.11: The velocity field is given by:

$$\begin{aligned} u_1 &= -\tfrac{\omega}{h} x_2 x_3 \\ u_2 &= +\tfrac{\omega}{h} x_1 x_3 \\ u_3 &= 0 \end{aligned}$$

(a) Determine the components of the velocity gradient tensor.
(b) Determine the components of the deformation tensor.
(c) Determine the components of the rotation tensor.

Problem 3.12: the Navier–Stokes equation is given by:

$$\frac{D\boldsymbol{V}}{Dt} = -\tfrac{1}{\varrho}\nabla p + \boldsymbol{V} \cdot \nabla \boldsymbol{V} + g, \text{ where } g = -e_3 g = -\nabla(gz)$$

(a) Give the index notation.
(b) Give the three components.

References

1. Aris R (1963) Vector, tensors and the basic equations of fluid mechanics. Prentice-Hall Inc., Englewood Cliffs, NJ
2. Vavra MH (1960) Aero-thermodynamics and flow in turbomachines. Wiley, Hoboken
3. Brand L (1947) Vector and tensor analysis. Wiley, New York
4. Klingbeil E (1966) Tensorrechnung für Ingenieure. Bibliographisches Institut, Mannheim
5. Lagally M (1944) Vorlesung über Vektorrechnung, dritte edn. Akademische Verlagsgeselschaft, Leipzig

Chapter 4
Tensors and Kinematics

In this chapter we discuss the role of tensors in explaining some kinematic aspects of continuum mechanics that deal with changes of the position and the shape of the material. The changes may include translation, rotation and deformation. Materials subjected to stresses experience deformation termed strain. As seen in previous chapter, there is a direct relationship between the stress and the deformation through the constitutive equation of material. This chapter provides the tools required for understanding the strain relation and its mathematical formulation.

4.1 Material and Spatial Description

4.1.1 Material Description

The motion of a particle or a material point with respect to a reference coordinate system is in general given by a time dependent position vector $x(t)$, Fig. 4.1.

To identify the motion of a *material point* at a certain instance of time $t = t_0 = 0$, we introduce the position vector $\xi = x(t_0)$. Thus, the motion of the fluid is described by the vector:

$$\boldsymbol{x} = \boldsymbol{x}(\boldsymbol{\xi}, t), x_i = x_i(\xi_j, t) \tag{4.1}$$

with x_i as the components of the vector $\boldsymbol{x}$. Equation (4.1) describes the path of a material point that has an initial position vector $\boldsymbol{\xi}$ that characterizes or better labels the material point at $t = t_0$. We refer to this description as the *material* description also called *Lagrangian* description. The material description deals with the motion of the individual particles of a continuum, and is used predominantly in *solid mechanics*. Considering another material points labeled ξ^i with different ξ-coordinates, their paths are similarly described by Eq. (4.1). The movement of a material point with

M. T. Schobeiri, *Tensor Analysis for Engineers and Physicists - With Application to Continuum Mechanics, Turbulence, and Einstein's Special and General Theory of Relativity*, https://doi.org/10.1007/978-3-030-35736-8_4

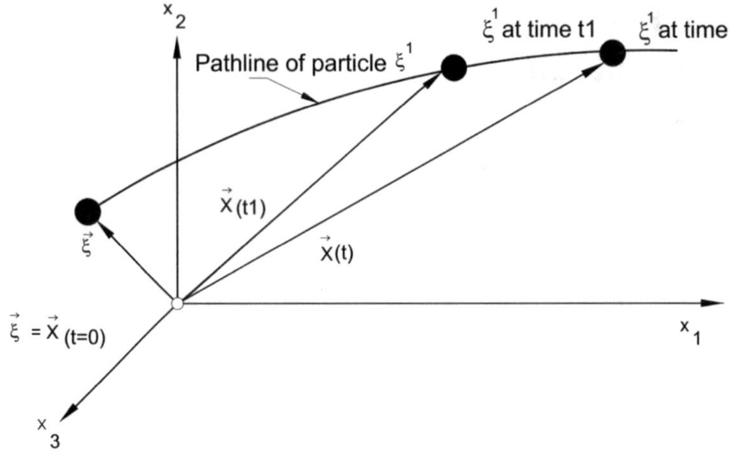

Fig. 4.1 Material description of a particle

the shape S_0 that started at $t = t_0$ and at the position vector ξ may move to another positions at another times and experience deformation and rotation such that its original shape changes from S_0 to S_1 to S_2 or generally to S_n. If we assume that the motion is continuous and single valued, then the inversion of Eq. (4.1) must give the initial position ξ of each material point which may be at any position $\boldsymbol{x}$ and any instant of time t; that is,

$$\boldsymbol{\xi} = \boldsymbol{\xi}(\boldsymbol{x}, \boldsymbol{t}),\ \xi_i = \xi_i(x_j, t). \tag{4.2}$$

The necessary and sufficient condition for an inverse function to exist is that the *Jacobian transformation function*

$$J = \left(\epsilon_{kmn} \frac{\partial x_k}{\partial \xi_1} \frac{\partial x_m}{\partial \xi_2} \frac{\partial x_n}{\partial \xi_3} \right) \tag{4.3}$$

does not vanish. With the above condition, the different shapes are related to each other via the Jacobian transformation function. Because of the significance of the Jacobian transformation function to continuum mechanics, which includes solid mechanics and fluid mechanics, we derive this function in Sect. 4.1.3.

4.1.2 Spatial Description

The material description we discussed in the previous section deals with the motion of the individual material points of a continuum, and is used in *solid mechanics*. In fluid dynamics, we are primarily interested in determining the flow quantities such as velocity, acceleration, density, temperature, pressure, and etc., at fixed points in space. For this purpose, we introduce the *spatial description*, which is also called the *Euler description*. The independent variables for the spatial descriptions are the space characterized by the position vector $\boldsymbol{x}$ and the time t. Consider the transformation of Eq. (4.1), where ξ is solved in terms of $\boldsymbol{x}$:

$$\boldsymbol{\xi} = \boldsymbol{\xi}(\boldsymbol{x}, \boldsymbol{t}), \; \xi_i = \xi_i(x_j, t). \tag{4.4}$$

The velocity vector is expressed as a function of position and time, $\boldsymbol{V}(\boldsymbol{\xi}, t)$, and because of Eq. (4.4) the position vector ξ is replaced by:

$$\boldsymbol{V}(\boldsymbol{\xi}, \boldsymbol{t}) = \boldsymbol{V}(\boldsymbol{\xi}(\boldsymbol{x}, t) = \boldsymbol{V}(\boldsymbol{x}, t). \tag{4.5}$$

For a fixed $\boldsymbol{x}$, Eq. (4.5) exhibits the velocity at the spatially fixed position $\boldsymbol{x}$ as a function of time. On the other hand, for a fixed t, Eq. (4.5) describes the velocity at the time t. With Eq. (4.5), any quantity described in spatial coordinates can be transformed into material coordinates provided the Jacobian transformation function J, which we discussed in the previous section, does not vanish. If the velocity is known in a spatial coordinate system, the path of the particle can be determined as the integral solution of the differential equation with the initial condition $\boldsymbol{x}(t_0)$ along the path $\boldsymbol{x} = \boldsymbol{x}(\xi, t)$ from the following relation:

$$\frac{dx}{dt} = \boldsymbol{V}(\boldsymbol{x}, t), \; \frac{dx_i}{dt} = V_i(x_j, t). \tag{4.6}$$

4.1.3 *Jacobian Transformation Function*

The Jacobian transformation function maps the position of a particle that moves through space at time t to any other position at any other time including $t = 0$. We consider a differential *material volume* at the time $t = 0$, to which we attach the reference coordinate system ξ_1, ξ_2, ξ_3, as shown in Fig. 4.2. At the time $t = 0$, the reference coordinate system is fixed so that the *undeformed* differential material volume dV_0 (Figs. 4.2 and 4.3) can be described as:

$$dV_0 = (e_1 d\xi_1 \times e_2 d\xi_2) \cdot e_3 d\xi_3 = d\xi_1 d\xi_2 d\xi_3. \tag{4.7}$$

Moving through the space, the differential material volume may undergo certain deformation and rotation. As deformation takes place, the sides of the material volume initially given as $d\xi_i$ would be convected into a non-rectangular, or curvilinear form. The changes of the deformed coordinates are then:

$$d\boldsymbol{x} = \frac{\partial \boldsymbol{x}}{\partial \xi_1} d\xi_1 + \frac{\partial \boldsymbol{x}}{\partial \xi_2} d\xi_2 + \frac{\partial \boldsymbol{x}}{\partial \xi_3} d\xi_3 = \frac{\partial \boldsymbol{x}}{\partial \xi_i} d\xi_i. \tag{4.8}$$

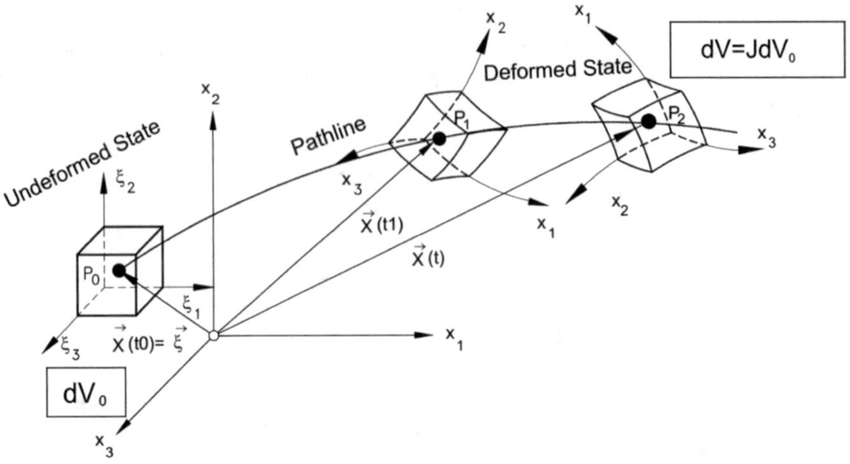

Fig. 4.2 Deformation of a differential volume at different instant of time

Using the index notation for the position vector $\boldsymbol{x} = e_k x_k$, Eq. (4.8) may be rearranged in the following way:

$$d\boldsymbol{x} = e_k \frac{\partial x_k}{\partial \xi_i} d\xi_i \equiv \boldsymbol{G}_i d\xi_i \tag{4.9}$$

with the vector $\boldsymbol{G}_i$ as

$$\boldsymbol{G}_i \equiv e_k \frac{\partial x_k}{\partial \xi_i} \tag{4.10}$$

where $\boldsymbol{G}_i$ is a is a base vector or a line element transformation vector that transforms the differential changes $d\xi_i$ into dx_i.[1] As we will see in Chap. 6, using base vectors such as $\boldsymbol{G}_i$, a metric tensor can be constructed as the scalar product of $\boldsymbol{G}_i$ and $\boldsymbol{G}_j$ such as $G_{ij} = \boldsymbol{G}_i \cdot \boldsymbol{G}_j$. The expression $\frac{\partial x_k}{\partial \xi_i}$ is called the deformation gradient. Figure 4.3 illustrates the deformation of the material volume and the transformation mechanism. The new deformed differential volume is obtained by:

$$dV = (d\boldsymbol{x_1} \times d\boldsymbol{x_2}) \cdot d\boldsymbol{x_3}. \tag{4.11}$$

Introducing Eq. (4.9) into Eq. (4.11) leads to:

$$dV = (\boldsymbol{G_1} d\xi_1 \times \boldsymbol{G_2} d\xi_2) \cdot \boldsymbol{G_3} d\xi_3. \tag{4.12}$$

[1] Actually, during the deformation process the Cartesian configuration generally changes to a curvilinear configuration. This subject is treated in Chap. 6.

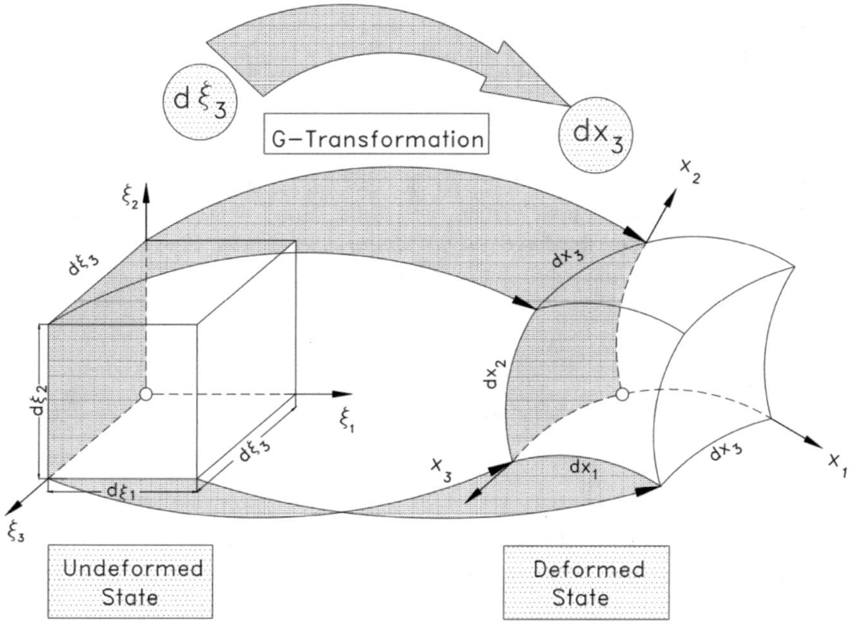

Fig. 4.3 Transformation of $d\xi_1, d\xi_2, d\xi_3$ into dx_1, dx_2, and dx_3 using G-transformation

Inserting $\boldsymbol{G}_i$ from Eq. (4.10) into Eq. (4.12) and considering the transformation described in Sect. 2.2.2 we arrive at the differential volume

$$dV = \left(e_k \frac{\partial x_k}{\partial \xi_1} \times e_m \frac{\partial x_m}{\partial \xi_2} \right) \cdot e_n \frac{\partial x_n}{\partial \xi_3} d\xi_1 d\xi_2 d\xi_3. \tag{4.13}$$

Now we replace the vector product and the following scalar product of the two unit vectors in Eq. (4.13) with the permutation symbol and the Kronecker delta:

$$dV = \epsilon_{kml} \delta_{\mathrm{ln}} \frac{\partial x_k}{\partial x_1} \frac{\partial x_m}{\partial \xi_2} \frac{\partial x_n}{\partial \xi_3} d\xi_1 d\xi_2 d\xi_3. \tag{4.14}$$

Applying the Kronecker delta to the terms with the indices l and n, we arrive at:

$$dV = \left(\epsilon_{kmn} \frac{\partial x_k}{\partial \xi_1} \frac{\partial x_m}{\partial \xi_2} \frac{\partial x_n}{\partial \xi_3} \right) (d\xi_1 d\xi_2 d\xi_3). \tag{4.15}$$

The expression in first parenthesis in Eq. (4.15) represents the *Jacobian function* J.

$$J = \left(\epsilon_{kmn} \frac{\partial x_k}{\partial \xi_1} \frac{\partial x_m}{\partial \xi_2} \frac{\partial x_n}{\partial \xi_3} \right). \tag{4.16}$$

The second parenthesis in Eq. (4.15) represents the initial infinitesimal material volume in the *undeformed state* at $t = 0$, described by Eq. (4.3). Using these terms, Eq. (4.15) is rewritten as:

$$dV = J dV_0 \tag{4.17}$$

where dV represents the differential volume in the *deformed state*, dV_0 has the same differential volume in the undeformed state at the time $t = 0$. The transformation function J is also called the *Jacobian functional determinant*. Performing the permutation in Eq. (4.16), the determinant is given as:

$$J = \det \begin{pmatrix} \frac{\partial x_1}{\partial \xi_1} & \frac{\partial x_2}{\partial \xi_1} & \frac{\partial x_3}{\partial \xi_1} \\ \frac{\partial x_1}{\partial \xi_2} & \frac{\partial x_2}{\partial \xi_2} & \frac{\partial x_3}{\partial \xi_2} \\ \frac{\partial x_1}{\partial \xi_3} & \frac{\partial x_2}{\partial \xi_3} & \frac{\partial x_3}{\partial \xi_3} \end{pmatrix}. \tag{4.18}$$

With the Jacobian functional determinant, we now have a necessary tool to directly relate any time dependent differential volume $dV = dV(t)$ to its fixed reference volume dV_0 at the reference time $t = 0$ as shown in Fig. 4.4. The Jacobian transformation function and its material derivative are the fundamental tools to understand the conservation laws using the integral analysis in conjunction with control volume method. To complete this section, we briefly discuss the material derivative of the Jacobian function. As the volume element dV follows the motion from $\boldsymbol{x} = \boldsymbol{x}(\boldsymbol{\xi}, t)$ to $\boldsymbol{x}(\boldsymbol{\xi}, t + dt)$ it changes and, as a result, the Jacobian transformation function undergoes a time change. To calculate this change, we determine the material derivative of J:

$$\frac{DJ}{Dt} = \frac{\partial J}{\partial t} + \boldsymbol{V} \cdot \nabla J. \tag{4.19}$$

Inserting Eq. (4.16) into Eq. (4.19), we obtain:

$$\frac{DJ}{Dt} = \frac{\partial}{\partial t}\left(\epsilon_{kmn} \frac{\partial x_k}{\partial \xi_1} \frac{\partial x_m}{\partial \xi_2} \frac{\partial x_n}{\partial \xi_3}\right) + V_j \frac{\partial}{\partial x_j}\left(\epsilon_{kmn} \frac{\partial x_k}{\partial \xi_1} \frac{\partial x_m}{\partial \xi_2} \frac{\partial x_n}{\partial \xi_3}\right). \tag{4.20}$$

Let us consider an arbitrary element of the Jacobian determinant, for example $\partial x_1 / \partial \xi_2$. Since the reference coordinate $\boldsymbol{\xi} \neq \boldsymbol{f}(\boldsymbol{t})$ is not a function of time and is fixed, the differentials with respect to t and ξ_2, can be interchanged resulting in:

$$\frac{\partial}{\partial t}\left(\frac{\partial x_1}{\partial \xi_2}\right) = \frac{\partial}{\partial \xi_2}\left(\frac{\partial x_1}{\partial t}\right) = \frac{\partial V_1}{\partial \xi_2}. \tag{4.21}$$

Performing similar operations for all elements of the Jacobian determinant and noting that the second expression on the right-hand side of Eq. (4.20) identically vanishes, we arrive at:

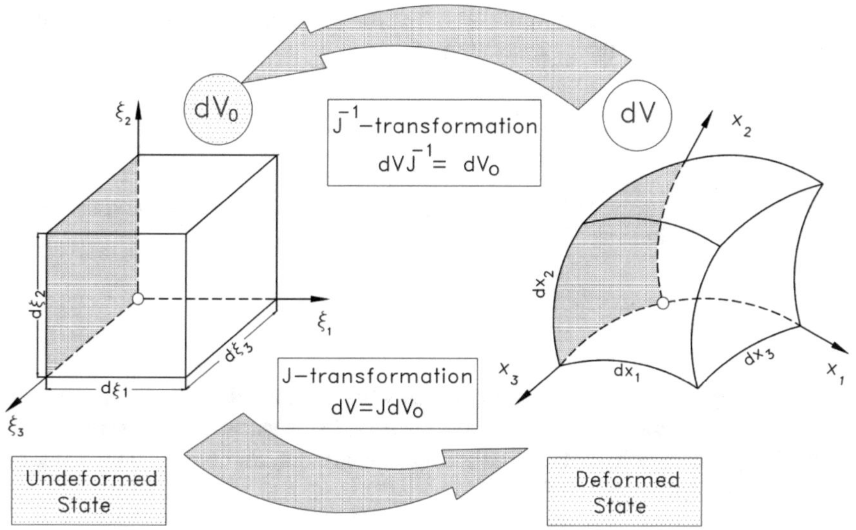

Fig. 4.4 Jacobian transformation of a material volume, change of states

$$\frac{DJ}{Dt} = \left(\frac{\partial V_1}{\partial x_1} + \frac{\partial V_2}{\partial x_2} + \frac{\partial V_3}{\partial x_3}\right) J. \tag{4.22}$$

The expression in the parenthesis of Eq. (4.22) is the well known divergence of the velocity vector. Using vector notation, Eq. (4.22) becomes:

$$\frac{DJ}{Dt} = (\nabla \cdot \boldsymbol{V})J. \tag{4.23}$$

4.2 Reynolds Transport Theorem

The conservation laws in integral form are, strictly speaking, valid for *closed systems*, where the *mass* does not cross the *system boundary*. In fluid mechanics, however, we are dealing with *open systems*, where the *mass flow* continuously crosses the system boundary. To apply the conservation laws to open systems, we briefly provide the necessary mathematical tools. In this section, we treat the volume integral of an arbitrary field quantity $f(\boldsymbol{X}, t)$ by deriving the *Reynolds transport theorem*. This is an important kinematic relation that we use in integral analysis of fluid mechanics, [1, 2].

The field quantity $f(\boldsymbol{X}, t)$ may be a zeroth, first or second order tensor valued function, such as mass, velocity vector, and stress tensor. The time dependent volume under consideration with a given time dependent surface moves through the flow field and may experience dilatation, compression and deformation. It is assumed to contain the same fluid particles at any time and therefore, it is called the material volume. The volume integral of the quantity $f(\boldsymbol{X}, t)$:

$$\boldsymbol{F}(t) = \int_{V(t)} f(\boldsymbol{X}, t) dv \tag{4.24}$$

is a function of time only. The integration must be carried out over the varying volume $v(t)$. The material change of the quantity $F(t)$ is expressed as:

$$\frac{DF(t)}{Dt} = \frac{D}{Dt}\int_{V(t)} f(\boldsymbol{X}, t) dv. \tag{4.25}$$

Since the shape of the volume $v(t)$ changes with time, the differentiation and integration cannot be interchanged. However, Eq. (4.25) permits the transformation of the time dependent volume $v(t)$ into the fixed volume v_0 at time $t = 0$ by using the Jacobian transformation function:

$$\frac{DF(t)}{Dt} = \frac{D}{Dt}\int_{v_0} f(\boldsymbol{X}, t) J dv_0. \tag{4.26}$$

With this operation in Eq. (4.26), it is possible to interchange the sequence of differentiation and integration:

$$\frac{DF(t)}{Dt} = \int_{v_0} \frac{D}{Dt}(f(\boldsymbol{X}, t) J) dv_0. \tag{4.27}$$

The chain differentiation of the expression within the parenthesis results in

$$\frac{DF(t)}{Dt} = \int_{v_0} \left(J\frac{D}{Dt} f(\boldsymbol{X}, t) + f(\boldsymbol{X}, t)\frac{DJ}{Dt} \right) dv_0. \tag{4.28}$$

Introducing the material derivative of the Jacobian function, Eq. (4.22) into Eq. (4.28) yields:

$$\frac{DF(t)}{Dt} = \int_{v_0} \left(\frac{D}{Dt} f(\boldsymbol{X}, t) + f(\boldsymbol{X}, t)\nabla \cdot \mathrm{V} \right) J dv_0. \tag{4.29}$$

Equation (4.29) permits the back transformation of the fixed volume integral into the time dependent volume integral:

$$\frac{DF(t)}{Dt} = \int_{v(t)} \left(\frac{D}{Dt} f(\boldsymbol{X}, t) + f(\boldsymbol{X}, t)\nabla \cdot \boldsymbol{V} \right) dv. \tag{4.30}$$

According to Eq. (4.1), the first term in the parenthesis can be written as:

$$\frac{Df}{Dt} = \frac{\partial f}{\partial t} + \boldsymbol{V} \cdot \nabla f. \tag{4.31}$$

Introducing Eq. (4.31) into Eq. (4.30) results in:

$$\frac{DF(t)}{Dt} = \int_{v(t)} \left(\frac{\partial}{\partial t} f(\boldsymbol{X}, t) + V \cdot \nabla f(\boldsymbol{X}, t) + f(X, t) \nabla \cdot \boldsymbol{V} \right) dv. \tag{4.32}$$

The chain rule applied to the second and third term in Eq. (4.32) yields:

$$\frac{DF(t)}{Dt} = \int_{v(t)} \left\{ \frac{\partial}{\partial t} f(\boldsymbol{X}, t) + \nabla \cdot (f(\boldsymbol{X}, t)\boldsymbol{V}) \right\} dv. \tag{4.33}$$

The second volume integral in Eq. (4.33) can be converted into a surface integral by applying the Gauss' divergence theorem:

$$\int_{v(t)} \nabla \cdot (f(\boldsymbol{X}, t)\boldsymbol{V}) dv = \int_{S(t)} f(\boldsymbol{X}, t)\boldsymbol{V} \cdot \boldsymbol{n} dS \tag{4.34}$$

where $\boldsymbol{V}$ represents the *flux velocity* and $\boldsymbol{n}$ the unit vector normal to the surface. Inserting Eq. (4.34) into Eq. (4.33) results in the following final equation, which is called the *Reynolds transport theorem*

$$\frac{DF(t)}{Dt} = \int_{v(t)} \frac{\partial}{\partial t} f(\boldsymbol{X}, t) dv + \int_{S(t)} f(\boldsymbol{X}, t)\boldsymbol{V} \cdot \boldsymbol{n} dS. \tag{4.35}$$

Equation (4.35) is valid for any system boundary with time dependent volume $V(t)$ and surface $S(t)$ at any time, including the time $t = t_0$, where the volume $V = V_C$ and the surface $S = S_C$ assume fixed values. We call V_C and S_C the *control volume* and *control surface*.

4.3 Translation, Deformation, Rotation

During a general three-dimensional motion, a fluid particle undergoes a translational and rotational motion which may be associated with deformation. The velocity of a particle at a given spatial, temporal position $(\boldsymbol{x} + d\boldsymbol{x}, t)$ can be related to the velocity at $(\boldsymbol{x}, t)$ by using the following Taylor expansion:

$$\boldsymbol{V}(\boldsymbol{x} + d\boldsymbol{x}, t) = \boldsymbol{V}(\boldsymbol{x}, t) + d\boldsymbol{V}. \tag{4.36}$$

Inserting in Eq. (4.36) for the differential velocity change $d\boldsymbol{V} = d\boldsymbol{x} \cdot \nabla \boldsymbol{V}$, Eq. (4.36) is re-written as:

$$\boldsymbol{V}(\boldsymbol{x} + d\boldsymbol{x}, t) = \boldsymbol{V}(\boldsymbol{x}, t) + d\boldsymbol{x} \cdot \nabla \boldsymbol{V}. \tag{4.37}$$

The first term on the right-hand side of Eq. (4.37) represents the translational motion of the fluid particle. The second expression is a scalar product of the differential displacement $d\boldsymbol{x}$ and the *velocity gradient* $\nabla \boldsymbol{V}$. We decompose the velocity gradient, which is a second order tensor, into two parts resulting in the following *identity*:

$$\nabla \boldsymbol{V} = \tfrac{1}{2}(\nabla \boldsymbol{V} + \nabla \boldsymbol{V}^{\boldsymbol{T}}) + \tfrac{1}{2}(\nabla \boldsymbol{V} - \nabla \boldsymbol{V}^{\boldsymbol{T}}). \tag{4.38}$$

The superscript T indicates that the matrix elements of the second order tensor $\nabla \boldsymbol{V}^T$ are the transpositions of the matrix elements that pertain to the second order tensor $\nabla \boldsymbol{V}$. The first term in the right-hand side represents the *deformation tensor*, which is a symmetric second order tensor:

$$\boldsymbol{D} = \tfrac{1}{2}(\nabla \boldsymbol{V} + \nabla \boldsymbol{V}^{\boldsymbol{T}}) = e_i e_j D_{ij} = \tfrac{1}{2} e_i e_j \left(\frac{\partial V_i}{\partial x_j} + \frac{\partial V_j}{\partial x_i} \right) \tag{4.39}$$

with components:

$$D_{ij} = \tfrac{1}{2} \left(\frac{\partial V_i}{\partial x_j} + \frac{\partial V_j}{\partial x_i} \right). \tag{4.40}$$

The second term of Eq. (4.38) is called the rotation or vorticity tensor, which is antisymmetric and is given by:

$$\boldsymbol{\Omega} = \tfrac{1}{2}(\nabla \boldsymbol{V} - \nabla \boldsymbol{V}^{\boldsymbol{T}}) = e_i e_j \Omega_{ij} = \tfrac{1}{2} e_i e_j \left(\frac{\partial V_j}{\partial x_i} - \frac{\partial V_i}{\partial x_j} \right). \tag{4.41}$$

The components are:

$$\Omega_{ij} = \tfrac{1}{2} \left(\frac{\partial V_j}{\partial x_i} = \frac{\partial V_i}{\partial x_j} \right). \tag{4.42}$$

Inserting Eqs. (4.39) and (4.41) into Eq. (4.37), we arrive at:

$$\boldsymbol{V}(\boldsymbol{x} + d\boldsymbol{x}, t) = \boldsymbol{V}(\boldsymbol{x}, t) + d\boldsymbol{x} \cdot \boldsymbol{D} + d\boldsymbol{x} \cdot \boldsymbol{\Omega}. \tag{4.43}$$

Equation (4.43) describes the kinematics of the fluid particle, which has a combined translational and rotational motion and undergoes a deformation.

Figure 4.5 illustrates the geometric representation of the rotation and deformation, [3]. Consider the fluid particle with a square-shaped cross section in the $x_1 - x_2$ plane at the time t. The position of this particle is given by the position vector $\boldsymbol{x} = \boldsymbol{x}(t)$. By moving through the flow field, the particle experiences translational motion to a new position $\boldsymbol{x} + d\boldsymbol{x}$. This motion may be associated with a rotational motion and a deformation. The deformation is illustrated by the initial and final state of diagonal $A - C$, which is stretched to $A' - C'$ and the change of the angle 2γ to $2\gamma'$. The rotational motion can be appropriately illustrated by the rotation of the diagonal by the angle $d\varphi_3 = \gamma' + d\alpha - \gamma$, where γ' can be eliminated using the relation $2\gamma' + d\alpha + d\beta = 2\gamma$. As a result, we obtain the infinitesimal rotation angle:

$$d\varphi_3 = \tfrac{1}{2}(d\alpha - d\beta) \tag{4.44}$$

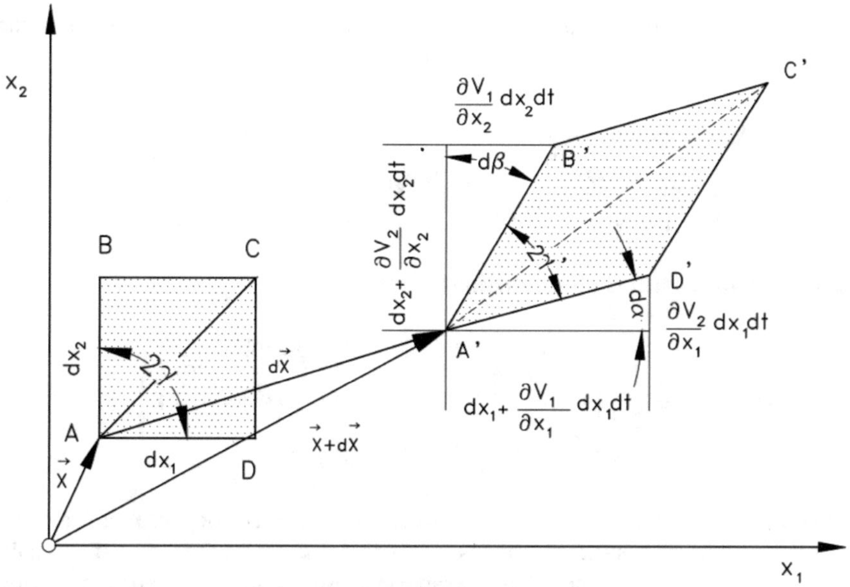

Fig. 4.5 Translation, rotation and deformation details of a fluid particle

where the subscript 3 denotes the direction of the rotation axis, which is parallel to x_3, Fig. 4.4. Referring to Fig. 4.5, direct relationships between $d\alpha$, $d\beta$ and the velocity gradients can be established by:

$$d\alpha \approx \tan(d\alpha) = \frac{\frac{\partial V_2}{\partial x_1} dx_1 dt}{dx_1 + \frac{\partial V_1}{\partial x_1} dx_1 dt} \approx \frac{\partial V_2}{\partial x_1} dt. \tag{4.45}$$

A similar relationship is given for the angle change $d\beta$:

$$d\beta \approx \tan(d\beta) = \frac{\frac{\partial V_1}{\partial x_2} dx_2 dt}{dx_2 + \frac{\partial V_2}{\partial x_2} dx_2 dt} \approx \frac{\partial V_1}{\partial x_2} dt. \tag{4.46}$$

Substituting Eqs. (4.45) and (4.46) into Eq. (4.44), the *rotation rate* in the x_3-direction is found:

$$\frac{d\varphi_3}{dt} = \tfrac{1}{2}\left(\frac{\partial V_2}{\partial x_1} - \frac{\partial V_1}{\partial x_2}\right). \tag{4.47}$$

Executing the same procedure, the other two components are:

$$\frac{d\varphi_1}{dt} = \tfrac{1}{2}\left(\frac{\partial V_3}{\partial x_2} - \frac{\partial V_2}{\partial x_3}\right), \frac{d\varphi_2}{dt} = \tfrac{1}{2}\left(\frac{\partial V_1}{\partial x_3} - \frac{\partial V_3}{\partial x_1}\right). \tag{4.48}$$

The above three terms in Eqs. (4.47) and (4.48) may be recognized as one-half of the three components of the vorticity vector $\boldsymbol{\omega}$, which is:

$$\boldsymbol{\omega} \equiv \nabla \times \boldsymbol{V} = \epsilon_{ijk} e_k \frac{\partial V_j}{\partial x_i}$$
$$\boldsymbol{\omega} = e_1 \left(\frac{\partial V_3}{\partial x_2} - \frac{\partial V_2}{\partial x_3} \right) + e_2 \left(\frac{\partial V_1}{\partial x_3} - \frac{\partial V_3}{\partial x_1} \right) + e_3 \left(\frac{\partial V_2}{\partial x_1} - \frac{\partial V_1}{\partial x_2} \right). \quad (4.49)$$

Examining the elements of the rotation matrix,

$$\Omega_{ij} = \begin{pmatrix} 0 & \frac{1}{2}\left(\frac{\partial V_2}{\partial x_1} - \frac{\partial V_1}{\partial x_2}\right) & \frac{1}{2}\left(\frac{\partial V_3}{\partial x_1} - \frac{\partial V_1}{\partial x_3}\right) \\ \frac{1}{2}\left(\frac{\partial V_1}{\partial x_2} - \frac{\partial V_2}{\partial x_1}\right) & 0 & \frac{1}{2}\left(\frac{\partial V_3}{\partial x_2} - \frac{\partial V_2}{\partial x_3}\right) \\ \frac{1}{2}\left(\frac{\partial V_1}{\partial x_3} - \frac{\partial V_3}{\partial x_1}\right) & \frac{1}{2}\left(\frac{\partial V_2}{\partial x_3} - \frac{\partial V_3}{\partial x_2}\right) & 0 \end{pmatrix} \quad (4.50)$$

we notice that the diagonal elements of the above antisymmetric tensor are zero and only three of the six non-zero elements are distinct. Except for the factor of one-half, these three distinct components are the same as those making up the vorticity vector. Comparing the components of the vorticity vector $\boldsymbol{\omega}$ given by Eq. (4.49) and the three distinct terms of Eq. (4.50), we conclude that the components of the vorticity vector, except for the factor of one-half, are identical with the *axial vector* of the antisymmetric tensor, Eq. (4.50). The axial vector of the second order tensor $\boldsymbol{\Omega}$ is the double scalar product of the third order permutation tensor $\boldsymbol{\epsilon} = e_i e_j e_k \epsilon_{ijk}$ with $\boldsymbol{\Omega}$:

$$\boldsymbol{\epsilon} : \boldsymbol{\Omega} = \epsilon_{ijk} e_i e_j e_k : e_m e_n \Omega_{mn} = e_i e_j \frac{1}{2} \left(\frac{\partial V_j}{\partial x_k} - \frac{\partial V_k}{\partial x_j} \right). \quad (4.51)$$

Expanding Eq. (4.51) results in:

$$\boldsymbol{\epsilon} : \boldsymbol{\Omega} = \frac{1}{2} \left[(e_1 \left(\frac{\partial V_2}{\partial x_3} - \frac{\partial V_3}{\partial x_2} \right) + e_2 \left(\frac{\partial V_3}{\partial x_1} - \frac{\partial V_1}{\partial x_3} \right) + e_3 \left(\frac{\partial V_1}{\partial x_2} - \frac{\partial V_2}{\partial x_1} \right) \right]. \quad (4.52)$$

Comparing Eq. (4.52) to Eq. (4.49) shows that the right-hand side of Eq. (4.52) multiplied with a negative sign is exactly equal the right-hand side of Eq. (4.49). This indicates that the axial vector of the rotation tensor is equal to the negative rotation vector and can be expressed as:

$$\nabla \times \boldsymbol{V} = -2\boldsymbol{\epsilon} : \boldsymbol{\Omega}. \quad (4.53)$$

The existence of the vorticity vector $\boldsymbol{\omega}$ and therefore, the rotation tensor Ω, is a characteristic of viscous flows that in general undergoes a *rotational motion*. This is particularly true for boundary layer flows, where the fluid particles move very close to the solid boundaries. In this region, the wall shear stress forces (friction forces) cause a combined deformation and rotation of the fluid particle. In contrast, for*inviscid flows*, or the flow regions, where the viscosity effect may be neglected, the rotation vector $\boldsymbol{\omega}$ may vanish if the flow can be considered isentropic. This ideal case is called *potential flow*, where the rotation vector $\nabla \times \boldsymbol{V} = 0$ in the entire potential flow field.

4.4 Strain Tensor

Equations (4.16) and (4.18) show that the Jacobian transformation function is a tool that enables the change of configuration from the initial undeformed to a deformed state. To quantify the change, we use the strain-displacement relation and derive the strain tensor. Strain is defined as the state of an elastic continuum to which forces are applied. Under the strain condition the volume and the shape of the continuum will change.

Considering the position vector $\boldsymbol{x} = \boldsymbol{x}(\boldsymbol{\xi}, \boldsymbol{t})$ as the sum of the vector $\boldsymbol{\xi}$ that is embedded in the time independent coordinate system (ξ_1, ξ_2, ξ_3) and the displacement vector $\boldsymbol{u}$, which is a function of time:

$$\boldsymbol{x}(t) = \boldsymbol{\xi} + \boldsymbol{u}(t). \tag{4.54}$$

As seen, the displacement vector $\boldsymbol{u}(t)$ is assumed to be a function of time. This assumption implies that one is interested in determining the transition from an undeformed state to the deformed one. This requires introducing the time as a forth dimension. In what follows, we assume that transition from one state to the next is completed, thus the independent variable time is not involved anymore. Taking the derivative of Eq. (4.54) with respect to ξ_i we have:

$$d\boldsymbol{x} = d\boldsymbol{\xi} + d\boldsymbol{u} = e_j\left(\frac{\partial \xi_j}{\partial \xi_i} + \frac{\partial u_j}{\partial \xi_i}\right)d\xi_i = e_j\left(\delta_{ji} + \frac{\partial u_j}{\partial \xi_i}\right)d\xi_i. \tag{4.55}$$

The changes of the position vector $\boldsymbol{x}$ in deformed state with respect to changes of $\boldsymbol{\xi}$ in undeformed state is:

$$d\boldsymbol{x} = \frac{\partial \boldsymbol{x}}{\partial \xi_i}d\xi_i = \boldsymbol{G}_i d\xi_i. \tag{4.56}$$

with $\boldsymbol{G}_i$ as the transformation vector function as shown in Fig. 4.3. Equating (4.56) with (4.55) we arrive at:

$$\boldsymbol{G}_i = e_i\left(\delta_{ij} + \frac{\partial \boldsymbol{u}_j}{\partial \xi_i}\right). \tag{4.57}$$

Now consider an infinitesimal line segment in the undeformed coordinate system, and the deformed one:

$$
\begin{aligned}
ds_\xi^2 &= d\boldsymbol{\xi} \cdot d\boldsymbol{\xi} = e_i \cdot e_j d\xi_i d\xi_j = \delta_{ij} d\xi_i d\xi_j = d\xi_i d\xi_i \\
ds_x^2 &= d\boldsymbol{x} \cdot d\boldsymbol{x} = \boldsymbol{G_i} \cdot \boldsymbol{G}_j x_i dx_j = g_{ij} x_i x_j
\end{aligned} \tag{4.58}
$$

with ds_ξ and ds_x as infinitesimal changes in undeformed and deformed coordinates, respectively. With Eqs. (4.55) and (4.57), we obtain

$$
ds_x^2 = G_{ij} d\xi_i d\xi_j, \text{ with } \boldsymbol{G}_i \cdot \boldsymbol{G}_j = G_{ij} \tag{4.59}
$$

with Eqs. (4.59) and (4.58) the scalar product becomes:

$$
\boldsymbol{G_i} \cdot \boldsymbol{G_j} = G_{ij} = e_j \left(\delta_{ji} + \frac{\partial u_j}{\partial \xi_i} \right) \cdot e_m \left(\delta_{mn} + \frac{\partial u_m}{\partial \xi_n} \right). \tag{4.60}
$$

The scala multiplication results in a modified version of Eq. (4.60)

$$
G_{ij} = \left(\delta_{ji} + \frac{\partial u_j}{\partial \xi_i} \right) \left(\delta_{jn} + \frac{\partial u_j}{\partial \xi_n} \right). \tag{4.61}
$$

The multiplication of the two parentheses in Eq. (4.61) gives:

$$
G_{ij} = \delta_{ij} + \frac{\partial u_i}{\partial \xi_j} + \frac{\partial u_j}{\partial \xi_i} + \frac{\partial u_m}{\partial \xi_i} \frac{\partial u_m}{\partial \xi_j}. \tag{4.62}
$$

With Eqs. (4.58), (4.59) and (4.62), we arrive at:

$$
ds_x^2 - ds_\xi^2 = (G_{ij} - \delta_{ij}) d\xi_i d\xi_j \equiv 2\gamma_{ij} d\xi_i d\xi_j \tag{4.63}
$$

with:

$$
\gamma_{ij} = \tfrac{1}{2}(G_{ij} - \delta_{ij}). \tag{4.64}
$$

Equation (4.64) is the strain tensor.

References

1. Schobeiri MT (2014) Engineering applied fluid mechanics. Graduate textbook. McGraw Hill, New York
2. Schobeiri MT (2010) Fluid mechanics for engineers. Graduate textbook. Springer, New York. ISBN 978-642-1193-6
3. White FM (1974) Viscous fluid flow. McGraw-Hill, New York
4. Klingbeil E (1966) Tensorrechnung Fuer Ingeneure. Hochschultaschenbuecher, Bibliographische Institut
5. Chung TJ (1988) Continuum mechanics. Prentice Hall, Englewood Cliffs

Chapter 5
Differential Balances in Continuum Mechanics

In this and the following chapter, we apply the tensor analysis to the fundamental laws of continuum mechanics. Continuum mechanics has two distinctive branches, the solid mechanics and the fluid mechanics.

The subject of solid mechanics is the study of the physical behavior of continuous materials with defined shapes. Once the material undergoes a certain level of stress its shape will experience a deformation that might be associated with rotation. If the stress is removed and the material returns to its initial shape, it is called elastic material. If, however, after removal of the stress, the previous undeformed state is not restored, then the material is called plastic material.

The second distinctive branch of continuum mechanics is fluid mechanics. It deals with the motion of *fluid particles* and describes their behavior under any dynamic condition where the particle velocity may range from low subsonic to hypersonic. It also includes the special case termed fluid statics, where the fluid velocity approaches zero. Fluids are encountered in various forms including homogeneous liquids, unsaturated, saturated, and superheated vapors, polymers and inhomogeneous liquids and gases. As we will see in the this chapters, only a few equations govern the motion of a fluid that consists of molecules. At microscopic level, the molecules continuously interact with each other moving with random velocities. The degree of interaction and the mutual exchange of momentum between the molecules increases with increasing temperature, thus, contributing to an intensive and random molecular motion. In this chapter we present the conservation laws of fluid mechanics that are necessary to understand the basics of flow physics from a unified point of view. The main subject of this chapter is the differential treatment of the conservation laws of fluid mechanics, namely conservation law of mass, linear momentum, angular momentum, and energy.

In many engineering applications, the fluid particles change the frame of reference from a *stationary frame* followed by a *moving frame*. The moving frame may

M. T. Schobeiri, *Tensor Analysis for Engineers and Physicists - With Application to Continuum Mechanics, Turbulence, and Einstein's Special and General Theory of Relativity*, https://doi.org/10.1007/978-3-030-35736-8_5

be translational, rotational or the combination of both. The absolute frame of reference is rigidly connected with the stationary parts, whereas the relative frame is attached to the rotating shaft, thereby turning with certain angular velocity about the machine axis. By changing the frame of reference from an absolute frame to a relative one, certain flow quantities remain unchanged, such as normal stress tensor, shear stress tensor, and deformation tensor. These quantities are indifferent with regard to a change of frame of reference. However, there are other quantities that undergo changes when moving from a stationary frame to a rotating one. Velocity, acceleration, and rotation tensor are a few. We first apply these laws to the stationary or absolute frame of reference.

The differential analysis is of primary significance to all engineering applications. A complete set of independent conservation laws exhibits a system of partial differential equations that describes the motion of a fluid particle. Once this differential equation system is defined, its solution delivers the detailed information about the flow quantities within the computational domain with given initial and boundary conditions.

5.1 Mass Flow Balance in Stationary Frame of Reference

The conservation law of mass requires that the mass contained in a material volume $v = v(t)$, must be constant:

$$m = \int_{v(t)} \rho dv. \tag{5.1}$$

Consequently, Eq. (5.1) requires that the substantial changes of the above mass must disappear:

$$\frac{Dm}{Dt} = \frac{D}{Dt} \int_{v(t)} \rho dv = 0. \tag{5.2}$$

Using the Reynolds transport theorem (see Chap. 4), the conservation of mass, Eq. (4.35), results in:

$$\frac{D}{Dt} \int_{v(t)} \rho dv = \int_{v(t)} \left(\frac{\partial \rho}{\partial t} + \nabla \cdot (\rho \boldsymbol{V}) \right) dv = 0. \tag{5.3}$$

Since the integral in Eq. (5.3) is zero, the integrand in the parentheses must vanish identically. As a result, we may write the continuity equation for unsteady and compressible flow as:

$$\frac{\partial \rho}{\partial t} + \nabla \cdot (\rho \boldsymbol{V}) = 0. \tag{5.4}$$

Equation (5.4) is a coordinate invariant equation. Its index notation in the Cartesian coordinate system given is:

$$\frac{\partial \rho}{\partial t} + \frac{\partial (\rho V_i)}{\partial x_i} = 0. \tag{5.5}$$

Expanding Eq. (5.5), we get:

$$\frac{\partial \rho}{\partial t} + \frac{\partial (\rho V_1)}{\partial x_1} + \frac{\partial (\rho V_2)}{\partial x_2} + \frac{\partial (\rho V_3)}{\partial x_3} = 0. \tag{5.6}$$

5.1.1 Incompressibility Condition

The condition for a working medium to be considered as incompressible is that the substantial change of its density along the flow path vanishes. This means that:

$$\frac{D\rho}{Dt} = \frac{\partial \rho}{\partial t} + \boldsymbol{V} \cdot \nabla \rho = 0. \tag{5.7}$$

Inserting Eq. (5.7) into Eq. (5.4) and performing the chain differentiation of the second term in Eq. (5.4) namely, $\nabla \cdot (\rho \boldsymbol{V}) = \rho \nabla \cdot \boldsymbol{V} + \boldsymbol{V} \cdot \nabla \rho$, the continuity equation for an incompressible flow reduces to:

$$\nabla \cdot \boldsymbol{V} = 0. \tag{5.8}$$

In a Cartesian coordinate system, Eq. (5.8) can be expanded as written in Eq. (5.9):

$$\frac{\partial V_1}{\partial x_1} + \frac{\partial V_2}{\partial x_2} + \frac{\partial V_3}{\partial x_3} = 0. \tag{5.9}$$

In an orthogonal, curvilinear coordinate system, the continuity balance for an incompressible fluid treated in Chap. 6.

5.2 Momentum Balance in Stationary Frame

In addition to the continuity equation we treated above, the detailed calculation of the entire flow field through different engineering devices and components requires the equation of motion in differential form. In the following, we provide the equation of motion in differential form in a four-dimensional time-space coordinate. We start from Newton's second law of motion and apply it to an infinitesimal fluid element shown in Fig. 5.1, with the mass dm for which the equilibrium condition is written as:

$$dm\boldsymbol{A} = d\boldsymbol{F}. \tag{5.10}$$

The acceleration vector $\boldsymbol{A}$ is the well known material derivative (see Chap. 3, 3.1:

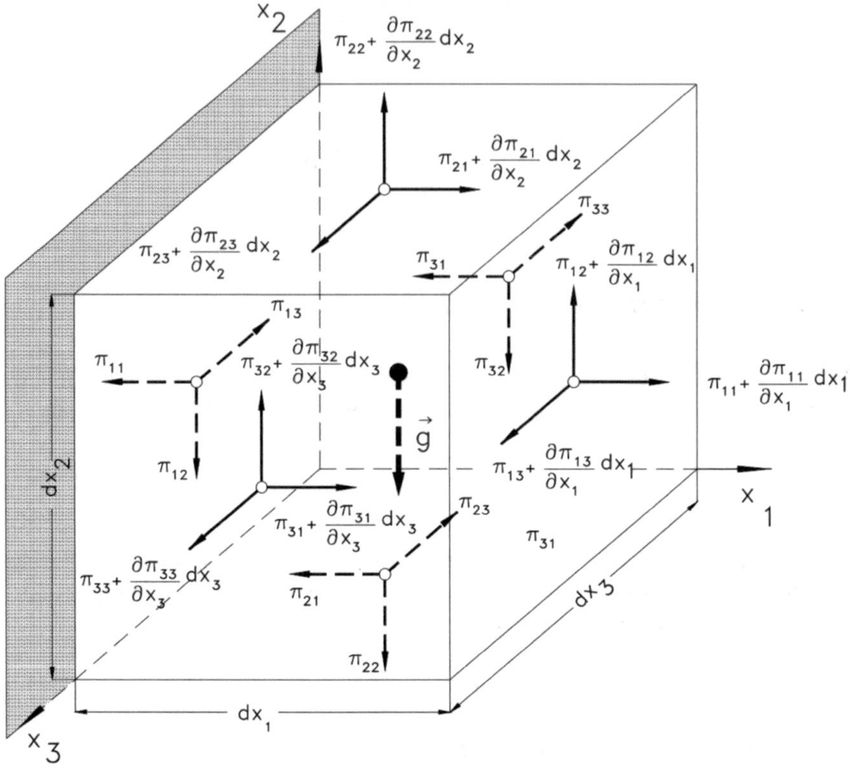

Fig. 5.1 Surface and gravitational forces acting on a volume element

$$A = \frac{DV}{Dt} = \frac{\partial V}{\partial t} + V \cdot \nabla V. \tag{5.11}$$

In Eq. (5.10), $\boldsymbol{A}$ is the acceleration vector and $d\boldsymbol{F}$ the vector sum of all forces exerting on the fluid element. In the absence of magnetic, electric or other extraneous effects, the force $d\boldsymbol{F}$ is equal to the vector sum of the surface force $d\boldsymbol{F}_s$ acting on the particle surface and the gravity force $dm\boldsymbol{g}$ as shown in Fig. 5.1. Inserting Eq. (5.11) into Eq. (5.10), we arrive at:

$$dm\left(\frac{\partial \boldsymbol{V}}{\partial t} + \boldsymbol{V} \cdot \nabla \boldsymbol{V}\right) = d\boldsymbol{F}_s + dm\boldsymbol{g}. \tag{5.12}$$

Consider the fluid element shown in Fig. 5.1 with sides dx_1, dx_2, dx_3 parallel to the axis of a Cartesian coordinate system. The stresses acting on the surfaces of this element are represented by the stress tensor $\boldsymbol{\Pi}$ which has the components π_{ij} that produce surface forces. The first index i refer to the axis, on which the fluid element surface is perpendicular, whereas the second index j indicates the direction of the stress component. Considering the stress situation in Fig. 5.1, the following resultant forces are acting on the surface dx_2dx_3 perpendicular to the x_1 axis:

$$e_1\frac{\partial\pi_{11}}{\partial x_1}dx_1dx_2dx_3,\ e_2\frac{\partial\pi_{12}}{\partial x_1}dx_1dx_2dx_3,\ e_3\frac{\partial\pi_{13}}{\partial x_1}dx_1dx_2dx_3. \tag{5.13}$$

The resulting forces acting on the other two surfaces are found in similar manner. Thus, the total resulting forces acting on the entire surface of the element are obtained by adding the nine components that result in Eq. (5.14):

$$\frac{d\boldsymbol{F}_s}{dv} = e_1\left(\frac{\partial\pi_{11}}{\partial x_1}+\frac{\partial\pi_{21}}{\partial x_2}+\frac{\partial\pi_{31}}{\partial x_3}\right)+e_2\left(\frac{\partial\pi_{12}}{\partial x_1}+\frac{\partial\pi_{22}}{\partial x_2}+\frac{\partial\pi_{32}}{\partial x_3}\right) + e_3\left(\frac{\partial\pi_{13}}{\partial x_1}+\frac{\partial\pi_{23}}{\partial x_2}+\frac{\partial\pi_{33}}{\partial x_3}\right). \tag{5.14}$$

Since the stress tensor $\boldsymbol{\Pi}$ and the volume of the fluid element are written as:

$$\boldsymbol{\Pi} = e_ie_j\pi_{ij}, \qquad dv = dx_1dx_2dx_3. \tag{5.15}$$

It can be easily shown that Eq. (5.14) is the divergence of the stress tensor expressed in Eq. (5.15)

$$\frac{d\boldsymbol{F}_s}{dv} = \nabla\cdot\boldsymbol{\Pi}. \tag{5.16}$$

The expression $\nabla\cdot\boldsymbol{\Pi}$ is the scalar differentiation of the second order tensor $\boldsymbol{\Pi}$ and is called divergence of the tensor field $\boldsymbol{\Pi}$ which is a first order tensor or a vector. Inserting Eq. (5.16) into Eq. (5.12) and dividing both sides by dm, results in the following **Cauchy equation of motion**:

$$\frac{\partial\boldsymbol{V}}{\partial t}+\boldsymbol{V}\cdot\nabla\boldsymbol{V} = \frac{1}{\rho}\nabla\cdot\boldsymbol{\Pi}+\boldsymbol{g}. \tag{5.17}$$

The stress tensor in Eq. (5.17) can be expressed in terms of deformation tensor, as we will see in the following section.

5.2.1 Relationship Between Stress Tensor and Deformation Tensor

Since the surface forces resulting from the stress tensor cause a deformation of fluid particles, it is obvious that one might attempt to find a functional relationship between the stress tensor and the velocity gradient:

$$\boldsymbol{\Pi} = f(\nabla\boldsymbol{V}). \tag{5.18}$$

As we saw in Chap. 2, the velocity gradient in Eq. (5.18) can be decomposed into an symmetric part called deformation tensor and an anti-symmetric part, called rotation or vorticity tensor:

$$\nabla \boldsymbol{V} = \frac{1}{2}(\nabla \boldsymbol{V} + \nabla \boldsymbol{V}^{\boldsymbol{T}}) + \frac{1}{2}(\nabla \boldsymbol{V} - \nabla \boldsymbol{V}^{\boldsymbol{T}}) = \boldsymbol{D} + \boldsymbol{\Omega}. \tag{5.19}$$

Consequently, the stress tensor may be set:

$$\boldsymbol{\Pi} = f(\nabla \boldsymbol{V}) = f(\boldsymbol{D}, \boldsymbol{\Omega}) \tag{5.20}$$

with the deformation tensor as:

$$\boldsymbol{D} = e_i e_j D_{ij} = \frac{1}{2} e_i e_j \left(\frac{\partial V_i}{\partial x_j} + \frac{\partial V_j}{\partial x_i} \right) \tag{5.21}$$

and the rotation tensor, which is antisymmetric, and is given by Eq. (2.27):

$$\boldsymbol{\Omega} = e_i e_j \Omega_{ij} = \frac{1}{2} e_i e_j \left(\frac{\partial V_j}{\partial x_i} - \frac{\partial V_i}{\partial x_j} \right). \tag{5.22}$$

Since the stress tensor $\boldsymbol{\Pi}$ in Eq. (5.20) is a *frame indifferent quantity*, it remains unchanged or invariant under any changes of frame of reference. Moving from an absolute frame into a relative one exhibits such a change in frame of reference. In order to understand the mathematical structure behind the principle of *frame indifference* or *objectivity*, we start with the *deformation gradient* $\boldsymbol{G}$ that we already discussed in Chap. 4. There, we used the vector $\boldsymbol{G}_i$ of the second order tensor $\boldsymbol{G}$ as a transformation vector to establish the Jacobian function that relates the undeformed volume element to the deformed one. Here we use the tensor $\boldsymbol{G}$ to establish a relationship between the non-deformed line element $d\xi_i$ and the corresponding deformed line element dx_i:

$$\begin{aligned} d\boldsymbol{x} &= \boldsymbol{G} \cdot d\boldsymbol{\xi} \\ dx_i &= G_{ik} d\xi_k \text{ with } G_{ik} \text{ as} \\ G_{ik} &= \frac{\partial x_k(\xi_1, \xi_2, \xi_3, t)}{\partial \xi_i}. \end{aligned} \tag{5.23}$$

As seen from the first equation in (5.23), the scalar operation between the second order tensor $\boldsymbol{G}$ and the first order tensor $d\boldsymbol{\xi}$ has resulted is a first order tensor $d\boldsymbol{x}$. Since the result of this operation is quite obvious, from now on, we may omit the *dot* (.) between the tensors. Similarly whenever we have a product of two second order tensors with a dot between them, the result is a second order tensor, thus we may omit the dot between them also. The same is true for the product of two second order tensor with a *double dot* (:) between them, the result is a scalar quantity. The inverse transformation of Eq. (5.23) is:

$$d\boldsymbol{\xi} = \boldsymbol{G}^{-1} d\boldsymbol{x}$$
$$d\xi_i = (G_{ik})^{-1} d\xi_k \text{ with } (G_{ik})^{-1} \text{ as}$$
$$(G_{ik})^{-1} = \frac{\partial \xi_i(x_1, x_2, x_3, t)}{\partial x_k}. \tag{5.24}$$

Thus with Eqs. (5.23) and (5.24) we have:

$$\boldsymbol{G}\boldsymbol{G}^{-1} = \boldsymbol{G}^{-1}\boldsymbol{G} = \boldsymbol{I} \tag{5.25}$$

with $\boldsymbol{I}$ as the unitary tensor. Differentiating Equation (5.25) with respect to time leads to:

$$\dot{\boldsymbol{G}}\boldsymbol{G}^{-1} + \boldsymbol{G}\dot{\boldsymbol{G}}^{-1} = 0. \tag{5.26}$$

Differentiating the deformation tensor G_{ik} with respect to time, we obtain:

$$\dot{G}_{ik} = \frac{\partial^2 x_i(\boldsymbol{\xi}, t)}{\partial \xi_k \partial t} = \frac{\partial V_i}{\partial x_l} \frac{\partial x_l}{\partial \xi_k}. \tag{5.27}$$

In Eq. (5.27) the first fraction is the matrix of the velocity gradient

$$e_l e_i \frac{\partial V_i}{\partial x_l} = \nabla \boldsymbol{V} \tag{5.28}$$

and the second fraction is the matrix of the deformation gradient, thus the derivative of the deformation gradient is:

$$\dot{\boldsymbol{G}} = (\nabla \boldsymbol{V}) \cdot \boldsymbol{G}. \tag{5.29}$$

Note that in Eq. (5.29) the velocity gradient $\nabla \boldsymbol{V}$, the deformation gradient $\boldsymbol{G}$ as well as its derivative $\dot{\boldsymbol{G}}$ are second order tensors. The latter is the result of the contraction between $\nabla \boldsymbol{V}$ and $\boldsymbol{G}$ resulting in second order tensor $\dot{\boldsymbol{G}}$. As elaborated above, since the result of Eq. (5.29), in what follows, we also omit the dot between $\nabla \boldsymbol{V}$ and $\boldsymbol{G}$ as we already did in Eqs. (5.24) and (5.25). Furthermore, we revisit Chap. 2, Eq. (2.17) where the components of the position vector $\boldsymbol{R}$ is transformed from a stationary frame into a rotating one through the transformation matrix $\boldsymbol{Q}$. In an analogous way, one could be tempted to establish a similar relation for the deformation gradient such as:

$$\boldsymbol{G}^* = \boldsymbol{Q}\boldsymbol{G}. \tag{5.30}$$

The fact that $\boldsymbol{G}^* \neq \boldsymbol{G}'$ underlines that $\boldsymbol{G}^*$ does not follow the transformation rules as $\boldsymbol{G}'$ does.

With the mathematical preparation given above we now turn our attention to the question of frame indifference property of the stress tensor. Given the causality principle on which Newtonian physics is based, whenever a material point is subjected to a stress, it undergoes a deformation. Thus there must be a unique relation between the stress and the deformation. This interaction is described by

$$\boldsymbol{\Pi} = f(\boldsymbol{G}) \tag{5.31}$$

with $\boldsymbol{\Pi}$ as the total stress tensor and $\boldsymbol{G}$ as the tensor of deformation gradient. The stress tensor can be transformed from an absolute (stationary) frame into a relative(moving) frame that includes translation and rotation. While its components follow the transformation rule discussed in Chap. 2 and given below:

$$\begin{aligned}[\boldsymbol{\Pi}] &= [\boldsymbol{Q}][\boldsymbol{\Pi}'][\boldsymbol{Q}]^T \\ [\boldsymbol{\Pi}'] &= [\boldsymbol{Q}]^T[\boldsymbol{\Pi}][\boldsymbol{Q}]\end{aligned} \tag{5.32}$$

the stress tensor itself will not change, meaning that it remains frame indifferent or *objective*. Applying this statement to Eq. (5.31), it follows that the argument of the functional f, namely the deformation gradient $\boldsymbol{G}$ must also follow the same transformation rule as the total stress tensor $\boldsymbol{\Pi}$ in Eq. (5.32). This, although quite intuitive from mathematical point of view, requires some physical explanation: Given is a system consisting of an absolute (stationary) frame and a relative (moving) frame that includes translation and rotation. An observer, A, located in the absolute frame measures the deformation gradient $\boldsymbol{G}$ of a material point using any technique to measure the dimensions dx_1, dx_2, dx_3 of the material point. The observer A calculates the total stress $\boldsymbol{\Pi}$ by virtue of Eq. (5.31). Another observer, B, situated in the relative frame measures the deformation gradient $\boldsymbol{G}^*$ in the relative frame and likewise calculates the stress $\boldsymbol{\Pi}^*$ using Eq. (5.31). Both observers start the observation of the motion at the same time $t = \tau$. Comparing the results, both observers come to the conclusion that $\boldsymbol{\Pi}^* \neq \boldsymbol{\Pi}$. To explain the difference, we resort to simple fluids such as liquids and gases. The stress tensor of such fluid can be expressed as:

$$\boldsymbol{\Pi} = -p\boldsymbol{I} + \boldsymbol{T} \tag{5.33}$$

with $-p\boldsymbol{I}$ as the pressure tensor, p the thermodynamic pressure and $\boldsymbol{T}$ the *friction stress tensor* to be determined. When the fluid is at rest, the friction tensor vanishes and $\boldsymbol{\Pi}$ reduces to the pressure tensor, which is a spherically symmetric second order tensor. For the friction stress tensor $\boldsymbol{T}$ (also called *extra stress tensor*) we introduce the following ansatz:

$$\boldsymbol{T} = f(\dot{\boldsymbol{G}}) = f((\nabla \boldsymbol{V}) \cdot \boldsymbol{G}) \tag{5.34}$$

with $\dot{\boldsymbol{G}}$ from Eq. (5.29). From Eq. (5.23) with $\boldsymbol{G}$ we have for an elastic medium in general a function that connects the elements of an undeformed reference configuration to the elements of a deformed configuration. In fluids, however, there is no specific undeformed reference configuration. Consequently, any configuration can be taken as the reference configuration meaning that $\boldsymbol{G} = \boldsymbol{I}$. As a result Eq. (5.29) can be rearranged as:

$$\dot{G}_{G=I} = \nabla V. \tag{5.35}$$

Considering Eq. (5.34), the friction stress tensor Equation (5.33) is reduced to:

$$T = f(\nabla V) \equiv f(L). \tag{5.36}$$

For the sake of simplifying the following steps, in Eq. (5.35) we have renamed $L \equiv \nabla V = \dot{G}$. Since the friction stress tensor T as an objective tensor follows the transformation rules, the tensor L and also $\dot{G}$ must follow the transformation rule. This, however, is not the case as we find out below. To proof this statement, we first differentiate Equation (5.30) with respect to time:

$$(G^*)^{\cdot} = \dot{Q}G + Q\dot{G}. \tag{5.37}$$

Using Eq. (5.29), where ∇V was replaced by L, we have:

$$(G^*)^{\cdot} = \dot{Q}G + Q\dot{G} = \dot{Q}G + QLG = (\dot{Q} + QL)G. \tag{5.38}$$

Rearranging Eq. (5.38),

$$(G^*)^{\cdot} = (\dot{Q} + QL)G = (\dot{Q} + QL)\,Q^T G^*. \tag{5.39}$$

Following Eq. (5.34) with $L \equiv \nabla V = \dot{G}|_{F=I}$ and $L^* = (\dot{G})^*|_{F^*=I}$ we conclude from Eq. (5.39) that:

$$L^* \equiv \dot{Q}\,Q^T + QL\,Q^T. \tag{5.40}$$

Now, if L is supposed to follow the transformation rule, meaning that $L^* \equiv L$, the first expression on the left hand side of Eq. (5.40), namely $\dot{Q}\,Q^T$ must disappear. To find out the circumstance under which $\dot{Q}\,Q^T$ becomes zero, we decompose the velocity gradient tensor ∇V into a symmetric part D and an antisymmetric part Ω:

$$L = \nabla V = \frac{1}{2}(\nabla V + \nabla V^T) + \frac{1}{2}(\nabla V - \nabla V^T) = D + \Omega. \tag{5.41}$$

Now we insert in Eq. (5.40) the decomposition from Eq. (5.41) and obtain

$$L^* \equiv \dot{Q}\,Q^T + Q(D + \Omega)\,Q^T = \dot{Q}\,Q^T + QD\,Q^T + Q\Omega\,Q^T \tag{5.42}$$

which is re-arranged as:

$$L^* = QD\,Q^T + \dot{Q}\,Q^T + Q\Omega\,Q^T. \tag{5.43}$$

Considering Eqs. (2.21) and (2.27) the last two terms in Eq. (5.43) cancel each other out resulting in:

$$L^* = QD\,Q^T. \tag{5.44}$$

With Eq. (5.44) as a frame indifferent tensor, the original ansatz in Eq. (5.36) must be revised as:

$$\boldsymbol{S} = f(\boldsymbol{L}^*) = f(\boldsymbol{D}). \tag{5.45}$$

Equation (5.45) satisfies the objectivity requirement. As a consequence, the total stress tensor is a function of deformation tensor $\boldsymbol{D}$ only.

$$\boldsymbol{\Pi} = f(\boldsymbol{D}). \tag{5.46}$$

A general form of Eq. (5.46) may be a polynomial in $\boldsymbol{D}$ as suggested in [1]:

$$\boldsymbol{\Pi} = f_1\boldsymbol{I} + f_2\boldsymbol{D} + f_3(\boldsymbol{D}\cdot\boldsymbol{D}) \tag{5.47}$$

with $\boldsymbol{I} = e_i e_j \delta_{ij}$ as the unit Kronecker tensor. To fulfill the frame indifference requirement, the functions f_i must be invariant. This means they depend on either the thermodynamic quantities, such as pressure, or the following three-principal invariant of the deformation tensor:

$$I_{1D} = \text{Tr}\boldsymbol{D} = \nabla\cdot\boldsymbol{V} = D_{ii} \tag{5.48}$$

$$I_{3D} = \det D_{ij} \tag{5.49}$$

$$I_{2D} = \frac{1}{2}(I_{1D}^2 - \boldsymbol{D}:\boldsymbol{D}) = \frac{1}{2}(D_{ii}D_{jj} - D_{ij}D_{ij}). \tag{5.50}$$

Of particular interest is the category of those fluids for which there is a linear relationship between the stress tensor and the deformation tensor. Many working fluids used in engineering applications, such as air, steam, and combustion gases, belong to this category. They are called the **Newtonian fluids** for which Eq. (5.47) is reduced to:

$$\boldsymbol{\Pi} = f_1\boldsymbol{I} + f_2\boldsymbol{D} \tag{5.51}$$

where the functions f_1 and f_2 are given by:

$$f_1 = (-p + \lambda\nabla\cdot\boldsymbol{V}),\ f_2 = 2\mu,\ \bar{\mu} = \lambda + \frac{2}{3}\mu \tag{5.52}$$

with μ as the absolute viscosity and $\bar{\mu}$ as the bulk viscosity. Introducing Eq. (5.52) into Eq. (5.47) results in the **Cauchy–Poisson law**:

$$\boldsymbol{\Pi} = (-p + \lambda\nabla\cdot\boldsymbol{V})\boldsymbol{I} + 2\mu\boldsymbol{D} = -pI + \lambda\nabla\cdot\boldsymbol{V}\boldsymbol{I} + 2\mu\boldsymbol{D} = -p\boldsymbol{I} + \boldsymbol{T}. \tag{5.53}$$

In Eq. (5.53), the terms with the coefficients involving viscosity are grouped together leading to a pressure tensor $-p\boldsymbol{I}$ and a friction stress tensor $\boldsymbol{T}$ that reads:

$$\boldsymbol{T} = \lambda(\nabla \cdot \boldsymbol{V})\boldsymbol{I} + 2\mu \boldsymbol{D}. \tag{5.54}$$

The first term on the right-hand side of Eq. (5.53) associated with the unit Kronecker tensor, $p\boldsymbol{I}$, represents the contribution of the thermodynamic pressure to the normal stress. The second term, $(\lambda\nabla \cdot \boldsymbol{V})\boldsymbol{I}$, exhibits a normal stress contribution caused by a volume dilatation or compression due to the compressibility of the working medium. For an incompressible medium, this term identically vanishes. The coefficient λ related to the coefficient of shear viscosity μ and the bulk viscosity $\bar{\mu}$ is given in Eq. (5.52) as $\bar{\mu} = \lambda + 2/3\mu$. For most of the fluids used in engineering applications, the bulk viscosity may be approximated as $\bar{\mu} = \lambda + 2/3\mu = 0$ leading to $\lambda = -2/3\mu$. This relation, frequently called the Stokes' relation, is valid for monoatomic gases [2]. Finally, the last term expresses a direct relationship between the shear stress tensor and the deformation tensor. For an incompressible fluid, Eq. (5.53) reduces to:

$$\boldsymbol{\Pi} = -p\boldsymbol{I} + 2\mu \boldsymbol{D}. \tag{5.55}$$

5.2.2 *Navier–Stokes Equation of Motion in Stationary Frame*

Inserting Eq. (5.53) into Eq. (5.17) we obtain:

$$\rho\frac{\partial V}{\partial t} + \rho \boldsymbol{V} \cdot \nabla \boldsymbol{V} = \nabla \cdot [(-p + \lambda\nabla \cdot \boldsymbol{V})\boldsymbol{I} + 2\mu \boldsymbol{D}] + \rho \boldsymbol{g}. \tag{5.56}$$

This is referred to as the **Navier–Stokes equation for compressible fluids**. In Eq. (5.56), the coefficient λ can be expressed in terms of shear viscosity μ. This, however, requires rearranging the second and third term in the bracket by using the index notation. Assuming that the viscosities μ, and thus λ are not varying spatially, for $\nabla \cdot (\lambda\nabla \cdot \boldsymbol{V}\boldsymbol{I})$ we may write:

$$\nabla \cdot (\lambda\nabla \cdot \boldsymbol{V}\boldsymbol{I}) = \left(e_i\frac{\partial}{\partial x_i}\right) \cdot \left(\lambda\frac{\partial V_j}{\partial x_j}e_k e_i \delta_{kl}\right) = \lambda\delta_{ik}\delta_{kl}e_l\frac{\partial^2 V_j}{\partial x_i \partial x_j}$$

$$\nabla \cdot (\lambda\nabla \cdot \boldsymbol{V}\boldsymbol{I}) = \lambda e_i \frac{\partial^2 V_j}{\partial x_i \partial x_j} = \lambda e_i \frac{\partial}{\partial x_i}\left(\frac{\partial V_j}{\partial x_j}\right) = \lambda\nabla(\nabla \cdot \boldsymbol{V}). \tag{5.57}$$

We apply the same procedure to $\nabla \cdot (2\mu \boldsymbol{D})$:

$$\nabla \cdot (2\mu \boldsymbol{D}) = 2\mu\left(e_i\frac{\partial}{\partial x_i}\right) \cdot \left[\frac{1}{2}e_j e_k\left(\frac{\partial V_j}{\partial x_k} + \frac{\partial V_k}{\partial x_j}\right)\right]$$

$$\nabla \cdot (2\mu \boldsymbol{D}) = \mu\left[e_k\frac{\partial}{\partial x_k}\left(\frac{\partial V_i}{\partial x_i}\right) + e_k\frac{\partial^2 V_k}{\partial x_i \partial x_i}\right] = \mu[\nabla(\nabla \cdot \boldsymbol{V}) + \Delta \boldsymbol{V}]. \tag{5.58}$$

Introducing Eqs. (5.57) and (5.58) into Eq. (5.56), we arrive at

$$\rho\left(\frac{\partial \boldsymbol{V}}{\partial t}+\boldsymbol{V}\cdot\nabla\boldsymbol{V}\right)=-\nabla p+(\lambda+\mu)\nabla(\nabla\cdot\boldsymbol{V})+\mu\Delta\boldsymbol{V}+\rho\boldsymbol{g}. \tag{5.59}$$

For $\lambda=-2/3\mu$, Eq. (5.59) results in:

$$\rho\left(\frac{\partial \boldsymbol{V}}{\partial t}+\boldsymbol{V}\cdot\nabla\boldsymbol{V}\right)=-\nabla p+\frac{\mu}{3}\nabla(\nabla\cdot\boldsymbol{V})+\mu\Delta\boldsymbol{V}+\rho\boldsymbol{g}. \tag{5.60}$$

For incompressible flows with constant shear viscosity, Eq. (5.56) reduces to:

$$\rho\frac{\partial \boldsymbol{V}}{\partial t}+\rho\boldsymbol{V}\cdot\nabla\boldsymbol{V}=\nabla\cdot(-p\boldsymbol{I}+2\mu\boldsymbol{D})+\rho\boldsymbol{g}. \tag{5.61}$$

Performing the differentiation on the right-hand side and dividing by ρ leads to:

$$\frac{\partial \boldsymbol{V}}{\partial t}+\boldsymbol{V}\cdot\nabla\boldsymbol{V}=-\frac{1}{p}\nabla p+\nu\nabla\boldsymbol{V}+\boldsymbol{g} \tag{5.62}$$

with $\nu=\mu/\rho$ as the kinematic viscosity and $\Delta=\nabla\cdot\nabla=\nabla^2$ as the *Laplace operator*. Equation (5.56) or its special case, Eq. (5.62) with the equation of continuity and energy, exhibits a system of partial differential equations. This system describes the flow field completely. Its solution yields the detailed distribution of flow quantities. In many engineering applications, with the exception of hydro power generation, the contribution of the gravitational term $\boldsymbol{g}=e_i g_i=e_3 g_3$ compared to the other terms is negligibly small. Equation (5.62) in Cartesian index notation is written as:

$$\frac{\partial V_i}{\partial t}+V_j\frac{\partial V_i}{\partial x_j}=-\frac{1}{p}\frac{\partial p}{\partial x_i}+\nu\frac{\partial^2 V_i}{\partial x_j\partial x_j}+g_i. \tag{5.63}$$

Using the Einstein summation convention, the three components of Eq. (5.63) are:

$$\begin{aligned}
\frac{\partial V_1}{\partial t}+V_1\frac{\partial V_1}{\partial x_1}+V_2\frac{\partial V_1}{\partial x_2}+V_3\frac{\partial V_1}{\partial x_3}&=-\frac{1}{\rho}\frac{\partial p}{\partial x_1}+\nu\left(\frac{\partial^2 V_1}{\partial x_1^2}+\frac{\partial^2 V_1}{\partial x_2^2}+\frac{\partial^2 V_1}{\partial x_3^2}\right)\\
\frac{\partial V_2}{\partial t}+V_1\frac{\partial V_2}{\partial x_1}+V_2\frac{\partial V_2}{\partial x_2}+V_3\frac{\partial V_2}{\partial x_3}&=-\frac{1}{\rho}\frac{\partial p}{\partial x_2}+\nu\left(\frac{\partial^2 V_2}{\partial x_1^2}+\frac{\partial^2 V_2}{\partial x_2^2}+\frac{\partial^2 V_2}{\partial x_3^2}\right)\\
\frac{\partial V_3}{\partial t}+V_1\frac{\partial V_3}{\partial x_1}+V_2\frac{\partial V_3}{\partial x_2}+V_3\frac{\partial V_3}{\partial x_3}&=-\frac{1}{\rho}\frac{\partial p}{\partial x_3}+\nu\left(\frac{\partial^2 V_3}{\partial x_1^2}+\frac{\partial^2 V_3}{\partial x_2^2}+\frac{\partial^2 V_3}{\partial x_3^2}\right)+.
\end{aligned} \tag{5.64}$$

To obtain the components of the Navier–Stokes equation in an orthogonal curvilinear coordinate system, we use metric coefficients, Christoffel symbols, and the index notation outlined in Chap. 6:

$$\boldsymbol{g}_i\left(\frac{\partial V^i}{\partial t}+V^jV^i_j+V^jV^k\Gamma^i_{kj}\right)=-\tfrac{1}{\rho}\boldsymbol{g}_ig^{ji}p_j+\nu\boldsymbol{g}_m[V^m_{ik}+$$
$$V^n_i\Gamma^m_{nk}+V^n_k\Gamma^m_{ni}-V^m_j\Gamma^j_{ik}+$$
$$V^p(\Gamma^n_{pi}\Gamma^m_{nk}-\Gamma^j_{ik}\Gamma^m_{ik}\Gamma^m_{pj}+\Gamma^m_{pi,k})]g^{ik}. \tag{5.65}$$

Using the Christoffel symbols and the physical components for a cylindrical coordinate system as specified in Chap. 6, we arrive at the component of Navier–Stokes equation in r-direction:

$$\frac{\partial V_r}{\partial t}+V_r\frac{\partial V_r}{\partial r}+\frac{V_\Theta}{r}\frac{\partial V_r}{\partial\theta}+V_z\frac{\partial V_r}{\partial z}-\frac{V_\Theta^2}{r}=-\frac{1}{\rho}\frac{\partial p}{\partial r}+$$
$$\nu\left(\frac{\partial^2 V_r}{\partial r^2}+\frac{1}{r^2}\frac{\partial^2 V_r}{\partial\Theta^2}+\frac{\partial^2 V_r}{\partial z^2}-2\frac{\partial V_\Theta}{r^2\partial\Theta}+\frac{\partial V_r}{r\partial r}-\frac{V_r}{r^2}\right) \tag{5.66}$$

in θ-direction,

$$\frac{\partial V_\Theta}{\partial t}+V_r\frac{\partial V_\Theta}{\partial r}+\frac{V_\Theta}{r}\frac{\partial V_\Theta}{\partial\Theta}+V_z\frac{\partial V_\Theta}{\partial z}+\frac{V_rV_\Theta}{r}=-\frac{1}{\rho}\frac{\partial p}{\partial r\partial\Theta}+$$
$$\nu\left(\frac{\partial^2 V_\Theta}{\partial r^2}+\frac{1}{r^2}\frac{\partial^2 V_\Theta}{\partial\Theta^2}+\frac{\partial^2 V_\Theta}{\partial z^2}+\frac{2}{r^2}\frac{\partial V_r}{\partial\Theta}+\frac{1}{r}\frac{\partial V_\Theta}{\partial r}-\frac{V_\Theta}{r^2}\right) \tag{5.67}$$

and in z-direction:

$$\frac{\partial V_z}{\partial t}+V_r\frac{\partial V_z}{\partial r}+\frac{V_\Theta}{r}\frac{\partial V_z}{\partial z}+V_z\frac{\partial V_z}{\partial z}=-\frac{1}{\rho}\frac{\partial p}{\partial z}+$$
$$\nu\left(\frac{\partial^2 V_z}{\partial r^2}+\frac{\partial^2 V_z}{\partial r^2\partial\Theta^2}+\frac{\partial^2 V_z}{\partial z^2}+\frac{1}{r}\frac{\partial V_z}{\partial r}\right). \tag{5.68}$$

5.2.3 Special Case: Euler Equation of Motion

For the special case of steady inviscid flow (no viscosity), Eq. (5.62) is reduced to:

$$\frac{\partial\boldsymbol{V}}{\partial t}+\boldsymbol{V}\cdot(\nabla\boldsymbol{V})=-\frac{1}{\rho}\nabla p+\boldsymbol{g}. \tag{5.69}$$

This equation is called *Euler equation of motion*. Its index notation is:

$$\frac{\partial V_i}{\partial t}+V_j\frac{\partial V_i}{\partial x_j}=-\frac{1}{\rho}\frac{\partial p}{\partial x_i}+g_i. \tag{5.70}$$

Replacing the convective term in Eq. (5.69) by the following vector identity:

$$\boldsymbol{V} \cdot \nabla \boldsymbol{V} = \nabla(\boldsymbol{V} \cdot \boldsymbol{V})/2 - \boldsymbol{V} \times (\nabla \times V) \tag{5.71}$$

we find that the convective acceleration is expressed in terms of the gradient of the kinetic energy $\nabla(\boldsymbol{V} \cdot \boldsymbol{V})/2 = \nabla(V^2/2)$, and a second term which is a vector product of the velocity and the vorticity vector $\nabla \times \boldsymbol{V}$. If the flow field under investigation allows us to assume a zero vorticity within certain flow regions, then we may assign a *potential* to the velocity field that significantly simplifies the equation system. This assumption is permissible for the flow region outside the boundary layer and is discussed more in detail in Chap. 6.

Before proceeding with the conservation of energy, in context of the Euler equation that describes the motion of inviscid flows, it is appropriate to present the *Bernoulli* equation, which exhibits a special integral form of Euler equations. For this purpose, we first rearrange the gravitational acceleration vector by introducing a scalar surface potential z, whose gradient ∇z has the same direction as the unit vector in x_3-direction. Furthermore, it has only one component that points in the negative x_3-direction. As a result, we may write $\boldsymbol{g} = -e_3 g = -g\nabla(z)$. Thus, the Euler equation of motion assumes the following form:

$$\frac{\partial \boldsymbol{V}}{\partial t} + \nabla\left(\frac{V^2}{2} + gz\right) + \frac{1}{\rho}\nabla p = \boldsymbol{V} \times (\nabla \times V). \tag{5.72}$$

Equation (5.72) shows that despite the inviscid flow assumption, it contains vorticities that are inherent in viscous flows and cause additional entropy production. This can be expressed in terms of the second law of thermodynamics, $Tds = dh = -dp/\rho$, where the changes of entropy, enthalpy, and static pressure, ds, dh, dp, or other thermodynamic properties are expressed in terms of the product of their gradients and a differential displacement as shown by Eq. (2.24) $dQ = d\boldsymbol{X} \cdot \nabla Q$. Replacing the quantity Q by the following properties, we obtain:

$$ds = d\boldsymbol{X} \cdot \nabla s, \quad dh = d\boldsymbol{X} \cdot \nabla h, \quad dp = d\boldsymbol{X} \cdot \nabla p \tag{5.73}$$

with s as the specific entropy, h as the specific static enthalpy, and p the static pressure. Inserting the above property changes into the first law of thermodynamics, $Tds = dh - dp/\rho$, we find:

$$\frac{\partial \boldsymbol{V}}{\partial t} + \nabla\left(h + \frac{V^2}{2} + gz\right) = \boldsymbol{V} \times (\nabla \times V) + T\nabla s. \tag{5.74}$$

As we comprehensively discuss in Chap. 5, the expression in the parentheses on the left-hand side of Eq. (5.74) is the total enthalpy $H = (h + V^2/2 + gz)$. In the absence of mechanical or thermal energy addition or rejection, H remains constant meaning that its gradient ∇H vanishes. Furthermore, for steady flow cases, Eq. (5.74) reduces to:

$$\boldsymbol{V} \times (\nabla \times V) = -T\nabla s. \tag{5.75}$$

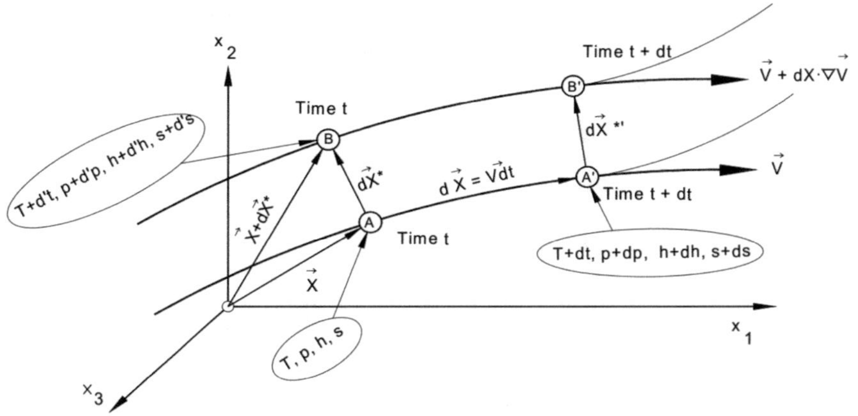

Fig. 5.2 Fluid particles at different thermodynamic state conditions

Equation (5.75) is an important result that establishes a direct relation between the vorticity and the entropy production in inviscid flows. In a flow field with discontinuities as a result of the presence of shock waves, there are always jumps in velocities across the shock front causing vorticity production and therefore, changes in entropy.

The Bernoulli equation can be obtained as a scalar product of the Euler differential equation (5.69) and a differential displacement vector. Figure 5.2 shows different displacement vectors that, in principle, may be used. The vector $d\boldsymbol{X}^*$ shows the differential distance between two neighboring fluid particles located at positions A and B at the same time t with the thermodynamic states shown in Fig. 5.2. The particles move along their flow paths and at $t + dt$, they occupy the positions □′ and B'. The distance □□′ is denoted by $d\boldsymbol{X}$. The thermodynamic conditions at $\boldsymbol{A}'$, denoted by $T + dT$, $p + dp$, $h + dh$, $s + ds$, indicate that the changes this particle has undergone are different from those of particle B. Thus, the vector $d\boldsymbol{X} = \boldsymbol{V}dt$ is the appropriate vector which we choose to multiply with the Euler equation of motion. For steady flow cases, the differential distance $d\boldsymbol{X}$ along the particle path is identical with a distance along a streamline. Thus, the multiplication of Euler equation (5.74) with the differential displacement $d\boldsymbol{X} = \boldsymbol{V}dt$, gives:

$$\boldsymbol{V} \cdot \left(\frac{\partial \boldsymbol{V}}{\partial t}\right) dt + d\boldsymbol{X} \cdot \nabla \left(\frac{V^2}{2} + gz\right) + d\boldsymbol{X} \cdot \frac{\nabla p}{\rho} = \boldsymbol{V} \cdot (\boldsymbol{V} \times \nabla \times V)dt. \quad (5.76)$$

The terms in Eq. (5.76) must be rearranged as follows. The first term starts with a scalar product of two vectors, eliminating the Kronecker delta and utilizing the Einstein summation convention results in:

$$\boldsymbol{V} \cdot \left(\frac{\partial \boldsymbol{V}}{\partial t}\right) dt = V_i \frac{\partial V_i}{\partial t} dt = \frac{1}{2}\frac{\partial (V_i V_i)}{\partial t} dt = \frac{1}{2}\frac{\partial V^2}{\partial t} dt = V\frac{\partial V}{\partial t} dt = \frac{\partial V}{\partial t} dX. \quad (5.77)$$

For rearrangement of the second and third term, we use Eq. (2.24) $dQ = d\boldsymbol{X} \cdot \nabla Q$. The fourth term in Eq. (5.76) identically vanishes because the vector V is perpendicular to the vector $\boldsymbol{V} \times \nabla \times \boldsymbol{V}$. As a result, we obtain:

$$\left(\frac{\partial V}{\partial t}\right) dX + d\left(\frac{V^2}{2} + gz\right) + \frac{dp}{\rho} = 0. \tag{5.78}$$

Integrating Eq. (5.78) results in:

$$\int \frac{\partial V}{\partial t} dX + \frac{V^2}{2} + gz + \int \frac{dp}{\rho} = C. \tag{5.79}$$

Integrating Eq. (5.78) from the initial point B to the final point E, we arrive at:

$$\int_B^E \frac{\partial V}{\partial t} dX + \frac{V_B^2}{2} + gz_B + \int_B^E \frac{dp}{\rho} = \frac{V_E^2}{2} + gz_E. \tag{5.80}$$

For an unsteady, incompressible flow, the integration of Equation (5.78) delivers:

$$\rho \int \frac{\partial V}{\partial t} dX + \rho \frac{V^2}{2} + \rho g z + p = C. \tag{5.81}$$

And finally, for a steady, incompressible flow, Eq. (5.81) is reduced to:

$$p + \rho \frac{V^2}{2} + \rho g z = C \tag{5.82}$$

which is the Bernoulli equation.

5.3 Mass Flow Balance in Rotating Frame

The mass flow balance in rotating frame is obtained form the one in stationary frame where the absolute velocity $\boldsymbol{V}$ is replaced by $\boldsymbol{V} = \boldsymbol{W} + \boldsymbol{\omega} \times \boldsymbol{R}$ in Eq. (5.4). With this replacement Eq. (5.4) reads:

$$\frac{\partial \rho}{\partial t} + \nabla \cdot [\rho(\boldsymbol{W} = \boldsymbol{\omega} \times \boldsymbol{R})] = 0. \tag{5.83}$$

Equation (5.83) is rearranged as:

$$\frac{\partial \rho}{\partial t} + (\boldsymbol{\omega} \times \boldsymbol{R}) \cdot \nabla \rho + \boldsymbol{W} \cdot \nabla \rho + \rho \nabla \cdot \boldsymbol{W} + \rho \nabla \cdot (\boldsymbol{\omega} \times \boldsymbol{R}) = 0. \tag{5.84}$$

The second expression in Eq. (5.84) is the convective change of density in circumferential direction $\boldsymbol{U} = \boldsymbol{\omega} \times \boldsymbol{R}$ so that can be written as:

$$(\boldsymbol{\omega} \times \boldsymbol{R}) \cdot \nabla \rho = \boldsymbol{U} \cdot \nabla \rho = \frac{d\rho}{dt}|_U \tag{5.85}$$

thus we can define the local change of density at a fixed position in relative system as:

$$\frac{\partial_R \rho}{\partial t} = \frac{\partial \rho}{\partial t} + \frac{d\rho}{dt}|_U = \frac{\partial \rho}{\partial t} + (\boldsymbol{\omega} \times \boldsymbol{R}) \cdot \nabla \rho. \tag{5.86}$$

Equation (5.86) implies that the local change of density in absolute frame can be replaced by:

$$\frac{\partial \rho}{\partial t} \equiv \frac{\partial_R \rho}{\partial t} - (\boldsymbol{\omega} \times \boldsymbol{R}) \cdot \nabla \rho. \tag{5.87}$$

Inserting Eq. (5.87) into Eq. (5.84) we have:

$$\frac{\partial_R \rho}{\partial t} + \boldsymbol{W} \cdot \nabla \rho + \rho \nabla \cdot \boldsymbol{W} + \rho \nabla \cdot (\boldsymbol{\omega} \times \boldsymbol{R}) = 0. \tag{5.88}$$

It can easily be shown that $\nabla \cdot (\boldsymbol{\omega} \times \boldsymbol{R}) = 0$ and Eq. (5.88) is reduced to

$$\frac{\partial_R \rho}{\partial t} + \nabla \cdot (\rho \boldsymbol{W}) = 0. \tag{5.89}$$

Equations (5.89) and (5.91) are the local change of density within relative frame of reference. Setting $\boldsymbol{\omega} = 0$ in Eq. (5.87) and subsequently in Eq. (5.88) results in continuity equation in absolute frame of reference.

5.4 Momentum Balance in Rotating Frame

The equation of motion for particles moving in a rotating frame is obtained by transforming the Cauchy equation of motion from the absolute frame into the rotating relative one. Decomposing the Cauchy total stress tensor into a pressure tensor and a friction tensor, we arrived at the Navier–Stokes equation with a detailed derivation given below.

5.4.1 Navier–Stokes Equation of Motion in Rotating Frame

With Eq. (2.34) that describes the acceleration in rotating frame that includes the centrifugal and Coriolis force and Eq. (5.53) that describes the frame indifferent nature of the total stress tensor, we have the complete mathematical tool at our disposal to perform the transformation of Navier–Stokes equation from a stationary frame into a rotating one. To do so, we need to rearrange first the acceleration term in Eq. (2.34) by replacing the acceleration in stationary frame $\ddot{\boldsymbol{X}}$ by DV/Dt with $\boldsymbol{V}$ as the absolute velocity, the relative acceleration $\boldsymbol{A_R}$ by DW/Dt and $\dot{\omega}$ by $\boldsymbol{d}\omega/dt$, and the translation acceleration $\ddot{\boldsymbol{C}}$ by $\boldsymbol{A}_{\text{Tr}}$. The translational acceleration vanishes if the origin of the relative frame does not move relative to the absolute frame. It vanishes also, if the relative frame moves with a constant velocity. Considering the above, the rearranged acceleration in relative frame reads:

$$\frac{DV}{Dt} = \frac{DW}{Dt} + \frac{d\omega}{dt} \times \boldsymbol{R} + \omega \times (\omega \times \boldsymbol{R}) + 2(\omega \times \boldsymbol{W}) + \boldsymbol{A}_{\text{Trans}}. \tag{5.90}$$

To arrive at the Navier–Stokes equation in rotating frame we first revisit the total stress tensor in Eq. (5.16) and decompose it in a pressure and a friction part:

$$\nabla \cdot \boldsymbol{\Pi} = \nabla \cdot (-p\boldsymbol{I} + \boldsymbol{T}) = -\nabla p + \nabla \cdot \boldsymbol{T}. \tag{5.91}$$

Now we consider the frame indifference argument that has lead to Eq. (5.53) and set Eq. (5.92) equal to the right hand side of Eq. (5.53).

$$\frac{DV}{Dt} = \frac{D_R \boldsymbol{W}}{Dt} + \frac{\partial \omega}{\partial t} \times \boldsymbol{R} + \boldsymbol{\omega} \times (\omega \times \boldsymbol{R}) + 2(\omega \times \boldsymbol{W}) + \boldsymbol{A}_{\text{Tr}} = -\frac{\nabla p}{\rho} + \nabla \cdot \boldsymbol{T} + \boldsymbol{g}. \tag{5.92}$$

In engineering cases the translational acceleration does not exist, meaning that $\boldsymbol{A}_{\text{Tr}} = 0$, furthermore, the order of magnitude of the gravitational acceleration compared to the other can be neglected. Considering the above and expanding the first term in Eq. (5.92), we obtain:

$$\frac{\partial \boldsymbol{W}}{\partial t} + \boldsymbol{W} \cdot \nabla \boldsymbol{W} + \left(\frac{\partial \omega}{\partial t} \times \boldsymbol{R} + \omega \times (\omega \times \boldsymbol{R}) + 2(\omega \times \boldsymbol{W}) \right) = -\frac{\nabla p}{\rho} + \nabla \cdot \boldsymbol{T}. \tag{5.93}$$

Equation (5.92) shows that the acceleration in an absolute frame DV/Dt is obtained from the relative acceleration vector $\boldsymbol{A_R} = D_R \boldsymbol{W}/Dt$ by adding the expressions in the large parentheses in Eq. (5.93). These expressions represent the additional terms associated with the transformation of the equation of motion from an absolute (non rotating) frame into a rotating relative frame. Reversing the transformation direction by setting $\omega = 0$ results in the identity $\boldsymbol{W} \equiv \boldsymbol{V}$ leading to the identity of Eq. (5.93) with Cauchy equation of motion Eq. (5.17). The centrifugal force $\omega \times \omega \times \boldsymbol{R}$ and the Coriolis force $2\omega \times \boldsymbol{W}$ in Eq. (5.93) cease to exist when ω approaches zero. Therefore they are called *apparent* or *fictitious forces* that do not exist in an absolute frame of reference. Transformation of the equation of motion from the absolute frame into the relative frame of reference and vice versa is encountered in turbomachinery. Figure 5.3 shows the interaction of the force vectors with a simplified impeller.

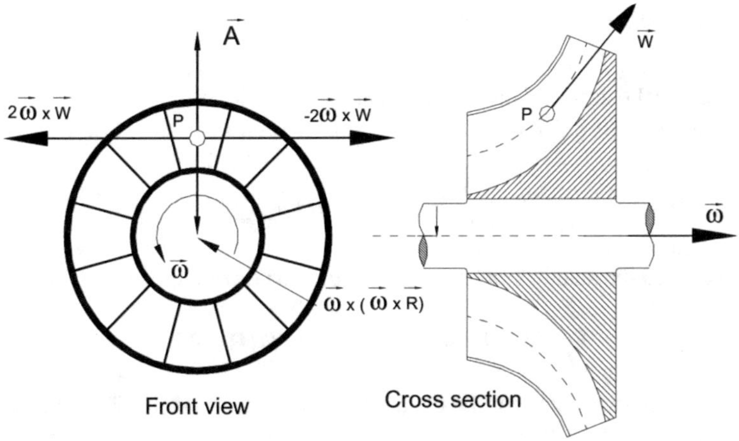

Fig. 5.3 Schematic of a centrifugal impeller with individual force vectors

5.5 Some Discussions on Navier–Stokes Equations

The flow in engineering applications in general is characterized by a three-dimensional, highly unsteady motion with random fluctuations due to the interactions between stationary and rotating components. Considering such the flows three distinctly different flow patterns are identified: (1) laminar flow (or non-turbulent flow) characterized by the absence of stochastic motions, (2) turbulent flow, where flow pattern is determined by a fully stochastic motion of fluid particles, and (3) transitional flow characterized by intermittently switching from laminar to turbulent at the same spatial position. Of the three patterns, the predominant one is the transitional flow pattern. The Navier–Stokes equations presented in this section generally describe the steady and unsteady flows through a variety of engineering components. Using a direct numerical simulation (DNS) approach delivers the most accurate results [3]. However, the application of DNS, for the time being, is restricted to simple flows at low Reynolds numbers. For calculating the complex flow field within a reasonable time, the Reynolds averaged Navier–Stokes (RANS) can be used. This issue is treated extensively in [4], where the laminar-turbulent transition, Reynolds- and ensemble averaging and the implementation of transition models into Navier–Stokes is treated extensively.

5.6 Energy Balance in Stationary Frame of Reference

For the complete description of flow process, the total energy equation is presented. This equation includes mechanical and thermal energy balances.

5.6.1 Mechanical Energy

The mechanical energy balance is obtained by the scalar multiplication of the equation of motion, Eq. (5.17), with the local velocity vector:

$$\rho \boldsymbol{V} \cdot \frac{D\boldsymbol{V}}{Dt} = \boldsymbol{V} \cdot (\nabla \cdot \boldsymbol{\Pi}) + \rho \boldsymbol{g} \cdot \boldsymbol{V}. \tag{5.94}$$

The first expression on the right hand side of Eq. (5.65) is modified using the following vector identity:

$$\boldsymbol{V} \cdot (\nabla \cdot \boldsymbol{\Pi}) = \nabla \cdot (\boldsymbol{\Pi} \cdot \boldsymbol{V}) = \mathrm{Tr}(\boldsymbol{\Pi} \cdot \nabla \boldsymbol{V}). \tag{5.95}$$

The velocity gradient ∇V can be decomposed into deformation $\boldsymbol{D}$ and rotation $\boldsymbol{\Omega}$ part as shown in Eq. (5.19):

$$\nabla \boldsymbol{V} = \frac{1}{2}(\nabla \boldsymbol{V} + \nabla \boldsymbol{V}^T) + \frac{1}{2}(\nabla \boldsymbol{V} - \nabla \boldsymbol{V}^T) = \boldsymbol{D} + \boldsymbol{\Omega}. \tag{5.96}$$

With this operation, the trace of the second order tensor in Eq. (5.95) is calculated from:

$$\mathrm{Tr}(\boldsymbol{\Pi} \cdot \nabla \boldsymbol{V}) = \boldsymbol{\Pi} : \boldsymbol{D} + \boldsymbol{\Pi} : \boldsymbol{\Omega}. \tag{5.97}$$

Since the second term on the right-hand side of Eq. (5.97) vanishes identically, Eq. (5.95) reduces to:

$$\boldsymbol{V} \cdot (\nabla \cdot \boldsymbol{\Pi}) = \nabla \cdot (\boldsymbol{\Pi} \cdot \boldsymbol{V}) - \boldsymbol{\Pi} : \boldsymbol{D}. \tag{5.98}$$

As a result, the mechanical energy balance, Eq. (5.94), becomes:

$$\rho \frac{D}{Dt}\left(\frac{V^2}{2}\right) = \nabla \cdot (\boldsymbol{V} \cdot \boldsymbol{\Pi}) - \boldsymbol{\Pi} : \boldsymbol{D} + \rho \boldsymbol{g} \cdot \boldsymbol{V}. \tag{5.99}$$

Incorporating Eq. (5.53) for Newtonian fluids into Eq. (5.99) leads to:

$$\begin{aligned} \rho \frac{D}{Dt}\left(\frac{V^2}{2}\right) =& \nabla \cdot [\boldsymbol{V} \cdot (-p + \lambda \nabla \cdot \boldsymbol{V}) I + 2\mu \boldsymbol{V} \cdot \boldsymbol{D}] \\ & - [(-p + \lambda \nabla \cdot \boldsymbol{V}) \nabla \cdot \boldsymbol{V} + 2\mu \boldsymbol{D} : \boldsymbol{D}] + \rho \boldsymbol{V} \cdot \boldsymbol{g} \end{aligned} \tag{5.100}$$

with $\boldsymbol{I} : \boldsymbol{D} = \nabla \cdot \boldsymbol{V}$. For incompressible flow, Eq. (5.100) is reduced to

$$\rho \frac{D}{Dt}\left(\frac{V^2}{2}\right) = \nabla \cdot [\boldsymbol{V} \cdot (-p I) + 2\mu \boldsymbol{V} \cdot \boldsymbol{D}] - 2\mu \boldsymbol{D} : \boldsymbol{D} + \rho \boldsymbol{V} \cdot \boldsymbol{g}. \tag{5.101}$$

The index notation of Eq. (5.101) is:

$$\frac{\partial(V_j V_j/2)}{\partial t} + V_k \frac{\partial(V_j V_j/2)}{\partial x_k} = -\frac{\partial}{\partial x_i}\left(\frac{p}{\rho} V_i\right) + \nu \frac{\partial}{\partial x_i} V_j \left(\frac{\partial V_i}{\partial x_j} + \frac{\partial V_j}{\partial x_i}\right)$$
$$- \nu \left(\frac{\partial V_i}{\partial x_j} + \frac{\partial V_j}{\partial x_i}\right)\frac{\partial V_j}{\partial x_i} + \varrho V_i g_i. \tag{5.102}$$

The sum of the last two terms in the second bracket of Eq. (5.100) is called the *dissipation function*:

$$\Phi = \lambda(\nabla \cdot \boldsymbol{V})(\nabla \cdot \boldsymbol{V}) + 2\mu \boldsymbol{D} : \boldsymbol{D}. \tag{5.103}$$

The dissipation function Eq. (5.103) is identical with the double scalar product between the friction stress tensor and the deformation tensor:

$$\Phi = \boldsymbol{T} : \boldsymbol{D}. \tag{5.104}$$

The index notation of Eq. (5.103) is

$$\Phi = \lambda \left(\frac{\partial V_i}{\partial x_i}\right)^2 + \frac{2}{4}\mu \left(\frac{\partial V_i}{\partial x_j} + \frac{\partial V_j}{\partial x_i}\right)\left(\frac{\partial V_i}{\partial x_j} + \frac{\partial V_j}{\partial x_i}\right). \tag{5.105}$$

Expanding Eq. (5.105) results in:

$$\Phi = \lambda \left(\frac{\partial V_1}{\partial x_1} + \frac{\partial V_2}{\partial x_2} + \frac{\partial V_3}{\partial x_3}\right)^2 + 2\mu \left[\left(\frac{\partial V_1}{\partial x_1}\right)^2 + \left(\frac{\partial V_2}{\partial x_2}\right)^2 + \left(\frac{\partial V_3}{\partial x_3}\right)^2\right]$$
$$\mu \left[\left(\frac{\partial V_1}{\partial x_2} + \frac{\partial V_2}{\partial x_1}\right)^2 + \left(\frac{\partial V_1}{\partial x_3} + \frac{\partial V_3}{\partial x_1}\right)^2 + \left(\frac{\partial V_2}{\partial x_3} + \frac{\partial V_3}{\partial x_2}\right)^2\right]. \tag{5.106}$$

In Eq. (5.106) the coefficient λ can be replaced by $\lambda = \bar{\mu} - 2/3\mu$ from Eq. (5.52). For an incompressible flow, Eq. (5.103) reduces to:

$$\Phi = 2\mu \boldsymbol{D} : \boldsymbol{D} \tag{5.107}$$

and as a result, Eq. (5.106) is written as:

$$\Phi = 2\mu \left[\left(\frac{\partial V_1}{\partial x_1}\right)^2 + \left(\frac{\partial V_2}{\partial x_2}\right)^2 + \left(\frac{\partial V_3}{\partial x_3}\right)^2\right] +$$
$$\mu \left[\left(\frac{\partial V_1}{\partial x_2} + \frac{\partial V_2}{\partial x_1}\right)^2 + \left(\frac{\partial V_1}{\partial x_3} + \frac{\partial V_3}{\partial x_1}\right)^2 + \left(\frac{\partial V_2}{\partial x_3} + \frac{\partial V_3}{\partial x_2}\right)^2\right]. \tag{5.108}$$

The dissipation function indicates the amount of mechanical energy dissipated as heat, which is due to the deformation caused by viscosity. Consider a viscous flow along a flat plate, an aircraft wing, a compressor stator or turbine rotor blade or any other engineering surfaces exposed to a flow. Close to the wall in the *boundary layer region*, the velocity experiences a high deformation because of the no-slip condition. By moving outside the boundary layer, the rate of deformation decreases leading to lower dissipation. To analyze the individual terms in the equation of energy and to demonstrate the role of shear stress and its effect on the dissipation of mechanical energy, we introduced the friction stress tensor, Eq. (5.54)

$$\boldsymbol{T} = \lambda(\nabla \cdot \boldsymbol{V})\boldsymbol{I} + 2\mu \boldsymbol{D}. \tag{5.109}$$

The off-diagonal elements of this tensor represent the shear stress components and characterize the *shear-deformative behavior* of this tensor. The diagonal elements of this tensor

$$T_{ii} = \lambda \frac{\partial V_i}{\partial x_i} + 2\mu D_{ii} \tag{5.110}$$

exhibit additional contributions caused by the volume dilatation or compression due to the compressibility of the working medium as mentioned before. This contribution is added to the normal stress components of the pressure tensor $p\boldsymbol{I} = e_i e_j \delta_{ij} p$. For an incompressible flow with $\nabla \cdot \boldsymbol{V} = D_{ii} = 0$, these terms identically disappear. Inserting Eq. (5.109) into Eq. (5.100), we arrive at:

$$\rho \frac{D}{Dt}\left(\frac{V^2}{2}\right) = -\boldsymbol{V} \cdot \nabla p + \nabla \cdot (\boldsymbol{T} \cdot \boldsymbol{V}) - \boldsymbol{T} : \boldsymbol{D} + \rho \boldsymbol{V} \cdot \boldsymbol{g}. \tag{5.111}$$

Equation (5.111) exhibits the mechanical energy balance in differential form. The first term on the right-hand side represents the rate of mechanical energy due to the change of pressure acting on the volume element. The second term is the rate of work done by viscous forces on the fluid particle. The third term represents the rate of irreversible mechanical energy due to the friction stress. It dissipates as heat and increases the internal energy of the system. This term corresponds to the dissipation function defined by Eq. (5.107). Finally, the forth term represents the mechanical energy necessary to overcome the gravity force acted on the fluid particle. Equation (5.111) exhibits the general differential form of mechanical energy balance for a

$$\boldsymbol{V} \cdot \nabla \left(p + \frac{1}{2}\rho V^2 + \rho g z\right) \equiv d\left(p + \frac{1}{2}\rho V^2 + \rho g z\right) = 0 \tag{5.112}$$

viscous flow. For a steady, incompressible, inviscid flow, Eq. (5.111) is simplified as: where the vector $\boldsymbol{g}$ is replaced by $\boldsymbol{g} = -g\nabla z$. Integration of the above equation leads to the Bernoulli equation of energy:

$$p + \frac{1}{2}\rho V^2 + \rho g z = \text{Constant}. \tag{5.113}$$

This equation is easily derived by multiplying the Euler equation of motion with a differential displacement.

5.6.2 Thermal Energy Balance

The thermal energy balance is described by the first law of thermodynamics which is postulated for a closed "thermostatic system". For this system, properties, such as temperature, pressure, entropy, internal energy, etc., have no spatial gradients. Since in an open system the thermodynamic properties undergo temporal and spatial changes, the classical first law must be formulated under open system conditions. To do so, we start from an open system within which a steady flow process takes place and replaces the differential operator, d, from classical thermodynamics by the substantial differential operator D. This operation implies the requirement that the thermodynamic system under consideration be at least in a locally stable equilibrium state. Starting from the first law for an internally irreversible system:

$$du = \delta Q - p dv + |\delta w_f| \tag{5.114}$$

where u is the internal energy, and Q[1] the specific thermal energy added to or removed from the system, p the thermodynamic pressure, v the specific volume, and δw_f the part of mechanical energy dissipated as heat by the internal friction. The subscript f refers to the irreversible nature of the process caused by internal friction. Applying the differential operator D:

$$\frac{Du}{Dt} = \delta \dot{Q} - p\frac{Dv}{Dt} + \delta \dot{w}_f \tag{5.115}$$

where $\delta \dot{Q}$ is the rate of thermal energy added (or removed) to or from the open system per unit mass and time. It can be expressed as the divergence of the thermal energy flux vector $\delta \dot{Q} = -\nabla \cdot \dot{\boldsymbol{q}}/\rho$. The rate of the mechanical energy dissipated as heat $\delta \dot{w}_f$ is identical to the third term $\boldsymbol{T} : \boldsymbol{D}/\rho$ in Eq. (5.111):

$$\frac{Du}{Dt} = -\frac{\nabla \cdot \dot{\boldsymbol{q}}}{\rho} - p\frac{Dv}{Dt} + \frac{\boldsymbol{T} : \boldsymbol{D}}{\rho}. \tag{5.116}$$

The negative sign of $-\nabla \cdot \dot{\boldsymbol{q}}$ is introduced to account for a positive heat transfer to the system. Furthermore, since the first term on the left-hand side is per unit mass and time, the divergence of the heat flux vector $\nabla \cdot \dot{\boldsymbol{q}}$ as well as the dissipation term $\boldsymbol{T} : \boldsymbol{D}$,

[1] Usually Q(kJ) represents the thermal energy and q(kJ/kg) the specific thermal energy. However, in order to avoid confusion that may arise using the same symbol for specific thermal energy and the heat flux vector $\dot{\boldsymbol{q}}$, we use Q for the specific thermal energy. We will use q, wherever there is no reason for confusion.

had to be divided by the density to preserve the dimensional integrity. Multiplying both sides of Eq. (5.116) with ρ, we obtain

$$\rho\frac{Du}{Dt} = -\nabla \cdot \dot{\boldsymbol{q}} - \rho p\frac{Dv}{Dt} + \boldsymbol{T} : \boldsymbol{D}. \tag{5.117}$$

In Eq. (5.117), first we replace the specific volume v by $1/\rho$ and consider the continuity equation

$$\frac{D\rho}{Dt} = -\rho\nabla \cdot \boldsymbol{V} \tag{5.118}$$

then we insert Eq. (5.118) into Eq. (5.117) and arrive at:

$$\rho\frac{Du}{Dt} = -\nabla \cdot \dot{\boldsymbol{q}} - p\nabla \cdot \boldsymbol{V} + \boldsymbol{T} : \boldsymbol{D}. \tag{5.119}$$

In Eq. (5.119) the internal energy can be related to the temperature by the thermodynamic relation $u = c_v T$ with c_v as the specific heat at constant volume. The heat flux vector $\dot{\boldsymbol{q}}$ can also be expressed in terms of temperature using the *Fourier heat conduction law*. For an *isotropic medium*, the Fourier law of heat conduction is written as:

$$\dot{\boldsymbol{q}} = -k\nabla T \tag{5.120}$$

with k (kJ/msec K) as the thermal conductivity. Introducing Eq. (5.120) into Eq. (5.119), for an incompressible fluid we get:

$$\rho C_v\frac{DT}{Dt} = k\nabla^2 T + 2\mu\boldsymbol{D} : \boldsymbol{D}. \tag{5.121}$$

For a steady flow, Eq. (5.121) can be simplified as:

$$C_v\boldsymbol{V} \cdot \nabla T = \frac{k}{\rho}\nabla^2 T + 2\nu\boldsymbol{D} : \boldsymbol{D}. \tag{5.122}$$

The thermal energy equation can equally be expressed in terms of enthalpy

$$dh = \delta Q + vdp + |\delta w_f|. \tag{5.123}$$

Following exactly the same procedure that has lead to Eq. (5.119), we find

$$\rho\frac{Dh}{Dt} = -\nabla \cdot \dot{\boldsymbol{q}} + \frac{Dp}{Dt} + \boldsymbol{T} : \boldsymbol{D} = k\nabla^2 T + \frac{Dp}{Dt} + \boldsymbol{T} : \boldsymbol{D}. \tag{5.124}$$

Introducing the temperature in terms of $h = c_p T$:

$$c_p\frac{DT}{Dt} = \tfrac{k}{\varrho}\nabla^2 T : \frac{1}{\varrho}\frac{Dp}{Dt} + \boldsymbol{T} : \boldsymbol{D}. \tag{5.125}$$

The index notation of Eq. (5.125) reads:

$$c_p\left(\frac{\partial T}{\partial t}+V_i\frac{\partial T}{\partial x_i}\right)=\frac{\kappa}{\rho}\left(\frac{\partial^2 T}{\partial x_i\partial x_i}\right)+\frac{1}{\rho}\left(\frac{\partial p}{\partial t}+V_i\frac{\partial p}{\partial x_i}\right)+\frac{1}{\rho}\Phi \tag{5.126}$$

and taking Φ from Eq. (5.108), we can expand Eq. (5.126) to arrive at:

$$c_p\left(\frac{\partial T}{\partial t}+V_1\frac{\partial T}{\partial x_2}+V_3\frac{\partial T}{\partial x_3}\right)=\frac{\kappa}{\rho}\left(\frac{\partial^2 T}{\partial x_1^2}+\frac{\partial^2 T}{\partial x_2^2}+\frac{\partial^2 T}{\partial x_3^2}\right)+$$
$$\frac{1}{\rho}\left(\frac{\partial p}{\partial t}+V_1\frac{\partial p}{\partial x_1}+V_2\frac{\partial p}{\partial x_2}+V_3\frac{\partial p}{\partial x_3}\right)+\frac{1}{\rho}\Phi. \tag{5.127}$$

5.6.3 *Total Energy*

The combination of the mechanical and thermal energy balances, Eqs. (5.111) and (5.119), results in the following *total energy equation*:

$$\rho\frac{D}{Dt}\left(u+\frac{V^2}{2}\right)=-\nabla\cdot\dot{\boldsymbol{q}}-\nabla\cdot(p\boldsymbol{V})+\nabla\cdot(\boldsymbol{T}\cdot\boldsymbol{V})+\rho\boldsymbol{V}\cdot\boldsymbol{g}. \tag{5.128}$$

We may rearrange the second and third term on the right-hand side of Eq. (5.128)

$$\rho\frac{D}{Dt}\left(u+\frac{V^2}{2}\right)=-\nabla\cdot\dot{\boldsymbol{q}}+\nabla\cdot[\boldsymbol{V}\cdot(-p\boldsymbol{I}+\boldsymbol{T})]+\rho\boldsymbol{V}\cdot\boldsymbol{g}. \tag{5.129}$$

The argument inside the parenthesis within the bracket exhibits the stress tensor

$$\rho\frac{D}{Dt}\left(u+\frac{V^2}{2}\right)=-\nabla\cdot\dot{\boldsymbol{q}}+\nabla\cdot(\boldsymbol{V}\cdot\boldsymbol{\Pi})+\rho\boldsymbol{V}\cdot\boldsymbol{g} \tag{5.130}$$

and the second term on the right-hand side of Eq. (5.130) constitutes the mechanical energy necessary to overcome the surface forces. The heat flux vector in Eq. (5.130) can be replaced by the Fourier equation (5.120) that gives

$$\rho\frac{D}{Dt}\left(u+\frac{V^2}{2}\right)=k\nabla^2 T+\nabla\cdot(\boldsymbol{V}\cdot\boldsymbol{\Pi})+\rho\boldsymbol{V}\cdot\boldsymbol{g}. \tag{5.131}$$

Equation (5.130) may be written in different forms using different thermodynamic properties. Since in an open system enthalpy is used, we replace the internal energy by the enthalpy $h = u + pv$ and find

$$\rho\frac{D}{Dt}\left(h+\frac{V^2}{2}\right)=\frac{\partial p}{\partial t}-\nabla\cdot\dot{\boldsymbol{q}}+\nabla\cdot(\boldsymbol{T}\cdot\boldsymbol{V})+\rho\boldsymbol{V}\cdot\boldsymbol{g} \tag{5.132}$$

and with the Fourier equation (5.120)

$$\rho\frac{D}{Dt}\left(h+\frac{V^2}{2}\right)=\kappa\nabla^2 T+\frac{\partial p}{\partial t}+\nabla\cdot(\boldsymbol{T}\cdot\boldsymbol{V})+\rho\boldsymbol{V}\cdot\boldsymbol{g}. \tag{5.133}$$

The expression in the parenthesis on the left-hand side is called the *total enthalpy* which is defined as $H=h+V^2/2$. With this definition, the re-arrangement of Eq. (5.133) gives

$$\rho\frac{DH}{Dt}=\rho\left(\frac{\partial H}{\partial t}+\boldsymbol{V}\cdot\nabla H\right)=k\nabla^2 T+\frac{\partial p}{\partial t}+\nabla\cdot(\boldsymbol{T}\cdot\boldsymbol{V})+\rho\boldsymbol{V}\cdot\boldsymbol{g} \tag{5.134}$$

and its index notation reads

$$\rho\left(\frac{\partial H}{\partial t}+V_i\frac{\partial H}{\partial x_i}\right)=k\frac{\partial^2 T}{\partial x_i\partial x_i}+\frac{\partial p}{\partial t}+\frac{\partial}{\partial x_i}(T_{ij}V_j)+\rho V_i g_i. \tag{5.135}$$

The gravitational term is $\boldsymbol{V}\cdot\boldsymbol{g}$ can be expressed as $\boldsymbol{V}\cdot\boldsymbol{g}=-\boldsymbol{V}\cdot\nabla(gz)$ which can be written as $\boldsymbol{V}\cdot\nabla(gz)=d\boldsymbol{X}/dt\cdot\nabla(gz)=d(gz)/dt$.

5.6.4 Entropy Balance

The second law of thermodynamics expressed in terms of internal energy as

$$ds=\frac{\delta Q}{T}=\frac{du+pdv}{T}. \tag{5.136}$$

The infinitesimal heat δQ added to or rejected from the system may include the heat generated by the irreversible dissipation process. Replacing the differential d by the material differential operators, we arrive at:

$$T\frac{Ds}{Dt}=\frac{Du}{Dt}+p\frac{Dv}{Dt}. \tag{5.137}$$

The right-hand side of Eq. (5.137) is expressed by Eq. (5.119) as:

$$\frac{Du}{Dt}+p\frac{Dv}{Dt}=-\frac{1}{\rho}\nabla\cdot\dot{\boldsymbol{q}}+\frac{1}{\rho}\boldsymbol{T}:\boldsymbol{D} \tag{5.138}$$

replacing the left-hand side of Eq. (5.138) by Eq. (5.137) results in

$$\rho \frac{Ds}{Dt} = -\frac{1}{T}\nabla \cdot \dot{\boldsymbol{q}} + \frac{1}{T}\boldsymbol{T} : \boldsymbol{D}. \tag{5.139}$$

The second term on the right-hand side, which includes the second order friction tensor T, is the dissipation function Eq. (5.103)

$$\rho \frac{Ds}{Dt} = -\frac{1}{t}\nabla \cdot \dot{\boldsymbol{q}} + \frac{1}{T}\Phi. \tag{5.140}$$

This equation shows clearly that the total entropy change Ds/Dt generally consists of two parts. The first part is the entropy change due to a reversible heat supply to the system (addition or rejection) and may assume positive, zero, or negative values. The second term exhibits the entropy production due to the irreversible dissipation and is always positive. Thus, Eq. (5.140) may be modified as:

$$\rho \frac{Ds}{Dt} = \rho \left(\frac{Ds}{Dt}\right)_{\text{rev}} + \rho \left(\frac{Ds}{Dt}\right)_{\text{irr}} \tag{5.141}$$

with $\rho \left(\frac{Ds}{Dt}\right)_{\text{rev}} = -\frac{1}{T}\nabla \cdot \dot{\boldsymbol{q}}$ and $\rho \left(\frac{Ds}{Dt}\right)_{\text{irr}} = \frac{\Phi}{T}$. The reversible part exhibits the heat added/rejected reversibly to/from the system, thus the entropy change can assume positive or negative values, whereas, for the irreversible, the entropy change is always positive.

References

1. Truesdell C, Noll W (1965) Handbuch der Physik. In: Flügge S (ed). Springer, Berlin
2. Truesdell C (1952) J. Rat. Mech. Anal. 1:125
3. Ferziger HJ, Peric M (2001) Computational methods for fluid dynamics, 3rd edn. Springer, Berlin
4. Schobeiri MT (2010) Fluid mechanics for engineers. Graduate textbook. Springer, New York. ISBN 978-642-1193-6

Chapter 6
Tensor Operations in Orthogonal Curvilinear Coordinate Systems

6.1 Change of Coordinate System

The vector and tensor operations we have discussed in the foregoing chapters were performed solely in rectangular coordinate system. It should be pointed out that we were dealing with quantities such as velocity, acceleration, and pressure gradient that are independent of any coordinate system within a certain frame of reference. In Chap. 5, we saw how the acceleration vector changes if we change the frame of reference. In this connection it is necessary to distinguish between a coordinate system and a frame of reference. The following example should clarify this distinction. In an absolute frame of reference, the flow velocity vector may be described by the rectangular Cartesian coordinate x_i:

$$\boldsymbol{V} = \boldsymbol{V}(x_1, x_2, x_3) = \boldsymbol{V}(\boldsymbol{X}). \tag{6.1}$$

It may also be described by a cylindrical coordinate system, which is a non-Cartesian coordinate system:

$$\boldsymbol{V} = \boldsymbol{V}(x, r, \theta) \tag{6.2}$$

or generally by any other non-Cartesian or curvilinear coordinate ξ_i that describes the vector:

$$\boldsymbol{V} = \boldsymbol{V}(\xi_1, \xi_2, \xi_3). \tag{6.3}$$

By changing the coordinate system, the components of the vector $\boldsymbol{V}$ will change. However the vector remains invariant under any transformation of coordinates. This is true for any other quantities such as acceleration vector, force vector, pressure or temperature gradient. As we saw in Chap. 5, the concept of invariance, however, is generally no longer valid if we change the frame of reference. For example, if the flow particles leave the absolute frame of reference and enter the relative frame of

M. T. Schobeiri, *Tensor Analysis for Engineers and Physicists - With Application to Continuum Mechanics, Turbulence, and Einstein's Special and General Theory of Relativity*, https://doi.org/10.1007/978-3-030-35736-8_6

reference, for example in a moving or rotating frame, the velocity will experience a change. In this chapter, we will pursue the concept of quantity invariance and discuss the fundamentals that are needed for coordinate transformation.

6.2 Co- and Contravariant Base Vectors, Metric Coefficients

As we saw in the previous chapters, a vector quantity is described in Cartesian coordinate system x_i by its components:

$$\boldsymbol{V} = \boldsymbol{e}_i V_i = \boldsymbol{e}_1 V_1 + \boldsymbol{e}_2 V_2 + \boldsymbol{e}_3 V_3 \tag{6.4}$$

with e_i as orthonormal unit vectors (Fig. 6.1 left). The same vector transformed into the curvilinear coordinate system ξ_k (Fig. 6.1 right) is represented by:

$$\boldsymbol{V} = \boldsymbol{g}_k V^k = \boldsymbol{g}_1 V^1 + \boldsymbol{g}_2 V^2 + \boldsymbol{g}_3 V^3 \tag{6.5}$$

where $\boldsymbol{g}_k$ are the *covariant* base vectors and V^k the *contravariant* components of $\boldsymbol{V}$ with respect to the base $\boldsymbol{g}_k$ in a curvilinear coordinate system. As shown in Fig. 6.1, the *covariant base* vectors $\boldsymbol{g}_1$, $\boldsymbol{g}_2$ and $\boldsymbol{g}_3$ are tangent vectors to the mutually orthogonal curvilinear coordinates and ξ_1, ξ_2, and ξ_3. Figure 6.2 shows a second set of base vectors, the reciprocal base vectors $\boldsymbol{g}^1$, $\boldsymbol{g}^2$, $\boldsymbol{g}^3$ also called *contravariant* base vectors.

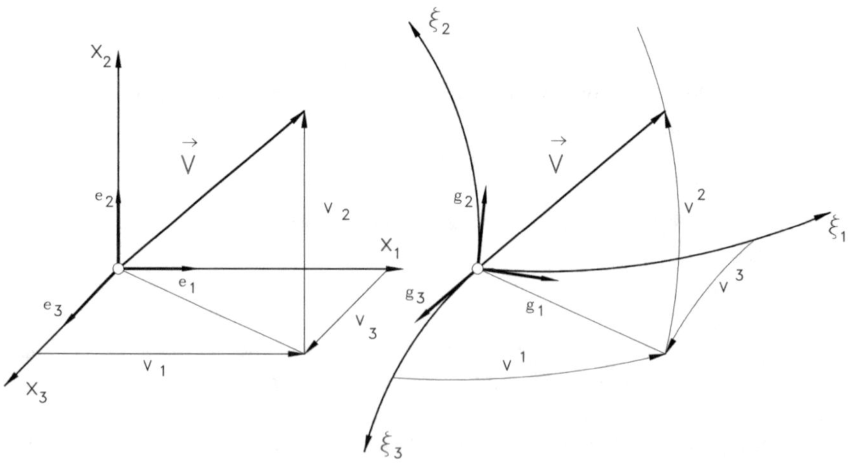

Fig. 6.1 Base vectors in a Cartesian (left) and in a generalized orthogonal curvilinear coordinate system (right)

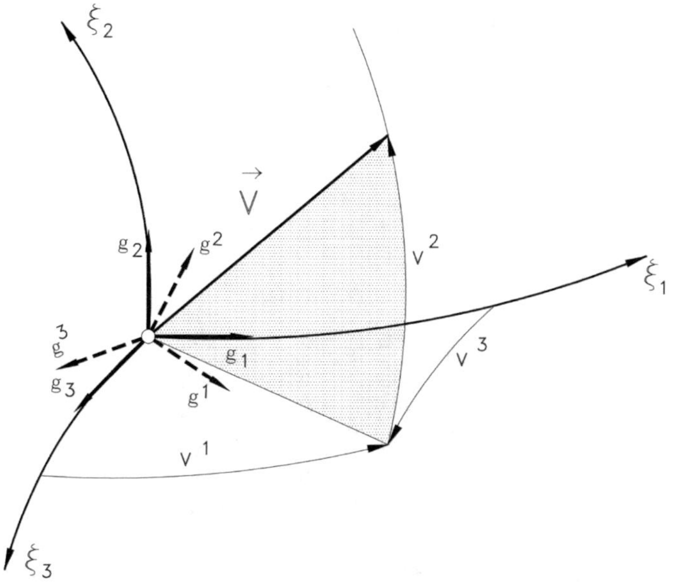

Fig. 6.2 Co- and contravariant base vectors

These base vectors are orthogonal to the planes described by $\boldsymbol{g}_1 - \boldsymbol{g}_2$, $\boldsymbol{g}_2 - \boldsymbol{g}_3$ and $\boldsymbol{g}_3 - \boldsymbol{g}_1$, respectively.

For curvilinear coordinate system, we place the indices diagonally for summing convenience. Unlike the Cartesian base vectors $\boldsymbol{e}_k$, that are orthonormal vectors (of unit length and mutually orthogonal), the base vectors $\boldsymbol{g}_k$ do not have unit lengths. The base vectors $\boldsymbol{g}_k$ represent the rate of change of the position vector $\boldsymbol{x}$ with respect to the curvilinear coordinates ξ_i.

$$\boldsymbol{g}_k = \frac{\partial \boldsymbol{x}}{\partial \xi_k} = \frac{\partial (\boldsymbol{e}_i x_i)}{\partial \xi_k} = \boldsymbol{e}_i \frac{\partial x_i}{\partial \xi_k}. \tag{6.6}$$

In Eq. (6.6) the unit vector $\boldsymbol{e}_i$, is not a functions of the coordinates x_i, therefor it was moved out of the parentheses.

We now define the *reciprocal base* vector $\boldsymbol{g}_k$ that we call the *contravariant base* vector:

$$\boldsymbol{g}^j = \boldsymbol{e}_m \frac{\partial \xi_j}{\partial x_m}. \tag{6.7}$$

The following examples should clarify how to calculate the base vectors. As the first example, we choose a cylindrical coordinate system with:

$$x_1 = r \cos \vartheta,\ x_2 = r \sin \vartheta,\ x_3 = z \tag{6.8}$$

with the curvilinear coordinate system:

$$\xi_1 \equiv r,\ \xi_2 \equiv \vartheta,\ \xi_3 \equiv z \tag{6.9}$$

that we insert into Eq. (6.8) to arrive at:

$$x_1 = r\cos\xi_2,\ x_2 = r\sin\xi_2,\ x_3 = z \tag{6.10}$$

expanding Eq. (6.6):

$$\begin{aligned}
\boldsymbol{g}_1 &= \boldsymbol{e}_1\frac{\partial x_1}{\partial \xi_1} + e_2\frac{\partial x_2}{\partial \xi_1} + e_3\frac{\partial x_3}{\partial \xi_1}\\
\boldsymbol{g}_2 &= e_1\frac{\partial x_1}{\partial \xi_2} + e_2\frac{\partial x_2}{\partial \xi_2} + e_3\frac{\partial x_3}{\partial \xi_2}\\
\boldsymbol{g}_3 &= e_1\frac{\partial x_1}{\partial \xi_3} + e_2\frac{\partial x_2}{\partial \xi_3} + e_3\frac{\partial x_3}{\partial \xi_3}
\end{aligned} \tag{6.11}$$

and performing the differentiation by using Eq. (6.10) yield:

$$\begin{aligned}
\boldsymbol{g}_1 &= e_1\cos\vartheta + e_2\sin\vartheta\\
\boldsymbol{g}_2 &= -e_1 r\sin\vartheta + e_2 r\cos\vartheta\\
\boldsymbol{g}_3 &= e_3.
\end{aligned} \tag{6.12}$$

As the second example we take a spherical coordinate system with:

$$x_1 = r\sin\vartheta\cos\Psi,\ x_2 = r\sin\varphi\sin\Psi,\ x_3 = r\cos\varphi. \tag{6.13}$$

with $xi_1 \equiv r,\ \xi_2 \equiv \varphi,\ \xi_3 \equiv \Psi$. Performing the same procedure as above, we find the covariant base vectors for a spherical coordinate system as:

$$\begin{aligned}
\boldsymbol{g}_1 &= e_1\sin\varphi\cos\Psi + e_2\sin\varphi\sin\Psi + e_3\cos\varphi\\
\boldsymbol{g}_2 &= e_1 r\cos\varphi\cos\Psi + e_2 r\cos\varphi\sin\Psi - e_3 r\cos\varphi\\
\boldsymbol{g}_3 &= -e_1 r\sin\varphi\sin\Psi + e_2 r\sin\varphi\cos\Psi.
\end{aligned} \tag{6.14}$$

6.2.1 *Transformation of Base Vectors*

Given two sets of curvilinear coordinate systems ξ_j and $\bar{\xi}_k$, where one coordinate system can be set as a function of the other coordinate system such as:

$$\xi_j = \xi_j(\bar{\xi}_k),\ \text{and}\ \bar{\xi}_k = \bar{\xi}_k(\xi_j) \tag{6.15}$$

from Eqs. (6.6) and (6.7) we further deduce that

$$\boldsymbol{g}_k = \boldsymbol{e}_i \frac{\partial x_i}{\partial \xi_k}, \; \bar{\boldsymbol{g}}_k = \boldsymbol{e}_i \frac{\partial x_i}{\partial \bar{\xi}_k}$$
$$\boldsymbol{g}^j = \boldsymbol{e}_m \frac{\partial \xi_j}{\partial x_m}, \; \bar{\boldsymbol{g}}^k = \boldsymbol{e}_m \frac{\partial \bar{\xi}_k}{\partial x_m} \tag{6.16}$$

dividing the equations in Eq. (6.16) and using a simple re-arrangement, we obtain the contravariant base vectors:

$$\boldsymbol{g}^j = \frac{\partial \xi_j}{\partial \bar{\xi}_k} \bar{\boldsymbol{g}}^k \text{ and } \bar{\boldsymbol{g}}^k = \frac{\partial \bar{\xi}_k}{\partial \xi_j} \boldsymbol{g}^j \tag{6.17}$$

and the corresponding covariant base vectors as:

$$\boldsymbol{g}_j = \frac{\partial \bar{\xi}_j}{\partial \xi_k} \bar{\boldsymbol{g}}_k \text{ and } \bar{\boldsymbol{g}}_k = \frac{\partial \xi_k}{\partial \bar{\xi}_j} \boldsymbol{g}_j. \tag{6.18}$$

The coefficients $\partial \xi_k / \partial \bar{\xi}_j$ and $\partial \xi_j / \partial \bar{\xi}_k$ correspond to the transformation matrices Q_{ij} and Q_{ji} we discussed in Chap. 2. Equations (6.17) and (6.18) show that the base vectors can be transformed from one curvilinear coordinate system to any other coordinate system using the transformation matrices.

6.2.2 *Transformation of Components*

The same procedure applied to the base vectors in Eqs. (6.17) and (6.18) can also be applied for transforming the components of a vector from one curvilinear coordinate system to the another. Consider the $\boldsymbol{V} = \boldsymbol{g}_i V^i = \boldsymbol{g}^i V_i$, $\boldsymbol{V} = \bar{\boldsymbol{g}}_i \bar{V}^i = \bar{\boldsymbol{g}}^i \bar{V}_i$

$$V^j = \frac{\partial \xi_j}{\partial \bar{\xi}_k} \bar{V}^k, \; \bar{V}^k = \frac{\partial \bar{\xi}_k}{\partial \xi_j} V^j$$
$$V_j = \frac{\partial \bar{\xi}_j}{\partial \xi_k} \bar{V}_k, \; \bar{V}_k = \frac{\partial \xi_k}{\partial \bar{\xi}_j} V_j. \tag{6.19}$$

Here again the transformation coefficients $\partial \xi_k / \partial \bar{\xi}_j$ and $\partial \bar{\xi}_j / \partial \xi_k$ correspond to the transformation matrices Q_{ij} and Q_{ji} we discussed in Chap. 2.

6.2.3 Metric Coefficients, Jacobian Determinants

The *covariant and covariant metric coefficients* are obtained by performing a scalar multiplication between the two covariant and two contravariant base vectors, respectively:

$$\boldsymbol{g}_i \cdot \boldsymbol{g}_j = g_{ij} = \boldsymbol{e}_m \cdot \boldsymbol{e}_n \frac{\partial x_m}{\partial \xi_i}\frac{\partial x_n}{\partial \xi_j} = \delta_{mn}\frac{\partial x_m}{\partial \xi_i}\frac{\partial x_n}{\partial \xi_j} = \frac{\partial x_m}{\partial \xi_i}\frac{\partial x_m}{\partial \xi_j}$$
$$\boldsymbol{g}^i \cdot \boldsymbol{g}^j = g^{ij} = \boldsymbol{e}_m \cdot \boldsymbol{e}_n \frac{\partial \xi_m}{\partial x_i}\frac{\partial \xi_n}{\partial x_j} = \delta_{mn}\frac{\partial \xi_m}{\partial x_i}\frac{\partial \xi_n}{\partial x_j} = \frac{\partial \xi_m}{\partial x_i}\frac{\partial \xi_m}{\partial x_j}. \tag{6.20}$$

Following the definition of the Jacobian function (4.16) and its expansion (4.18) the last expressions in Eq. (6.20) are written in matrix form:

$$[g_{ij}] = \begin{bmatrix} g_{11} & g_{12} & g_{13} \\ g_{21} & g_{22} & g_{23} \\ g_{31} & g_{32} & g_{33} \end{bmatrix}, \ [g^{ij}] = \begin{bmatrix} g^{11} & g^{12} & g^{13} \\ g^{21} & g^{22} & g^{23} \\ g^{31} & g^{32} & g^{33} \end{bmatrix}. \tag{6.21}$$

The corresponding determinants are

$$g = |g_{ij}| = \begin{vmatrix} g_{11} & g_{12} & g_{13} \\ g_{21} & g_{22} & g_{23} \\ g_{31} & g_{32} & g_{33} \end{vmatrix}, \ \frac{1}{g} = |g^{ij}| = \begin{vmatrix} g^{11} & g^{12} & g^{13} \\ g^{21} & g^{22} & g^{23} \\ g^{31} & g^{32} & g^{33} \end{vmatrix}. \tag{6.22}$$

In Eq. (6.22) g represents the determinant of the covariant metric coefficients. Using the definition of the Jacobian determinant from Eq. (6.20) we obtain:

$$J = \det\left[\frac{\partial x_m}{\partial \xi_j}\right] = \left|\frac{\partial x_m}{\partial \xi_j}\right|,$$
$$\frac{1}{J} = \det\left[\frac{\partial \xi_1}{\partial x_m}\right] = \left|\frac{\partial \xi_j}{\partial x_m}\right| \tag{6.23}$$

and in conjunction with Eq. (6.22) it follows that:

$$|g_{ij}| = \left|\frac{\partial x_m}{\partial \xi_j}\right|^2 = \begin{vmatrix} g_{11} & g_{12} & g_{13} \\ g_{21} & g_{22} & g_{23} \\ g_{31} & g_{32} & g_{33} \end{vmatrix} = J^2$$
$$|g^{ij}| = \left|\frac{\partial \xi_j}{\partial x_m}\right|^2 = \begin{vmatrix} g^{11} & g^{12} & g^{13} \\ g^{21} & g^{22} & g^{23} \\ g^{31} & g^{32} & g^{33} \end{vmatrix} = \frac{1}{J^2}. \tag{6.24}$$

From Eqs. (6.24) and (6.22), we find the relationship between g and J:

$$g \equiv |g_{ij}| = J^2, \ \sqrt{g} = J. \tag{6.25}$$

For the orthogonal curvilinear coordinate systems the non-diagonal elements in Eq. (6.21) disappear because for $\boldsymbol{g}_i \cdot \boldsymbol{g}_j = 0$ for $i \neq j$; the expansion of Eq. (6.20) delivers:

$$
\begin{aligned}
g_{ij} &= \frac{\partial x_1}{\partial \xi_i}\frac{\partial x_1}{\partial \xi_j} + \frac{\partial x_2}{\partial \xi_i}\frac{\partial x_2}{\partial \xi_j} + \frac{\partial x_3}{\partial \xi_i}\frac{\partial x_3}{\partial \xi_j} \\
g_{11} &= \frac{\partial x_1}{\partial \xi_1}\frac{\partial x_1}{\partial \xi_1} + \frac{\partial x_2}{\partial \xi_1}\frac{\partial x_2}{\partial \xi_1} + \frac{\partial x_3}{\partial \xi_1}\frac{\partial x_3}{\partial \xi_1} \\
g_{22} &= \frac{\partial x_1}{\partial \xi_2}\frac{\partial x_1}{\partial \xi_2} + \frac{\partial x_2}{\partial \xi_2}\frac{\partial x_2}{\partial \xi_2} + \frac{\partial x_3}{\partial \xi_2}\frac{\partial x_3}{\partial \xi_2} \\
g_{33} &= \frac{\partial x_1}{\partial \xi_3}\frac{\partial x_1}{\partial \xi_3} + \frac{\partial x_2}{\partial \xi_3}\frac{\partial x_2}{\partial \xi_3} + \frac{\partial x_3}{\partial \xi_3}\frac{\partial x_3}{\partial \xi_3}.
\end{aligned}
\tag{6.26}
$$

The co- and contra variant base vectors are interrelated by the *mixed Kronecker delta* δ_k^j as follows:

$$
\boldsymbol{g}_k \cdot \boldsymbol{g}^j \equiv g_k^j = \boldsymbol{e}_i \cdot \boldsymbol{e}_m \frac{\partial x_i}{\partial \xi_k}\frac{\partial \xi_j}{\partial x_m} = \delta_{im}\frac{\partial x_i}{\partial \xi_k}\frac{\partial \xi_j}{\partial x_m} = \frac{\partial \xi_j}{\partial \xi_k} \equiv \delta_k^j. \tag{6.27}
$$

The new mixed Kronecker delta δ_k^j from Eq. (6.27) has the values:

$$
\boldsymbol{g}_k \cdot \boldsymbol{g}^j \equiv g_k^j = \delta_k^j,\ \delta_k^j = 1 \text{ for } k = j,\ \delta_k^j = 0 \text{ for } k \neq j. \tag{6.28}
$$

In Eqs. (6.27) and (6.28) we have with g_k^j a *mixed* metric coefficient.

Continuing with the coordinates in Eq. (6.10), the co- and contravariant metric coefficients for cylindrical coordinate system are:

$$
(g_{ij}) = \begin{pmatrix} 1 & 0 & 0 \\ 0 & r^2 & 0 \\ 0 & 0 & 1 \end{pmatrix},\ (g^{ij}) = \begin{pmatrix} 1 & 0 & 0 \\ 0 & 1/r^2 & 0 \\ 0 & 0 & 1 \end{pmatrix} \tag{6.29}
$$

and for spherical coordinate system:

$$
(g_{ij}) = \begin{pmatrix} 1 & 0 & 0 \\ 0 & r^2 & 0 \\ 0 & 0 & r^2 \sin^2\varphi \end{pmatrix},\ (g^{ij}) = \begin{pmatrix} 1 & 0 & 0 \\ 0 & 1/r^2 & 0 \\ 0 & 0 & \frac{1}{r^2 \sin^2\varphi} \end{pmatrix}. \tag{6.30}
$$

As seen for both coordinate systems the metric coefficients $g_{ij} = g^{ij} = 0$ for $i \neq j$.

6.2.4 *Vectors, Components and Scalar Product*

As shown in Chap. 2, vectors are invariant with respect to coordinate systems transformation. However, their components undergo changes according to the coordinate system configuration. Consider the vector $\boldsymbol{V}$ decomposed in its co- and contravariant components:

$$\begin{aligned}\boldsymbol{V} &= \boldsymbol{g}^k V_k = \boldsymbol{g}^1 V_1 + \boldsymbol{g}^2 V_2 + \boldsymbol{g}^3 V_3 \\ \boldsymbol{V} &= \boldsymbol{g}_k V^k = \boldsymbol{g}_1 V^1 + \boldsymbol{g}_2 V^2 + \boldsymbol{g}_3 V^3.\end{aligned} \tag{6.31}$$

Given two vectors $\boldsymbol{U}$ and $\boldsymbol{V}$, their scalar product has the following configuration:

$$\begin{aligned}\boldsymbol{V} \cdot \boldsymbol{U} &= (g_i V^i) \cdot (g_j U^j) = g_{ig} V^i U^j \\ \boldsymbol{V} \cdot \boldsymbol{U} &= (g^i V_i) \cdot (g^j U_j) = g^{ig} V_i U_j \\ \boldsymbol{V} \cdot \boldsymbol{U} &= (g^i V_i) \cdot (g_j U^j) = g_i^j V_i U^j.\end{aligned} \tag{6.32}$$

The *mixed metric* coefficient defined in Eq. (6.27) appears in Eq. (6.32) as the product of the covariant and the contravariant base vector:

$$g_i^j = \boldsymbol{g}_i \cdot \boldsymbol{g}^j. \tag{6.33}$$

6.3 Relation Between, Contravariant and Covariant Base Vectors

Two methods are presented below that relate the contravariant base vectors to the covariant base vectors. In the first method we assume that each contravariant base vector is related to the covariant base through the following relation:

$$\begin{aligned}\boldsymbol{g}^1 &= A^{11}\boldsymbol{g}_1 + A^{12}\boldsymbol{g}_2 + A^{13}\boldsymbol{g}_3 \\ \boldsymbol{g}^2 &= A^{21}\boldsymbol{g}_1 + A^{22}\boldsymbol{g}_2 + A^{323}\boldsymbol{g}_3 \\ \boldsymbol{g}^3 &= A^{31}\boldsymbol{g}_1 + A^{32}\boldsymbol{g}_2 + A^{33}\boldsymbol{g}_3\end{aligned} \tag{6.34}$$

with the coefficients A^{ij} yet to be determined. The system of three equations in Eq. (6.34) is written as

$$\boldsymbol{g}^i = A^{ij}\boldsymbol{g}_j. \tag{6.35}$$

To find the coefficients A^{ij}, we multiply Eq. (6.35) with $\boldsymbol{g}^k$:

$$\boldsymbol{g}^i \cdot \boldsymbol{g}^k = A^{ij}\boldsymbol{g}_j \cdot \boldsymbol{g}^k. \tag{6.36}$$

Equation (6.36) leads to $g^{ik} = A^{ij}\delta^k_j$. The right hand side is different from zero only if $j = k$, that means:

$$g^{ik} = A^{ik}. \tag{6.37}$$

Thus we obtain from Eq. (6.34) that

$$\boldsymbol{g}^j = g^{ij}\boldsymbol{g}_j. \tag{6.38}$$

Similarly we find:

$$\boldsymbol{g}_i = g_{ij}\boldsymbol{g}^j. \tag{6.39}$$

A second alternative to find the contravariant base vectors is using the triple scalar product. To understand how, in this context, the use of triple scalar product functions, we resort to Cartesian coordinate system with vector operation $\boldsymbol{e}_1 = \boldsymbol{e}_2 \times \boldsymbol{e}_3$ and set for the curvilinear coordinate system $\alpha\boldsymbol{g}^1 = \boldsymbol{g}_2 \times \boldsymbol{g}_3$ with α to be determined. The scalar multiplication of $\alpha\boldsymbol{g}^1$ with $\boldsymbol{g}_1$ results in $\alpha = (\boldsymbol{g}_2 \times \boldsymbol{g}_3) \cdot \boldsymbol{g}_1$ which is the triple scalar product with the symbol that we will use in the following:

$$\alpha = (\boldsymbol{g}_2 \times \boldsymbol{g}_3) \cdot \boldsymbol{g}_1 \equiv [\boldsymbol{g}_1, \boldsymbol{g}_2, \boldsymbol{g}_3]. \tag{6.40}$$

Considering the sequence of indices we have:

$$\begin{aligned} &[\boldsymbol{g}_1, \boldsymbol{g}_2, \boldsymbol{g}_3] \neq 0 \\ &[\boldsymbol{g}_2, \boldsymbol{g}_2, \boldsymbol{g}_3] = 0 \\ &[\boldsymbol{g}_3, \boldsymbol{g}_2, \boldsymbol{g}_3] = 0 \end{aligned} \tag{6.41}$$

with the bracket $[\boldsymbol{g}_1, \boldsymbol{g}_2, \boldsymbol{g}_3]$ as the symbol for triple scalar product we obtain the contravariant base vectors:

$$\begin{aligned} \boldsymbol{g}^1 &= \frac{\boldsymbol{g}_2 \times \boldsymbol{g}_3}{[\boldsymbol{g}_1, \boldsymbol{g}_2, \boldsymbol{g}_3]} \\ \boldsymbol{g}^2 &= \frac{\boldsymbol{g}_3 \times \boldsymbol{g}_1}{[\boldsymbol{g}_1, \boldsymbol{g}_2, \boldsymbol{g}_3]} \\ \boldsymbol{g}^3 &= \frac{\boldsymbol{g}_1 \times \boldsymbol{g}_2}{[\boldsymbol{g}_1, \boldsymbol{g}_2, \boldsymbol{g}_3]} \text{ or generally:} \\ \boldsymbol{g}^k &= \frac{\boldsymbol{g}_l \times \boldsymbol{g}_m}{[\boldsymbol{g}_1, \boldsymbol{g}_2, \boldsymbol{g}_3]}. \end{aligned} \tag{6.42}$$

In the last equation of (6.42), klm undergoes cyclic positive permutations (123), (231), (312) and negative permutations (132), (321), (213).

6.3.1 Vector Product of Base Vectors in Curvilinear Coordinate System

The vector or cross product of two vectors in a curvilinear coordinate system is determined by the product of their co- and covariant base vectors. We can construct the vector product using the orthonormal unit vectors from Eq. (6.7):

$$\boldsymbol{g}_k \times \boldsymbol{g}_m = \left(\boldsymbol{e}_i \frac{\partial x_i}{\partial \xi_k}\right) \times \left(\boldsymbol{e}_j \frac{\partial x_j}{\partial \xi_m}\right) = \varepsilon_{ijn} e_n \frac{\partial x_i}{\partial \xi_k} \frac{\partial x_j}{\partial \xi_m}. \tag{6.43}$$

The permutation symbol ε_{ijk} in Eq. (6.43) is the Levi-Civita's symbol used in Cartesian coordinate system as shown in Eq. (1.14). For the curvilinear coordinate system we introduce modified permutation symbols ϵ_{ijk}, ϵ^{ijk} defined as follows:

$$\begin{aligned} \boldsymbol{g}_i \times \boldsymbol{g}_j &= \epsilon_{ijk} \boldsymbol{g}^k \\ \boldsymbol{g}^i \times \boldsymbol{g}^j &= \epsilon^{ijk} \boldsymbol{g}_k. \end{aligned} \tag{6.44}$$

In Eq. (6.44) the symbols ϵ_{ijk}, ϵ^{ijk} are related to the Cartesian permutation symbol ε_{ijk} through the determinant g:

$$\begin{aligned} \epsilon_{ijk} &= \varepsilon_{ijk}\sqrt{g}, \\ \epsilon^{ijk} &= \varepsilon_{ijk}\frac{1}{\sqrt{g}} \text{ with } g = \frac{1}{|g^{ij}|} = |g_{ij}|. \end{aligned} \tag{6.45}$$

As an example, a cross product of two vectors are written in contravariant version as:

$$\boldsymbol{A} \times \boldsymbol{B} = \epsilon_{ijk} A^i B^j \boldsymbol{g}^k = \sqrt{g} \begin{vmatrix} \boldsymbol{g}^1 & \boldsymbol{g}^2 & \boldsymbol{g}^3 \\ A^1 & A^2 & A^3 \\ B^1 & B^2 & B^3 \end{vmatrix}. \tag{6.46}$$

The same product in covariant version

$$\boldsymbol{A} \times \boldsymbol{B} = \epsilon^{ijk} A_i B_j \boldsymbol{g}_k = \frac{1}{\sqrt{g}} \begin{vmatrix} \boldsymbol{g}_1 & \boldsymbol{g}_2 & \boldsymbol{g}_3 \\ A_1 & A_2 & A_3 \\ B_1 & B_2 & B_3 \end{vmatrix}. \tag{6.47}$$

The covariant and contravariant permutation symbols ϵ_{ijk} and ϵ^{ijk} are antisymmetric tensor with respect to subscripts or superscripts.

$$\epsilon_{ijk} = -\epsilon_{jik} = -\epsilon_{ikj} = \epsilon_{kji}. \tag{6.48}$$

6.3.2 Raising and Lowering Indices

The operations presented in Eqs. (6.38) and (6.39) already show how to raise and lower the indices. Given a covariant base vector, the contravariant base is obtained by multiplying the covariant base vector with the contravariant metric coefficient:

$$\boldsymbol{g}^i = g^{ij}\boldsymbol{g}_j. \tag{6.49}$$

Similarly, we find the covariant base vector by multiplying the contravariant base vector with the covariant metric coefficient:

$$\boldsymbol{g}_i = g_{ij}\boldsymbol{g}^j. \tag{6.50}$$

The matrices of contra- and covariant metric coefficients are give below with the product:

$$\begin{bmatrix} g^{11} & g^{12} & g^{13} \\ g^{21} & g^{22} & g^{23} \\ g^{31} & g^{32} & g^{33} \end{bmatrix} \begin{bmatrix} g_{11} & g_{12} & g_{13} \\ g_{21} & g_{22} & g_{23} \\ g_{31} & g_{32} & g_{33} \end{bmatrix} = \begin{bmatrix} 1 & 0 & 0 \\ 0 & 1 & 0 \\ 0 & 0 & 1 \end{bmatrix}. \tag{6.51}$$

The matrix product (6.51) shows that the matrix (g^{ij}) is the inverse of (g_{ij}).

6.4 Physical Components of Tensors

As mentioned previously, the base vectors $\boldsymbol{g}_i$ or $\boldsymbol{g}^j$ are not unit vectors. Consequently the contravariant and covariant components of a vector such as V^i or V_j do not reflect the *physical components* of the vector $\boldsymbol{V}$ in a curvilinear coordinate system. The real length of a component in a curvilinear coordinate system is, on the one side absorbed by the magnitude of the base vector and on the other side by the component itself. In Cartesian coordinate system with orthonormal bases $\boldsymbol{e}_i \cdot \boldsymbol{e}_i = 1$ and $\boldsymbol{e}_i \cdot \boldsymbol{e}_j = 0$ for $i \neq j = 0$, the component of a vector represents the true length of that component. In curvilinear coordinate system, the true length of the component is represented by the physical component.

6.4.1 Physical Components of First Order Tensors

To obtain the physical components of first order tensors or vectors, first the corresponding unit vectors must be found. They are obtained by dividing the base vector by its magnitude, thus the covariant unit base vector is obtained from:

$$\boldsymbol{g}_i^* = \frac{\boldsymbol{g}_i}{|\boldsymbol{g}_i|} = \frac{\boldsymbol{g}_i}{\sqrt{\boldsymbol{g}_i \cdot \boldsymbol{g}_i}} = \frac{\boldsymbol{g}_i}{\sqrt{g_{(ii)}}} \tag{6.52}$$

and the contravariant unit base vector is:

$$\boldsymbol{g}^{*i} = \frac{\boldsymbol{g}^i}{|\boldsymbol{g}^i|} = \frac{\boldsymbol{g}^i}{\sqrt{\boldsymbol{g}^i \cdot \boldsymbol{g}^i}} = \frac{\boldsymbol{g}^i}{\sqrt{g^{(ii)}}} \tag{6.53}$$

where $\boldsymbol{g}_i^*$, represents the unit base vector and $|g^i|$ the absolute value of the base vector. The expression (ii) denotes that no summing is carried out, whenever the indices are enclosed within parentheses. The vector can now be expressed in terms of its unit base vectors and the corresponding physical components:

$$\boldsymbol{V} = \boldsymbol{g}_i V^i = \boldsymbol{g}_i^* V^{*i} = \frac{\boldsymbol{g}_i}{\sqrt{g_{(ii)}}} V^{*1}. \tag{6.54}$$

Thus the covariant and contravariant physical components can be obtained from:

$$\begin{aligned}
&V_i^* = \sqrt{g^{(ii)}}V_i, \; V^{*i} = \sqrt{g_{(ii)}}V^i, \text{ which is expanded as follows}\\
&V_1^* = \sqrt{g^{11}}V_1, \; V^{*1} = \sqrt{g_{11}}V^1\\
&V_2^* = \sqrt{g^{22}}V_2, \; V^{*1} = \sqrt{g_{22}}V^2\\
&V_3^* = \sqrt{g^{33}}V_3, \; V^{*3} = \sqrt{g_{33}}V^3.
\end{aligned} \tag{6.55}$$

In Cartesian coordinate system the physical components of the base vectors are identical with the unit vectors.

6.4.2 Physical Components of a Second Order Tensor

In the following we present the physical components of a second order tensor $\boldsymbol{T}$. The procedure shown bellow follows the same procedure used in Sect. 6.3.1. Expressing in its co-, contravariant and mixed components, we have:

$$\begin{aligned}
\boldsymbol{T} &= T^{kl}\boldsymbol{g}_k\boldsymbol{g}_l = T_{kl}\boldsymbol{g}^k\boldsymbol{g}^l = T_k^l\boldsymbol{g}^k\boldsymbol{g}_l\\
\boldsymbol{T} &= T^{*kl}\boldsymbol{g}_k^*\boldsymbol{g}_l^* = T_{kl}^*\boldsymbol{g}^{*k}\boldsymbol{g}^{*l} = T_k^{*l}\boldsymbol{g}^{*k}\boldsymbol{g}_l^*.
\end{aligned} \tag{6.56}$$

The procedure for obtaining the physical components of a second order tensor presented in Eq. (6.56) can be applied to a third or higher order tensors such as a fourth order one $\boldsymbol{T} = T^{klmn}\boldsymbol{g}_k\boldsymbol{g}_l\boldsymbol{g}_m\boldsymbol{g}_n$.

6.4.3 Derivatives of Base Vectors, Christoffel Symbols

In a curvilinear coordinate system, the base vectors are generally functions of the coordinates itself. This fact must be considered while differentiating the base vectors. Consider the partial derivative:

$$\boldsymbol{g}_{i,j} \equiv \frac{\partial \boldsymbol{g}_i}{\partial \xi_j} = \frac{\partial}{\partial \xi_j}\left(e_m \frac{\partial x_m}{\partial \xi_i}\right) = e_m \frac{\partial^2 x_m}{\partial \xi_j \partial \xi_i} = \left(\frac{\partial \xi_n}{\partial x_m}\boldsymbol{g}_n\right)\frac{\partial^2 x_m}{\partial \xi_j \partial \xi_i}. \tag{6.57}$$

The abbreviation "," (comma followed by an index) refers to the partial derivative of the base vector $\boldsymbol{g}_i$, with respect to the coordinate ξ_i. In Eq. (6.57), the unit vector $\boldsymbol{e}_m$ was replaced by Eq. (6.6). Rearranging Eq. (6.57) and introducing the Christoffel symbol of second kind, we get:

$$\boldsymbol{g}_{i,j} = \frac{\partial \xi_n}{\partial x_m}\frac{\partial^2 x_m}{\partial \xi_j \partial \xi_i}\boldsymbol{g}_n. \tag{6.58}$$

Introducing Eqs. (6.58) into Eq. (6.57) and defining the *Christoffel symbol of second kind* Γ^n_{ij}, we find the derivative of the covariant base vector $\boldsymbol{g}_{i,j}$ as:

$$\begin{aligned} \Gamma^n_{ij} &\equiv \frac{\partial \xi_n}{\partial x_k}\frac{\partial^2 x_k}{\partial \xi_j \partial \xi_i} \\ \boldsymbol{g}_{i,j} &= \Gamma^n_{ij}\boldsymbol{g}_n. \end{aligned} \tag{6.59}$$

Equation (6.58) can be expressed in terms of Christoffel symbol of *first kind*:

$$\begin{aligned} \Gamma_{ijn} &\equiv \frac{\partial x_m}{\partial \xi_n}\frac{\partial^2 x_m}{\partial \xi_j \partial \xi_i} \\ \boldsymbol{g}_{i,j} &= \Gamma_{ijn}\boldsymbol{g}^n \end{aligned} \tag{6.60}$$

with Γ_{ijn} as the Christoffel symbols of first kind. From Eqs. (6.59) and (6.60) it follows that the Christoffel symbols of the second kind is related to those of the first kind by:

$$\Gamma^k_{ij} = \Gamma_{ijm} g^{mk}. \tag{6.61}$$

The Christoffel symbols were introduced to relate the partial derivative of the base vector to the base vector itself. Also from Eqs. (6.59) and (6.60) we immediately conclude that the Christoffel symbols are symmetric at their lower indices.

$$\Gamma^k_{ij} = \Gamma^k_{ji}. \tag{6.62}$$

To relate the derivatives of the contravariant base vectors to their contravariant base vectors, we now introduce another symbol Λ^{j}_{ik} and assume that it satisfies the following relation:

$$\boldsymbol{g}^{j}_{,i} = \frac{\partial \boldsymbol{g}^{j}}{\partial \xi_i} \equiv \Lambda^{j}_{ik} \boldsymbol{g}^{k}. \tag{6.63}$$

To establish a relationship between Λ^{j}_{ik} and Γ^{j}_{ik} we resort to the mixed Kronecker delta $\boldsymbol{g}^{j} \cdot \boldsymbol{g}_{i} = \delta^{j}_{i}$ and take its partial derivative:

$$(\delta^{j}_{i})_{,k} = (\boldsymbol{g}^{j} \cdot \boldsymbol{g}_{i})_{,k} = \boldsymbol{g}^{j}_{,k} \cdot \boldsymbol{g}_{i} + \boldsymbol{g}^{j} \cdot \boldsymbol{g}_{i,_k} = 0 \tag{6.64}$$

and insert the results of Eqs. (6.60) and (6.63) into Eq. (6.64) and obtain:

$$\Lambda^{i}_{jk} = -\Gamma^{i}_{jk}. \tag{6.65}$$

With Eqs. (6.60) and (6.65) we have

$$\begin{aligned} \boldsymbol{g}_{k,l} &= \Gamma^{m}_{kl} \boldsymbol{g}_{m} \\ \boldsymbol{g}^{k}_{,l} &= -\Gamma^{k}_{lm} \boldsymbol{g}^{m}. \end{aligned} \tag{6.66}$$

The Christoffel symbols can be expressed in terms of metric tensors and their derivatives. For this purpose we first perform a scalar multiplication of Eq. (6.66) with the contravariant and covariant base vector:

$$\begin{aligned} \boldsymbol{g}_{k,l} \cdot \boldsymbol{g}^{n} &= \Gamma^{m}_{kl} \boldsymbol{g}_{m} \cdot \boldsymbol{g}^{n} = \Gamma^{m}_{kl} \delta^{n}_{m} = \Gamma^{n}_{kl} \\ \boldsymbol{g}^{k}_{,l} \cdot \boldsymbol{g}_{n} &= -\Gamma^{k}_{lm} \boldsymbol{g}^{m} \cdot \boldsymbol{g}_{n} = \Gamma^{k}_{lm} \boldsymbol{g}^{m} \delta^{n}_{m} = -\Gamma^{k}_{ln}. \end{aligned} \tag{6.67}$$

Now we consider the covariant base vector:

$$\boldsymbol{g}_{l} = g_{kl} \boldsymbol{g}^{k} \tag{6.68}$$

and differentiate it with respect to ξ_n:

$$\boldsymbol{g}_{l,n} = g_{kl,n} \boldsymbol{g}^{k} + g_{kl} \boldsymbol{g}^{k}{}_{,n} \tag{6.69}$$

the scalar multiplication of Eq. (6.69) with $\boldsymbol{g}^{p}$ and considering Eqs. (6.66) and (6.67), we arrive at:

$$\Gamma^{p}_{\text{ln}} = g^{kp} g_{\boldsymbol{kl},\boldsymbol{n}} - g_{kl} g^{pq} \Gamma^{k}_{nq}. \tag{6.70}$$

After multiplying both sides of Eq. (6.70) with g_{pm}, we find:

$$g_{pm} \Gamma^{p}_{\text{ln}} = g_{pm} g^{kp} g_{kl,n} - g_{pm} g_{kl} g^{pq} \Gamma^{k}_{nq} \tag{6.71}$$

the multiplication of covariant with contravariant metric coefficients in Eq. (6.71), the step-by-step simplification of Eq. (6.71) delivers:

$$
\begin{aligned}
g_{pm}\Gamma^{p}_{ln} &= g_{pm}g^{kp}g_{kl,n} - g_{pm}g_{kl}g^{pq}\Gamma^{k}_{nq} \\
g_{pm}\Gamma^{p}_{ln} &= \delta^{k}_{m}g_{kl,n} - \delta^{q}_{m}g_{kl}\Gamma^{k}_{nq} \\
g_{pm}\Gamma^{p}_{ln} &= g_{ml,n} - g_{kl}\Gamma^{k}_{nm}.
\end{aligned} \tag{6.72}
$$

Taking the last equation from (6.72) and its cyclic permutation of l, m, n gives

$$
\begin{aligned}
g_{pm}\Gamma^{p}_{ln} + g_{kl}\Gamma^{k}_{nm} &= g_{lm,n} \\
g_{pn}\Gamma^{p}_{ml} + g_{km}\Gamma^{k}_{nl} &= g_{mn,l} \\
g_{pl}\Gamma^{p}_{nm} + g_{kn}\Gamma^{k}_{lm} &= g_{nl,m}.
\end{aligned} \tag{6.73}
$$

Multiplying the last two equations in (6.73) with $\frac{1}{2}$ and the first with $-\frac{1}{2}$ and considering the symmetry of lower indices of $\Gamma^{p}_{mn} = \Gamma^{p}_{nm}$, we arrive at:

$$
g_{pm}\Gamma^{p}_{ln} = \frac{1}{2}(g_{mn,l} + g_{nl,m} - g_{lm,n}). \tag{6.74}
$$

Finally the multiplication of Eq. (6.74) with the contravariant metric coefficient g^{kn} results in:

$$
\Gamma^{k}_{lm} = \Gamma^{k}_{ml} = \frac{1}{2}g^{kn}(g_{mn,l} + g_{nl,m} - g_{lm,n}). \tag{6.75}
$$

Equation (6.75) directly relates the Christoffel symbol to the derivatives of the metric coefficient. For the flat space the metric coefficient g^{kn} becomes δ_{km} and its derivations become zero, leading to $\Gamma^{k}_{lm} = 0$. Since the Christoffel symbols of first kind can be converted into the second kind and vice versa using metric coefficient, for the sake of convenience, in what follows, we use the Christoffel symbols of second kind. It should be noted that the Christoffel symbols do not follow the transformation rules and therefore they are not tensors.

To present a few examples for calculating the Christoffel, we use the cylindrical and spherical coordinate systems given by Eqs. (6.10) and (6.13). For the cylindrical coordinate system:

$$
(\Gamma^{l}_{lm}) = \begin{pmatrix} 0 & 0 & 0 \\ 0 & -r & 0 \\ 0 & 0 & 0 \end{pmatrix}, \; (\Gamma^{2}_{lm}) = \begin{pmatrix} 0 & 1/r & 0 \\ 1/r & 0 & 0 \\ 0 & 0 & 0 \end{pmatrix}, \; (\Gamma^{3}_{lm}) = \begin{pmatrix} 0 & 0 & 0 \\ 0 & 0 & 0 \\ 0 & 0 & 0 \end{pmatrix} \tag{6.76}
$$

and for spherical coordinate system:

$$(\Gamma^l_{ij}) = \begin{pmatrix} 0 & 0 & 0 \\ 0 & -r & 0 \\ 0 & 0 & -r\sin^2\varphi \end{pmatrix}, \; (\Gamma^2_{ij}) = \begin{pmatrix} 0 & \frac{1}{r} & 0 \\ \frac{1}{r} & 0 & 0 \\ 0 & 0 & -r\sin\varphi\cos\varphi \end{pmatrix},$$

$$(\Gamma^3_{ij}) = \begin{pmatrix} 0 & 0 & \frac{1}{r} \\ 0 & 0 & \cot\varphi \\ \frac{1}{r} & \cot\varphi & 0 \end{pmatrix}. \tag{6.77}$$

6.4.4 Spatial Derivatives in Curvilinear Coordinate System

The differential operator ∇, *Nabla*, is the first spatial differential operator and was defined in Cartesian coordinate system by Eq. (3.4). The operator ∇ has a vector character and acts on the argument that immediately follows it. In curvilinear coordinate system ∇ is defined as:

$$\nabla = \left(\boldsymbol{g}^i \frac{\partial}{\partial \xi^i}\right) = \boldsymbol{g}^1 \frac{\partial}{\partial \xi^1} + \boldsymbol{g}^2 \frac{\partial}{\partial \xi^2} + \boldsymbol{g}^3 \frac{\partial}{\partial \xi^3}. \tag{6.78}$$

Since in Cartesian coordinate system the co- and covariant base vectors $\boldsymbol{e}^i$ and $\boldsymbol{e}_i$ are identical with the orthonormal base vectors $\boldsymbol{e}_i$, it was not necessary to call them co- and covariant base vectors. In curvilinear coordinate system, however, because we operate in co- and contravariant coordinate systems, we define the differential operator vector ∇ which has the direction of the contravariant base vector $\boldsymbol{g}^i$ and differentiates with respect to the coordinate ξ^i.

The second spatial differential operator is the *Laplace* operator $\boldsymbol{\Delta}$ defined as the scalar product of two Nabla operators $\boldsymbol{\Delta} = \nabla \cdot \nabla = \nabla^2$. Thus, it is a second order differential operator that acts on the argument that immediately follows it. The argument might be any tensor valued function that is twice differentiable. In the following $\boldsymbol{\Delta}$ is given in curvilinear coordinate systems:

$$\boldsymbol{\Delta} = \left(\boldsymbol{g}^i \frac{\partial}{\partial \xi^i}\right) \cdot \left(\boldsymbol{g}^j \frac{\partial}{\partial \xi^j}\right). \tag{6.79}$$

While in Cartesian coordinate system the differentiation of the orthonormal unit vectors identically vanishes, the differentiation of the base vectors in curvilinear coordinate system is nonzero. Performing the differentiation on $\boldsymbol{g}^j$, one has to consider that the Christoffel symbols that are always associated with the base vector that must also be differentiated. As a result, we are dealing with a complex outcome of the operation as seen below, where we apply $\boldsymbol{\Delta}$ to a velocity vector. In continuum mechanics the results of this operation is involved in determining the shear stress $\nu\boldsymbol{\Delta}\boldsymbol{V}$ of a Newtonian fluid. Applying $\boldsymbol{\Delta}$ to the velocity $\boldsymbol{V}$ vector as a first order tensor:

$$\boldsymbol{\Delta V} = \boldsymbol{g}_m g^{ik} \lfloor V^m_{,ik} + V^n_{,i}\Gamma^m_{nk} + V^n_{,k}\Gamma^m_{nl} - V^m_{,j}\Gamma^j_{ik} + V^p(\Gamma^n_{pi}\Gamma^m_{nk} - \Gamma^j_{ik}\Gamma^m_{pj} + \Gamma^m_{pi,k})\rfloor. \tag{6.80}$$

6.5 Application of ∇ to Tensor Functions

In this section, the operator ∇ will be applied to different arguments such as zeroth, first, second and higher order tensors. If the argument is a zeroth order tensor which is a scalar quantity such as pressure, temperature, entropy or any other thermo-fluid quantities, the results of the operation is the gradient of the scalar field which is a vector quantity:

$$\nabla p = \boldsymbol{g}^i \frac{\partial p}{\partial \xi^i} \equiv \boldsymbol{g}^i p_{,i} \tag{6.81}$$

If the argument is a first order tensor such as a velocity vector, the order of the resulting tensor depends on the operation character between the operator ∇ and the argument. Here, as in Chap. 3, we distinguish between three products: (1) Scalar, (2) vector (cross) and (3) tensor products.

6.5.1 *Scalar Product of ∇ and a First Order Tensor*

The scalar product of ∇ and a vector results in a scalar quantity. In case of a velocity $\boldsymbol{V}$, the result of the operation is called the *divergence* of the vector field (see also Chap. 3):

$$\nabla \cdot \boldsymbol{V} = \left(\boldsymbol{g}^i \frac{\partial}{\partial \xi^i}\right) \cdot (\boldsymbol{g}_k V^j) = \boldsymbol{g}^i \cdot \left(\frac{\partial \boldsymbol{g}_j}{\partial \xi^i} V^j + \frac{\partial V^j}{\partial \xi^i} \boldsymbol{g}_j\right). \tag{6.82}$$

As seen, the chain rule of differentiation is applied to the argument $\boldsymbol{g}_i V^j$. It should be noted that a scalar operation leads to a contraction of the order of tensor on which the operator is acting. The scalar operation in Eq. (6.82) leads to:

$$\nabla \cdot \boldsymbol{V} = V^i_{,i} + V^j \Gamma^i_{ij}. \tag{6.83}$$

For Cartesian coordinate system, Eq. (6.83) reduces to Eq. (3.26). These equations are identical with the equation of continuity for incompressible flow.

6.5.2 Scalar Product of ∇ and a Second Order Tensor

A scalar operation that involves ∇ and a second order tensor, such as the stress tensor $\mathbf{\Pi}$ or the deformation tensor $\boldsymbol{D}$, results in a first order tensor which is a vector:

$$\nabla \cdot \mathbf{\Pi} = \left(\boldsymbol{g}^m \frac{\partial}{\partial \xi^m} \right) \cdot (\boldsymbol{g}_i \boldsymbol{g}_j \pi^{ij}) = \boldsymbol{g}^m \cdot (\boldsymbol{g}_k \boldsymbol{g}_l)(\pi^{kl}_{,m} + \pi^{nl}\Gamma^k_{nm} + \pi^{kn}\Gamma^l_{nm}). \quad (6.84)$$

The argument in Eq. (6.84) consists of three elements, where the chain rule of differentiations is applied. Performing the differentiation, the right hand side of Eq. (6.84) is reduced to:

$$\nabla \cdot \mathbf{\Pi} = \boldsymbol{g}_j(\pi^{mj}_{,m} + \pi^{nj}\Gamma^m_{nm} + \pi^{mn}\Gamma^j_{mn}). \quad (6.85)$$

Is $\mathbf{\Pi}$ the stress tensor, so $\nabla \cdot \mathbf{\Pi}$ is its divergence which is directly related to the stress vector.

6.5.3 Vector Product of ∇ and a First Order Tensor

A vector or cross product of ∇ and a first order tensor such as the velocity vector results in a *rotation* or *curl* of the vector field. Again, in this case, the chain rule of differentiation is applied as shown below:

$$\nabla \times \boldsymbol{V} = \left(\boldsymbol{g}^i \frac{\partial}{\partial \xi^i} \right) \times (\boldsymbol{g}^j V_j) = \boldsymbol{g}^i \times \mathbf{g}^j V_{j,i} + \boldsymbol{g}^i \times \boldsymbol{g}^j_{,i} V_j. \quad (6.86)$$

Implementing the Christoffel symbol in the second expression of Eq. (6.86) and performing the vector operation we obtain in:

$$\nabla \times \boldsymbol{V} = \boldsymbol{g}^i \times \boldsymbol{g}^j (V_{j,i} - \Gamma^k_{ij} V_k) = \frac{1}{\sqrt{g}} \varepsilon^{ijk} \boldsymbol{g}_k (V_{j,i} - \Gamma^k_{ij} V_k) = \frac{1}{\sqrt{g}} \varepsilon^{ijk} \boldsymbol{g}_k (V_{j,i}) \quad (6.87)$$

with ε^{ijk} as the permutation symbol that functions similar to the one for Cartesian coordinate system and $\sqrt{g} = \sqrt{|g_{ii}|}$. As seen from the last expression in Eq. (6.87), the product of $\varepsilon^{ijk}\Gamma^k_{ii}$ identically vanishes because of the antisymmetric property of the permutation symbol ε^{ijk} and the symmetric property of Γ^k_{ii}.

6.5.4 Tensor Product of ∇ and a First Order Tensor

This product is the gradient of a first order tensor such as the velocity vector $\boldsymbol{V}$. In a curvilinear coordinate system it reads:

$$\nabla V = g^i g_j (V^j_{,i} + V^k \Gamma^j_{ik}). \tag{6.88}$$

In contrast to the scalar product, where the order of the first order tensor V was contracted, the tensor operation (6.88) the order of the product raises. Contracting the base vectors in Eq. (6.88) delivers the divergence of the velocity field as derived earlier.

6.6 Covariant Derivative

In Chap. 2 we have shown that in Cartesian coordinate system embedded in Euclidean space, tensors of order zero, first, second or higher order are invariant with respect to coordinate transformation. In Chap. 3 we have seen that the gradient of a scalar quantity (tensor of order zero) is a first order tensor that is invariant with respect to coordinate transformation. Generally all laws of physics are supposed to be invariant with respect to any change of frame of reference. Differentiating a tensor generally changes its order from a lower order to a higher one. The gradient of a zeroth order tensor (scalar quantity) and a first order tensor (vector quantity) delivered a first and a second order tensor. In this section, we want to investigate the validity of the above statement if we leave the Cartesian coordinate system and move into a curvilinear coordinate system. With other word: Is the differentiation of a tensor valued function in a curvilinear coordinate system a tensor or not? As we will see in what follows, the answer is no. But, by introducing the *covariant derivative* we will arrive at a tensor valued function.

6.6.1 Covariant Derivative of a First Order Tensor

We now consider a vector expressed in contravariant components:

$$V = g_k V^k \tag{6.89}$$

and obtain its gradient operation in a step-by-step process:

$$\begin{aligned}
\nabla V &= g^i \frac{\partial (g_j V^j)}{\partial \xi^i} = g^i \left(\frac{\partial g_j}{\partial \xi^i} V^j + \frac{\partial V^j}{\partial \xi^i} g_j \right) \\
\nabla V &= g^i \frac{\partial (g_j V^j)}{\partial \xi^i} = g^i g_j V^j_{,i} + g^i V^j g_{j,i} \\
\nabla V &= g^i g_j V^j_{,i} + g^i V^j \Gamma^n_{ji} g_n = g^i g_j (V^j_{,i} + V^m \Gamma^j_{mi}).
\end{aligned} \tag{6.90}$$

The result of differentiation of ∇V would be a second order tensor if the second term of the third equation in (6.90) namely $V^m \Gamma^j_{mi}$ vanishes. Since the component V^m is a non-zero quantity, the Christoffel symbol Γ^j_{mi} must disappear. This is only the case in a Cartesian coordinate system. In order to establish a relationship for derivatives of physical quantities such as the gradient of a vector field in a curvilinear coordinate system to be tensors, the expression in the parentheses of the third equation in (6.90) must follow the transformation rules. To this end, we introduce a new operator $\boldsymbol{\nabla}_i$ that is directly related to the Nabla operator as seen below:

$$\boldsymbol{\nabla} = \boldsymbol{g}^i \boldsymbol{\nabla}_i. \tag{6.91}$$

In contrast to $\boldsymbol{\nabla}$-operator that acts on both the vector components and the base vectors, the new spatial operator $\boldsymbol{\nabla}_i$ acts only on the components. The following example clarifies the statement:

$$\boldsymbol{\nabla V} = g^i \frac{\partial(\boldsymbol{g}_j V^j)}{\partial \xi^i} = \boldsymbol{g}^i \boldsymbol{\nabla}_i (\boldsymbol{g}_i V^j) = \boldsymbol{g}^i \boldsymbol{g}_j \boldsymbol{\nabla}_i (V^j). \tag{6.92}$$

In Eq. (6.92) the new operator ∇_i acts on components only, so the base vector can move outside the parentheses. From Eqs. (6.92) and (6.90) it follows that:

$$\boldsymbol{\nabla}_j V^i = \frac{\partial V^i}{\partial \xi_j} + V^k \Gamma^i_{jk} \text{ or } \boldsymbol{\nabla}_j V^i = V^i_{,j} + V^k \Gamma^i_{jk}. \tag{6.93}$$

Equation (6.93) is the covariant derivative of the contravariant components of a vector. To distinguish the covariant differentiation from the partial differentiation, three different symbols are encountered in the literature from which $\boldsymbol{\nabla}_i$ is already shown above. In the following example the three symbols are presented in conjunction with a vector:

$$\begin{aligned}
\boldsymbol{\nabla V} &= \boldsymbol{g}^i \boldsymbol{g}_j \boldsymbol{\nabla}_i (V^j) \\
\boldsymbol{\nabla V} &= \boldsymbol{g}^i \boldsymbol{g}_j V^j|_i \\
\boldsymbol{\nabla V} &= \boldsymbol{g}^i \boldsymbol{g}_j V^j_{;i}.
\end{aligned} \tag{6.94}$$

On a case by case basis, in this and the following chapters we may use "|" or $\boldsymbol{\nabla}_i$. The semicolon ";" in the third expression is an alternative symbol which is found in a few tensor books, but is not used in this and the following sections. Thus, the covariant derivative of the contravariant component of a vector is:

$$V^i|_j = V^i_{,j} + \Gamma^i_{jk} V^k. \tag{6.95}$$

In the same way, we construct the covariant derivative of the covariant components of a vector as:

$$V_i|_j = V_{i,j} - \Gamma^k_{ij} V_k. \tag{6.96}$$

In concluding this section, we present the partial differential operator ∇ and the covariant differential operator and to highlight their differences:

$$\nabla = \mathbf{g}^j \frac{\partial}{\partial \xi_j}$$
$$\nabla_j = \frac{\partial}{\partial \xi_j} + \Gamma_j. \tag{6.97}$$

The Christoffel symbol in Eq. (6.97) is a set of three matrices. If ∇_i acts on a vector, for example V^i, then the other missing indices in Christoffel symbols are linked to the component index as shown in the following equation:

$$\nabla_j V^i = \left(\frac{\partial}{\partial \xi_j} + \Gamma_j\right) V^i = \frac{\partial V^i}{\partial \xi^j} + \Gamma^i_{jk} V^k. \tag{6.98}$$

6.6.2 Covariant Derivative of a Second Order Tensor

For obtaining the covariant derivatives of tensors, the differentiation procedure applied to vectors can be extended to second and higher order tensors. Given second order stress tensor:

$$\mathbf{\Pi} = g_i g_j \pi^{ij} = g_i \mathbf{g}^j \pi^i_j = \mathbf{g}^i \mathbf{g}^j \pi_{ij} \tag{6.99}$$

its spatial differentiation is the gradient of a second order tensor:

$$\nabla \mathbf{\Pi} = \nabla(\mathbf{g}^i \mathbf{g}^j \pi_{ij}) = \mathbf{g}^k \frac{\partial}{\partial \xi^k} (\mathbf{g}^i \mathbf{g}^j \pi_{ij}) = \mathbf{g}^k \frac{\pi \mathbf{\Pi}}{\partial \xi^k} = \mathbf{g}^k \mathbf{\Pi}_{,k}. \tag{6.100}$$

Equation (6.100) is a third order tensor with the derivative of its nine components:

$$\mathbf{\Pi}_{,k} = \mathbf{g}^i \mathbf{g}^j \pi_{ij,k} + \mathbf{g}^i_{,k} \mathbf{g}^j \pi_{ij} + \mathbf{g}^i \mathbf{g}^j_{,k} \pi_{ij}. \tag{6.101}$$

Applying the covariant differentiation to covariant, contravariant and mixed components, we obtain:

$$\pi_{ij}|_k = \pi_{ij,k} - \Gamma^m_{ik} \pi_{mj} - \Gamma^m_{jk} \pi_{im}$$
$$\pi^{ij}|_k = \pi^{ij}_{,k} + \Gamma^i_{mk} \pi^{mj} + \Gamma^j_{mk} \pi^{im}$$
$$\pi^i_j|_k = \pi^i_{j,k} + \Gamma^i_{mk} \pi^m_j - \Gamma^m_{jk} \pi^i_m. \tag{6.102}$$

6.7 Application Example 1: Inviscid Incompressible Flow Motion

As the first application example, the equation of motion for an inviscid incompressible and steady flow is transformed into a cylindrical coordinate system, where it is decomposed in its three components r, **2**, z. The coordinate invariant version of the equation is written as:

$$\boldsymbol{V} \cdot \nabla \boldsymbol{V} = -\frac{1}{\rho}\nabla p. \tag{6.103}$$

The transformation and decomposition procedure is shown in the following steps.

6.7.1 Equation of Motion in Curvilinear Coordinate Systems

The second order tensor on the left hand side can be obtained using Eq. (6.91):

$$\nabla \boldsymbol{V} = \boldsymbol{g}^i \boldsymbol{g}_j (V^j_{,i} + V^k \Gamma^j_{ik}). \tag{6.104}$$

The scalar multiplication with the velocity vector $\boldsymbol{V}$ leads to:

$$\boldsymbol{V} \cdot \nabla \boldsymbol{V} = \boldsymbol{g}_m V^m \cdot \boldsymbol{g}^i \boldsymbol{g}_j (V^j_{,i} + V^k \Gamma^j_{ik}). \tag{6.105}$$

Introducing the mixed Kronecker delta:

$$\boldsymbol{V} \cdot \nabla \boldsymbol{V} = \delta^i_m \boldsymbol{g}_j V^m (V^j_{,i} + V^k \Gamma^j_{ik}). \tag{6.106}$$

For an orthogonal curvilinear coordinate system the mixed Kronecker delta is:

$$\begin{aligned} \delta^i_m &= 1 \text{ for } i = m \\ \delta^i_m &= 0 \text{ for } i \neq m. \end{aligned} \tag{6.107}$$

Taking Eq. (6.107) into account, Eq. (6.63) yields:

$$\boldsymbol{V} \cdot \nabla \boldsymbol{V} = \boldsymbol{g}_j V^i (V^j_{,i} + V^k \Gamma^j_{ik}) \tag{6.108}$$

and rearranging the indices, we find:

$$\boldsymbol{V} \cdot \nabla \boldsymbol{V} = \boldsymbol{g}_i (V^j V^i_j + V^j V^k \Gamma^i_{kj}). \tag{6.109}$$

The pressure gradient on the right hand side of Eq. (6.103) is calculated form:

$$\nabla p = \boldsymbol{g}^i \frac{\partial p}{\partial \xi^i} = \boldsymbol{g}^i p_{,i}. \tag{6.110}$$

Replacing the contravariant base vector with the covariant one using Eq. (6.110) leads to:

$$\nabla p = \boldsymbol{g}^i \frac{\partial p}{\partial \xi^i} = \boldsymbol{g}_i g^{ji} p_j. \tag{6.111}$$

Incorporating Eqs. (6.111) and (6.109) into Eq. (6.103) yields:

$$\boldsymbol{g}_i (V^j V^i_{,j} + V^j V^k \Gamma^i_{kj}) = -\frac{1}{\rho} \boldsymbol{g}_i g^{ji} p_j. \tag{6.112}$$

In i-direction, the equation of motion is:

$$V^j V^i_{,j} + V^j V^k \Gamma^i_{kj} = -\frac{1}{\rho} g^{ji} p^j. \tag{6.113}$$

6.7.2 Special Case: Cylindrical Coordinate System

To transfer equation (6.113) in any arbitrary curvilinear coordinate system, first the coordinate system must be specified. As an example, we consider the cylinder coordinate system. It is related to the Cartesian coordinate system by:

$$x_1 = r \cos \Theta, \; x_2 = r \sin \Theta, \; x_3 = z. \tag{6.114}$$

The curvilinear coordinate system is represented by:

$$\xi_1 = r, \; \xi_2 = \Theta, \; \xi_3 = z. \tag{6.115}$$

6.7.3 Base Vectors, Metric Coefficients

The base vectors are calculated from Eq. (6.7) that is decomposed in its components as:

$$\begin{aligned}
\boldsymbol{g}_1 &= \boldsymbol{e}_1 \frac{\partial x_1}{\partial \xi^1} + e_2 \frac{\partial x^2}{\partial \xi^1} + e_3 \frac{\partial x_3}{\partial \xi^1} \\
\boldsymbol{g}_2 &= e_1 \frac{\partial x_1}{\partial \xi^2} + e_2 \frac{\partial x_2}{\partial \xi^2} + e_3 \frac{\partial x_3}{\partial \xi^2} \\
\boldsymbol{g}_3 &= e_1 \frac{\partial x_1}{\partial \xi^3} + e_2 \frac{\partial x_2}{\partial \xi^3} + e_3 \frac{\partial x_3}{\partial \xi^3}.
\end{aligned} \tag{6.116}$$

The differentiation of the Cartesian coordinates yields:

$$\begin{aligned} \boldsymbol{g}_1 &= e_1 \cos\theta + e_2 \sin\theta \\ \boldsymbol{g}_2 &= -e_1 r \sin\theta + e_2 r \cos\theta \\ \boldsymbol{g}_3 &= e_3. \end{aligned} \tag{6.117}$$

The co- and contravariant metric coefficients are:

$$(g_{ij}) = \begin{pmatrix} 1 & 0 & 0 \\ 0 & r^2 & 0 \\ 0 & 0 & 1 \end{pmatrix}, \; (g^{ij}) = \begin{pmatrix} 1 & 0 & 0 \\ 0 & l/r^2 & 0 \\ 0 & 0 & 1 \end{pmatrix}. \tag{6.118}$$

The contravariant base vectors are obtained from:

$$\begin{aligned} \boldsymbol{g}^i &= g^{ij} \boldsymbol{g}_j \\ \boldsymbol{g}^1 &= g^{11} \boldsymbol{g}_1 + g^{12} \boldsymbol{g}_2 + g^{13} \boldsymbol{g}_3 \\ \boldsymbol{g}^2 &= g^{21} \boldsymbol{g}_1 + g^{22} \boldsymbol{g}_2 + g^{23} \boldsymbol{g}_3 \\ \boldsymbol{g}^3 &= g^{31} \boldsymbol{g}_1 + g^{32} \boldsymbol{g}_2 + g^{33} \boldsymbol{g}_3. \end{aligned} \tag{6.119}$$

Since the mixed metric coefficient are zero, Eq. (6.119) reduces to:

$$\boldsymbol{g}^1 = g^{11} \boldsymbol{g}_1, \; \boldsymbol{g}^2 = g^{22} \boldsymbol{g}_2, \; \boldsymbol{g}^3 = g^{33} \boldsymbol{g}_3. \tag{6.120}$$

6.7.4 Christoffel Symbols

The calculation procedure for obtaining the Christoffel symbols follows Eq. (6.59), where one zero- element and one non-zero element are calculated as an example.

$$\begin{aligned} \Gamma^1_{11} &= \frac{1}{2} g^{11} \left(\frac{\partial g_{11}}{\partial \xi_1} + \frac{\partial g_{11}}{\partial \xi_1} - \frac{\partial g_{11}}{\partial \xi_1} \right) = 0 \\ \Gamma^1_{22} &= \frac{1}{2} g^{11} \left(\frac{\partial g_{21}}{\partial \xi_2} + \frac{\partial g_{12}}{\partial \xi_2} - \frac{\partial g_{22}}{\partial \xi_1} \right) = -r. \end{aligned} \tag{6.121}$$

All other elements are calculated similarly. They are shown in the following matrices:

$$(\Gamma^1_{lm}) = \begin{pmatrix} 0 & 0 & 0 \\ 0 & -r & 0 \\ 0 & 0 & 0 \end{pmatrix}, \; (\Gamma^2_{lm}) = \begin{pmatrix} 0 & 1/r & 0 \\ l/r & 0 & 0 \\ 0 & 0 & 0 \end{pmatrix}, \; (\Gamma^3_{lm}) = \begin{pmatrix} 0 & 0 & 0 \\ 0 & 0 & 0 \\ 0 & 0 & 0 \end{pmatrix}. \tag{6.122}$$

Introducing the non-zero Christoffel symbols into Eq. (6.113), the components in g_l, g_2 and g_3 directions are:

$$V^1 V^1_{,1} + V^2 V^1_{,2} + V^3 V^1_{,3} + \Gamma^1_{22} V^2 V^2 = -\frac{1}{\rho} g^{11} p_{,1}$$
$$V^1 V^2_{,1} + V^2 V^2_{,2} + V^3 V^2_{,3} + 2\Gamma^2_{21} V^2 V^1 = -\frac{1}{\rho} g^{22} p_{,2}$$
$$V^1 V^3_{,1} + V^2 V^3_{,2} + V^3 V^3_{,3} = -\frac{1}{\rho} g^{33} p_{,3}. \tag{6.123}$$

6.7.5 Introduction of Physical Components

The physical components can be calculated from Eqs. (6.52) and (6.53):

$$V^*_i = \sqrt{g^{(ii)}} V_i, \; V^{*i} = \sqrt{g_{(ii)}} V^i$$
$$V^{*1} = \sqrt{g_{(11)}} V_1, \; V^{*2} = \sqrt{g_{(22)}} V^2; \; V^{*3} = \sqrt{g_{33}} V^3$$
$$V^{*1} = \sqrt{1} V^1; \; V^{*2} = \sqrt{r^2} V^2; \; V^{*3} = \sqrt{1} V^3. \tag{6.124}$$

The V^i-components expressed in terms of V^{*i} are:

$$V^1 = V^{*1}; \; V^2 = \frac{1}{r} V^{*2}; \; V^3 = V^{*3}. \tag{6.125}$$

Introducing Eq. (6.124) into Eq. (6.123) results in:

$$V^{*1} V^{*1}_{,1} + \frac{V^{*2}}{r} V^{*1}_{,2} + V^{*3} V^{*1}_{,3} + \Gamma^1_{22} \frac{V^{*2} V^{*2}}{r^2} = -\frac{1}{\rho} g^{11} p_{,1}$$
$$V^{*1} \frac{V^{*2}_{,1}}{r} - V^{*1} \frac{V^{*2}}{r^2} + \frac{V^{*2}}{r^2} V^{*2}_{,2} + V^{*3} \frac{V^{*2}_{,3}}{r} + \frac{2}{r} \Gamma^2_{21} V^{*2} V^{*1} - \frac{1}{\rho} g^{22} p_{,2}$$
$$V^{*1} V^{*3}_{,1} + \frac{V^{*2}}{r} V^{*3}_{,2} + V^{*3} V^{*3}_{,3} = -\frac{1}{\rho} g^{33} p_{,2}. \tag{6.126}$$

According to the definition:

$$\xi_1 = r; \; \xi_2 = \Theta; \; \xi_3 = z \tag{6.127}$$

the physical components of the velocity vectors are:

$$V^{*1} = V_r; \; V^{*2} = V_\Theta; \; V^{*3} = V_z \tag{6.128}$$

and insert these relations into Eq. (6.126), the resulting components in r, **1**, and z directions are:

$$V_r \frac{\partial V_r}{\partial r} + \frac{V_\Theta}{r} \frac{\partial V_r}{\partial \Theta} + V_z \frac{\partial V_r}{\partial z} - \frac{V_\Theta^2}{r} = -\frac{1}{\rho} \frac{\partial p}{\partial r}$$
$$V_r \frac{\partial V_\Theta}{\partial r} + \frac{V_\Theta}{r} \frac{\partial V_\Theta}{\partial \Theta} + V_z \frac{\partial V_\Theta}{\partial z} + \frac{V_r V_\Theta}{r} = -\frac{1}{\rho} \frac{\partial p}{r \partial \Theta}$$
$$V_r \frac{\partial V_z}{\partial r} + \frac{V_\Theta}{r} \frac{\partial V_z}{\partial \Theta} + V_z \frac{\partial V_z}{\partial z} = -\frac{1}{\rho} \frac{\partial p}{\partial z}. \quad (6.129)$$

6.8 Application Example 2: Viscous Flow Motion

As the second application example, the Navier–Stokes equation of motion for a viscous incompressible flow is transferred into a cylindrical coordinate system, where it is decomposed in its three components r, **2**, z. The coordinate invariant version of the equation is written as:

$$\boldsymbol{V} \cdot \nabla \boldsymbol{V} = -\frac{1}{\rho} \nabla p + \nu \nabla^2 \boldsymbol{V}. \quad (6.130)$$

The second term on the right hand side of Eq. (6.130) exhibits the shear stress force. It was treated in Sect. 6.6 and is the only term that has been added to the equation of motion for inviscid flow.

6.8.1 *Equation of Motion for Viscous Flow*

The transformation and decomposition procedure is similar to the example in Sect. 6.6. Therefore, a step by step derivation is not necessary to be presented.

$$\begin{aligned} \boldsymbol{g}_i (V^j V^i_{j} + V^j V^k \Gamma^i_{kj}) = -\frac{1}{\rho} \boldsymbol{g}_i g^{ji} p_j + \nu \boldsymbol{g}_m [& V^m_{,ik} + \\ & V^n_{,i} \Gamma^m_{nk} + V^n_{,k} \Gamma^m_{ni} - V^m_{,j} \Gamma^j_{ik} + \\ & V^p (\Gamma^n_{pi} \Gamma^m_{nk} - \Gamma^j_{ik} \Gamma^m_{pj} + \Gamma^m_{pi,k})] g^{ik}. \end{aligned} \quad (6.131)$$

6.8.2 *Special Case: Cylindrical Coordinate System*

Using the Christoffel symbols from Sect. 6.6.4, the physical components from Sect. 6.6.5, and inserting the corresponding relations into Eq. (6.131), the resulting components in r, **1**, and z directions are:

$$V_r \frac{\partial V_r}{\partial r} + \frac{V_\Theta}{r} \frac{\partial V_r}{\partial \theta} + V_z \frac{\partial V_r}{\partial z} - \frac{V_\Theta^2}{r} = -\frac{1}{\rho} \frac{\partial p}{\partial r} +$$
$$\nu \left(\frac{\partial^2 V_r}{\partial r^2} + \frac{1}{r^2} \frac{\partial^2 V_r}{\partial \Theta^2} + \frac{\partial^2 V_r}{\partial z^2} - 2 \frac{\partial V_\Theta}{r^2 \partial \Theta} + \frac{\partial V_r}{r \partial r} - \frac{V_r}{r^2} \right) \tag{6.132}$$

$$V_r \frac{\partial V_\Theta}{\partial r} + \frac{V_\Theta}{r} \frac{\partial V_\Theta}{\partial \Theta} + V_z \frac{\partial V_\Theta}{\partial z} + \frac{V_r V_\Theta}{r} = -\frac{1}{\rho} \frac{\partial p}{r \partial \Theta}$$
$$+\nu \left(\frac{\partial^2 V_\Theta}{\partial r^2} + \frac{1}{r^2} \frac{\partial^2 V_\Theta}{\partial \Theta^2} + \frac{\partial^2 V_\Theta}{\partial z^2} + \frac{2}{r^2} \frac{\partial V_r}{\partial \Theta} + \frac{1}{r} \frac{\partial V_\Theta}{\partial r} - \frac{V_\Theta}{r^2} \right)$$
$$V_r \frac{\partial V_z}{\partial r} + \frac{V_\Theta}{r} \frac{\partial V_z}{\partial z} + V_z \frac{\partial V_z}{\partial z} = -\frac{1}{\rho} \frac{\partial p}{\partial z}$$
$$+ \nu \left[\frac{\partial^2 V_z}{\partial r^2} + \frac{\partial^2 V_z}{r^2 \partial \Theta^2} + \frac{\partial^2 V_z}{\partial z^2} + \frac{1}{r} \frac{\partial V_z}{\partial r} \right]. \tag{6.133}$$

References

1. Aris R (1962) Vector, tensors and the basic equations of fluid mechanics. Prentice-Hall, Inc, Englewood Cliffs
2. Kaestner S (1964) Vektoren, Tensoren, Spinoren. Akademie-Verlag, Leipzig
3. Brand L (1947) Vector and tensor analysis. Wiley, New York
4. Klingbeil E (1966) Tensorrechnung für Ingenieure. A. Tensor operations in orthogonal curvilinear coordinate systems 495. Bibliographisches Institut, Mannheim
5. Lagally M (1944) Vorlesung über Vektorrechnung, dritte edn. Akademische Verlagsgeselschaft, Leipzig
6. Vavra MH (1960) Aero-thermodynamics and flow in turbomachines. Wiley Inc, Hoboken

Chapter 7
Tensor Application, Navier–Stokes Equation

Tensor analysis plays a substantive role in finding exact solutions for laminar flows through channels with curved walls. In this section, a set of exact solutions of the Navier–Stokes equations for two-dimensional *laminar* flow through several curved channels is presented. Generally in fluid mechanics, exact analytical solutions of the Navier–Stokes equations are available for a very small number of cases, where the flow is assumed to be *unidirectional*. This implies that the velocity vector has a component in longitudinal (or streamwise) direction only but it changes in lateral direction. In curved channels, where the velocity vector of a two-dimensional flow has generally two components, the coordinate system can be transformed such that the velocity vector has only one component in a streamwise direction. In providing the exact solutions, in the following sections, several cases are presented that are of fundamental significance for understanding the motion of viscous flows. In addition to the solution of Navier–Stokes equations, temperature distribution provides insight into the physics of viscous dissipation. In this section, first we find the exact solutions of the Navier–Stokes equation in terms of velocity distribution. The temperature distribution is obtained by introducing the energy equation that we combine with the Navier–Stokes equation.

7.1 Steady Viscous Flow Through a Curved Channel

In solving the Navier–Stokes equation, we investigate the influence of channel curvature and pressure gradient on the flow quantities such as velocity and temperature distributions. These distributions are primarily affected by the curvature and pressure gradient. To calculate the above quantities, conservation laws of fluid mechanics and thermodynamics are applied. For an incompressible Newtonian fluid, the Navier–Stokes equations describe the flow motion completely. These equations have exact

M. T. Schobeiri, *Tensor Analysis for Engineers and Physicists - With Application to Continuum Mechanics, Turbulence, and Einstein's Special and General Theory of Relativity*, https://doi.org/10.1007/978-3-030-35736-8_7

solutions for only a few special cases. For the major part of practical problems encountered in applied fluid mechanics, however, it is hardly possible to find any exact solutions. This deficiency is in part due to the complexity of the individual flow field and its geometry under consideration. Despite this fact, the existence of exact solutions of fluid mechanics problems including the velocity and temperature distribution within viscous flows are of particular interest to the computational fluid dynamics (CFD) community dealing with development of CFD codes. A comprehensive code assessment and validation requires both the experimental verification and theoretical confirmation. For the latter case, a comparison with existing exact solutions exhibits an appropriate procedure to demonstrate the code capability. For symmetric flows through channels with positive and negative pressure gradients exact solutions are found by Jeffery [1] and Hammel [2]. For asymmetric curved channels with convex and concave walls, exact solutions of the Navier–Stokes equation are found by Schobeiri [3, 4], where the influence of the wall curvature on the velocity and temperature distribution is discussed. Furthermore, a class of approximate solutions of Navier–Stokes is presented in [5–7]. This section treats the influence of curvature and pressure gradient on temperature and velocity distributions by solving the energy and momentum equations. Under the assumption that the flow is two dimensional, steady, incompressible, and has constant viscosity, the conservation laws of fluid mechanics and thermodynamics are transformed into a curvilinear coordinate system. The system describes the two-dimensional, symmetric and asymmetrically curved channels with convex and concave walls.

7.1.1 Description of the Curved Channel Geometry

To determine the influence of curvature and pressure gradient on temperature distribution, the velocity distribution must be known. This requires the solution of continuity equation (6.84) and the Navier–Stokes equations (6.132). For a two-dimensional flow, we prescribe that the velocity component normal to the flow direction must vanish. As a result, the integration of Eq. (6.132) must fulfill both the continuity and the Navier–Stokes equations. This is possible only if the Christoffel symbols Γ^i_{ki} are not functions of the coordinates themselves. The corresponding channel with the curvilinear coordinate is then obtained from the transformation:

$$w = -\frac{2}{a+ib} \ln z \text{ with } z = x + iy \text{ and } w = \xi_1 + i\xi_2 \tag{7.1}$$

with ξ_i as the orthogonal curvilinear coordinate system:

$$x = e^{-\frac{1}{2}(a\xi_1 - b\xi_2)} \cos\left(\frac{a\xi_1 + b\xi_2}{2}\right)$$
$$y = -e^{-\frac{1}{2}(a\xi_1 - b\xi_2)} \sin\left(\frac{a\xi_2 + b\xi_1}{2}\right) \tag{7.2}$$

with a and b as real constants that define the configuration of the channel and ξ_1 and ξ_2 as the orthogonal curvilinear coordinates. The corresponding metric coefficients and Christoffel symbols are:

$$g^{11} = g^{22} = \frac{4}{a^2 + b^2} e^{a\xi_1 - b\xi_2},\ g^{12} = g^{21} = 0 \tag{7.3}$$

$$\Gamma^1_{kl} = \frac{1}{2}\begin{pmatrix} -a & b \\ b & +a \end{pmatrix},\ \Gamma^2_{kl} = -\frac{1}{2}\begin{pmatrix} b & a \\ a & -b \end{pmatrix}. \tag{7.4}$$

With Eqs. (7.3) and (7.4) and the requirement that the velocity component in ξ_2 must vanish, the integration of the continuity equation (6.85) leads to:

$$V^1 = \frac{4\boldsymbol{v}}{a^2 + b^2} F(\xi_2) e^{a\xi_1 - b\xi_2} \tag{7.5}$$

where V^1 is the contravariant component of the velocity in the ξ_1-direction, v is the kinematic viscosity, and $F = F(\xi_2)$ is a function to be determined. Thus, the only physical component of the velocity vector is the one in the ξ_1-direction, for which we may omit the superscript 1 and set:

$$U \equiv V^{*1} = \frac{V^1}{\sqrt{g^{11}}} = \frac{2\boldsymbol{v}}{\sqrt{a^2 + b^2}} F(\xi_2) e^{\frac{1}{2}(a\xi_1 - b\xi_2)}. \tag{7.6}$$

Equation (7.6) must strictly satisfy the Navier–Stokes equations (6.133) in order to be an exact solution. Neglecting the body forces, for the two-dimensional flow with $V^3 = 0$ Eq. (6.132) leads in ξ_1-direction to:

$$\begin{aligned} &V^1 V^1_{,1} + (V^1)^2 \Gamma^1_{11} = -\frac{1}{\rho} p_{,1} g^{11} + \\ &\boldsymbol{\nu}[V^1_{,11} + V^1_{,22} + V^1_{,1}(\Gamma^1_{11} - \Gamma^1_{22}) + V^1_{,2}(2\Gamma^1_{12} - \Gamma^2_{11} - \Gamma^2_{22})] g^{11} \end{aligned} \tag{7.7}$$

and in the ξ_2-direction:

$$(V^1)^2 \Gamma^2_{11} = -\frac{1}{\rho} p_{,2} g^{22} + \boldsymbol{v}[2V^1_{,11}\Gamma^2_{11} + 2V^1_{,2}\Gamma^2_{12}] g^{22} \tag{7.8}$$

with ν as kinematic viscosity. Introducing the integration results of the continuity (7.6) into the system of differential equations (7.7) and (7.8) and eliminating the pressure terms, the result of the first integration is:

$$F'' - 2bF' + (a^2 + b^2)F - \frac{a}{2}F^2 + K_1 = 0 \tag{7.9}$$

with K_1 as the integration constant. Dividing Eq. (7.9) by its maximum value $F_{\max}$, the dimensionless velocity function is obtained from:

$$\Phi'' - 2b\Phi' + (a^2 + b^2)\Phi - \frac{a}{2}F_{\max}\Phi^2 + C_1 = 0 \tag{7.10}$$

where

$$\Phi = \frac{F}{F_{\max}} \text{ and } C_1 = \frac{K_1}{F_{\max}}. \tag{7.11}$$

The significant parameter affecting the flow within the curved channel is the Reynolds number, which is defined as:

$$\mathrm{Re} = \frac{\boldsymbol{\Delta}s \cdot U_m}{\nu} \tag{7.12}$$

where $\boldsymbol{\Delta}s$ and U_m are the distance and the maximum velocity in the ξ_1-direction. The latter is obtained by setting in Eq. (7.6), the coordinate ξ_2 equal to $\xi_{2\max}$ which is the coordinate of the maximum velocity:

$$U_m = \frac{2\nu}{\sqrt{a^2 + b^2}} F_{\max} e^{\frac{1}{2}(a\xi_1 - b\xi_{2\max})}. \tag{7.13}$$

With the distance $\boldsymbol{\Delta}s$:

$$\boldsymbol{\Delta}s = -\frac{\sqrt{a^2 + b^2}}{a} e^{-\frac{1}{2}(a\xi_1 - b\xi_{2\max})} \tag{7.14}$$

and Eq. (7.13), the Reynolds number becomes:

$$\mathrm{Re} = -\frac{2}{a}F_{\max}. \tag{7.15}$$

Introducing Eq. (7.15) into Eq. (7.10) leads to:

$$\Phi'' - 2b\Phi' + (a^2 + b^2)\Phi + \frac{a^2}{4}\mathrm{Re}\,\Phi^2 + C_1 = 0. \tag{7.16}$$

7.1.2 Case I: Solution of the Navier–Stokes Equation

Equation (7.16) describes the motion of a viscous flow through curved channels pertaining to the coordinate transformation (7.2) discussed in Sect. 7.1.1. It includes both the Navier–Stokes and continuity equations that are reduced to a single, ordinary, nonlinear, second-order differential equation. The solutions of Eq. (7.16) are functions of the coordinate ξ_2 and incorporate the Reynolds number as a parameter.

Special cases of Eq. (7.16) are the purely radial flow, where $a = -2$ and $b = 0$, and the flow through concentric cylinders with $a = 0$ and $b = 1$.

For the solution of Eq. (7.16), a numerical integration procedure is applied. Starting from the initial conditions specified below and the determination of constant C_1, an iteration method is developed that reduces the boundary-value problem to an initial value one. The solution of differential equation (7.16) must fulfill the governing initial and boundary conditions. The boundary conditions are given by the non-slip conditions at the channel walls:

$$\begin{aligned} \xi_2 &= \xi_{B1} \equiv 0.1, \quad \mathbf{\Phi} = \mathbf{\Phi}_{B1} \equiv 0 \\ \xi_2 &= \xi_{2_{B2}} \equiv 0.5, \quad \mathbf{\Phi} = \mathbf{\Phi}_{B2} \equiv 0 \end{aligned} \tag{7.17}$$

where the indices $B1$ and $B2$ refer to the convex and concave channel walls. The initial condition is described by the maximum value of the velocity distribution and its position $\xi_2 = \xi_{2\,\max}$, which is unknown for the time being:

$$\xi_2 = \xi_{2\,\max}, \; \mathbf{\Phi} = \mathbf{\Phi}_{\max} = \pm 1, \; \mathbf{\Phi}' = \mathbf{\Phi}'_{\max} = 0. \tag{7.18}$$

The positive sign of $\mathbf{\Phi}$ indicates an increase of the cross-section area in direction of decreasing ξ_1 which is associated with the positive pressure gradient. The negative sign characterizes the accelerated flow in direction of increasing ξ_1, where negative pressure gradient prevails. The constant C_1 in Eq. (7.16) specifies the solution of Eq. (7.16) and significantly affects the convergence speed. It must be determined so that the above boundary and initial conditions are identically fulfilled. Detailed solution is found in [5].

7.1.3 *Case I: Curved Channel, Negative Pressure Gradient*

Once the solution of Eq. (7.16) is found, the dimensionless velocity distribution is obtained from Eq. (7.13) as:

$$\frac{U}{U_m} \equiv \frac{V^*}{V^*_m} = \mathbf{\Phi} e^{\frac{1}{2} b(\xi_{\max} - \xi_2)}. \tag{7.19}$$

As seen earlier, the solution $\mathbf{\Phi} = \mathbf{\Phi}(\xi_2)$ is a function of the coordinate ξ_2 only and incorporates the Reynolds number as a parameter. Thus, the velocity distributions represented by Eq. (7.19) exhibit similar solutions. An asymmetrically curved channel with convex and concave walls is generated by choosing $a = -1$ and $b = 1$.

As shown in Fig. 7.1, the negative pressure gradient is established by an asymmetrically convergent channel with convex and concave walls. For Reynolds number $\mathrm{Re} = 500$ the velocity distributions at the coordinate $\xi_1 = 3.8$ exhibit an almost parabolic shape with the maximum close to $\xi_2 = 0.3$. For the similarity reasons

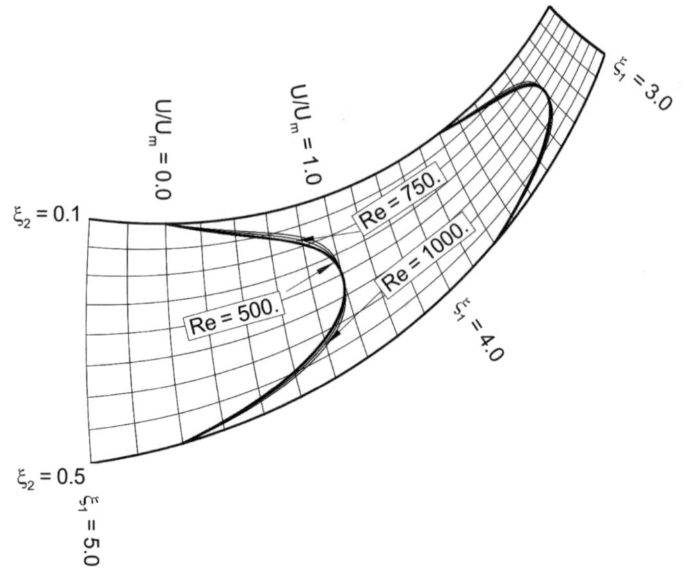

Fig. 7.1 Accelerated laminar flow through a two-dimensional curved channel at different Reynolds numbers

explained above, similar velocity distribution is plotted at $\xi_1 = 3.8$ for the same Reynolds number. Increasing the Reynolds number to Re = 750, 1000 respectively, results in steeper velocity slopes at both walls, Fig. 7.1. As a consequence, the velocity profile tends to become fuller, particularly for higher Reynolds numbers. As shown, the viscosity effect is restricted predominantly to the wall regions and continuously reduces by increasing the Reynolds number. For Reynolds numbers up to Re = 5000, velocity distributions can be calculated without convergence problems. Thus for an accelerated flow, the stability of the laminar flow and the transition from laminar into turbulent flow are apparently extended to higher Reynolds numbers as expected.

7.1.4 Case I: Curved Channel, Positive Pressure Gradient

The positive pressure gradient within the asymmetrically curved channel discussed above is created by reversing the flow direction. Figure 7.2 shows the flow at different Reynolds numbers. Figure 7.2, for Re = 500, the velocity distribution on the concave wall is fully attached. The fluid particles moving in streamwise direction are exposed to three different forces: (1) the wall shear stress force acting in opposite flow direction, decelerates the fluid particles; (2) the decelerating effect of the wall shear stress is intensified by the pressure forces which also act in opposite direction causing the flow to further decelerate; and (3) the centrifugal force caused by

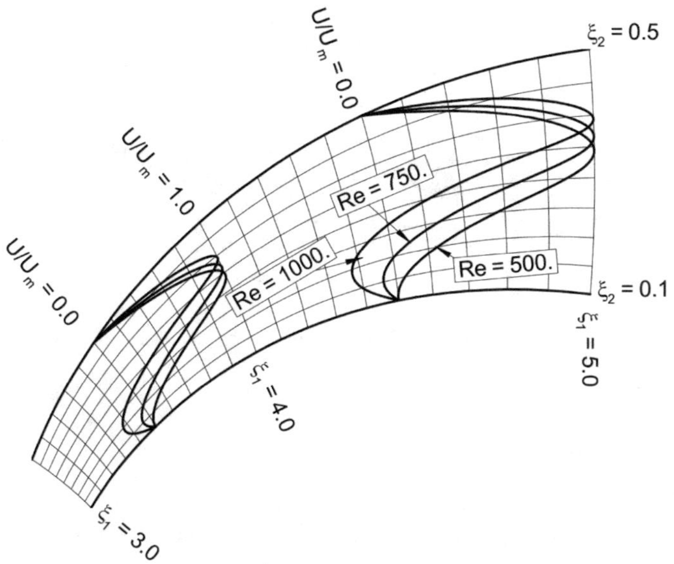

Fig. 7.2 Decelerated laminar flow through a two-dimensional channel at different Reynolds number

the channel curvature pushes the fluid particle away from the convex wall towards the concave one increasing the susceptibility of flow to separation. Increasing the Reynolds number to Re $= 1500$ causes the separation zone on the convex channel wall to become larger.

7.1.5 Case II: Radial Flow, Positive Pressure Gradient

We are now interested in determining the effect of pressure gradient on the velocity distribution in the absence of curvature. To investigate this, we generate a channel with straight wall geometry by setting $a = -2$ and $b = 0$. With these new constants, Eq. (7.16) reduces to:

$$\mathbf{\Phi}'' + 4\mathbf{\Phi} + \text{Re}\,\mathbf{\Phi}^2 + C_1 = 0. \tag{7.20}$$

This special case constitutes a purely radial laminar flow through a channel with straight walls and is known as the Hamel-flow, [3]. The results are shown in Fig. 7.3, where the velocity distributions are plotted for three different Reynolds numbers. Close to the wall at Re $= 500$, the flow exhibits a tendency for separation on both walls. Increasing the Reynolds number to Re $= 750$ and 1500 respectively causes the flow separation on both walls. A comparison with the results in Fig. 7.2 clearly indicates that the difference in velocity distributions is attributed to the nature of wall curvature.

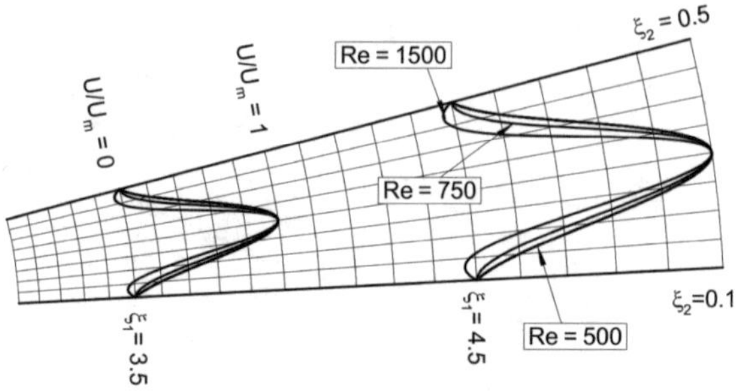

Fig. 7.3 Decelerated laminar flow through a two-dimensional channel with straight walls at different Reynolds numbers

7.2 Temperature Distribution

To determine the temperature distribution within the curved channel we combine the mechanical and thermal energy balances as we discussed in [7]:

$$\rho \frac{Du}{Dt} = -\nabla \cdot \dot{\boldsymbol{q}} - p(\nabla \cdot \boldsymbol{V}) + \boldsymbol{T} : \nabla \boldsymbol{V} \tag{7.21}$$

with u as the internal energy, $\dot{\boldsymbol{q}}$ the heat flux vector, and $\boldsymbol{T}$ the shear stress tensor. Considering the thermodynamic relationship, for the steady incompressible flow, Eq. (7.21) reduces to:

$$\rho c_v \frac{dT}{dt} = -\nabla \cdot \dot{\boldsymbol{q}} + \boldsymbol{T} : \nabla \boldsymbol{V} \tag{7.22}$$

where $du = c_v dT$, c_v is the specific heat capacity at constant volume and T is the temperature of the working medium. Using the decomposition of the velocity gradient, the Fourier equation of conduction, and the Stokes relation:

$$\nabla \boldsymbol{V} = \boldsymbol{D} + \boldsymbol{\Omega}, \quad \dot{\boldsymbol{q}} = -\boldsymbol{\kappa} \nabla T, \quad \boldsymbol{T} = 2\mu \boldsymbol{D} \tag{7.23}$$

with $\boldsymbol{\kappa}$ as the thermal conductivity, μ the absolute viscosity, and $\boldsymbol{D}$ and $\boldsymbol{\Omega}$ as second-order tensors of the deformation and rotation, respectively (see Eqs. (5.20) and (5.22)). Introducing Eq. (7.23) into Eq. (7.22) and considering the identity $\boldsymbol{D} : \boldsymbol{\Omega} = 0$, the result of this operation leads to the differential equation of the energy in terms of temperature:

$$c_v (\nabla T) \cdot \boldsymbol{V} = \frac{\kappa}{\rho} \nabla^2 T + 2\nu \boldsymbol{D} : \boldsymbol{D}. \tag{7.24}$$

The above differential equation is invariant with respect to the coordinate system transformation. Its index notation , however, takes into account the specific geometry of the coordinate system. We expand Eq. (7.24):

$$c_v T_{,i} V^i = \frac{k}{\rho} g^{ik}(T_{,ik} - T_{,j}\Gamma^j_{ik}) + \frac{1}{2}\nu(V_{k,l} + V_{l,k} - V_m\Gamma^m_{kl} - V_m\Gamma^m_{lk})$$
$$\times (g^{li}V^k_{,i} + g^{ki}V^l_{,i} + g^{li}V^m\Gamma^k_{mi} + g^{ki}V^m\Gamma^l_{mi}) \quad (7.25)$$

and insert all Γ^i_{jk} from Eq. (7.4) into Eq. (7.25). Considering the unidirectional assumption we made previously, Eq. (7.25) is reduced to:

$$c_v T_{,1} V^1 = \frac{\kappa}{\rho}(T_{,11} + T_{,22}) + \nu(aV_1V^1_{,1} - bV_1V^1_{,2} + V_{1,2}V^1_{,2})g^{11}. \quad (7.26)$$

7.2.1 Case I: Solution of Energy Equation

Equation (7.26) is a second order, nonlinear, partial differential equation, in which the temperature $T = T(\xi_1, \xi_2)$. It can be reduced to an ordinary differential equation by making the following ansatz:

$$T = T(\xi_1, \xi_2) = \frac{4}{a^2 + b^2} G e^{(a\xi_1 - b\xi_2)} + T_w \quad (7.27)$$

with $G = G(\xi_2)$ and T_w as the wall temperature. Without loss of generality we may assume a constant wall temperature. Incorporating Eqs. (7.27) and (7.5) into Eq. (7.26) we obtain:

$$ac_\nu \nu GF = \frac{\kappa}{\rho}[(a^2 + b^2)G - 2bG' + G''] + \nu^3[(a^2 + b^2)F^2 - 2bFF' + F'^2]. \quad (7.28)$$

Dividing Eq. (7.28) by $F^2_{\max}$ and introducing the Reynolds number from Eq. (7.15) we arrive at:

$$\mathbf{\Theta}'' - 2b\mathbf{\Theta}' + (a^2 + b^2)\mathbf{\Theta} + \frac{a^2}{2}\,\text{PrRe}\,\mathbf{\Phi\Theta} +$$
$$+ \frac{a^2}{4}\,\text{PrRe}\,[(a^2 + b^2)\mathbf{\Phi}^2 - 2b\mathbf{\Phi\Phi}' + \mathbf{\Phi}'^2] = 0. \quad (7.29)$$

In Eq. (7.29) the function $\mathbf{\Theta}$ is defined as $\mathbf{\Theta} = c_v G/v^2$ Re, with $\text{Pr} = \frac{\mu c_p}{\kappa}$, as the Prandtl number. For gases the Prandtl number is around 0.7 and for water around 7. Detailed distributions of the values for the absolute viscosity μ, the thermal conductivity $\boldsymbol{\kappa}$ and the Prandtl number for dry air can be taken from Fig. 7.4. These values change slightly if the humidity ratio $\boldsymbol{T} = m_{\text{water}}/m_{\text{air}}$ increases from 0 to 10%.

The terms $\mathbf{\Phi}$ and $\mathbf{\Phi}'$ are given as the solution of the Navier–Stokes equations (7.10). The solution of the ordinary, nonlinear, second-order differential (7.29) must satisfy the following boundary conditions:

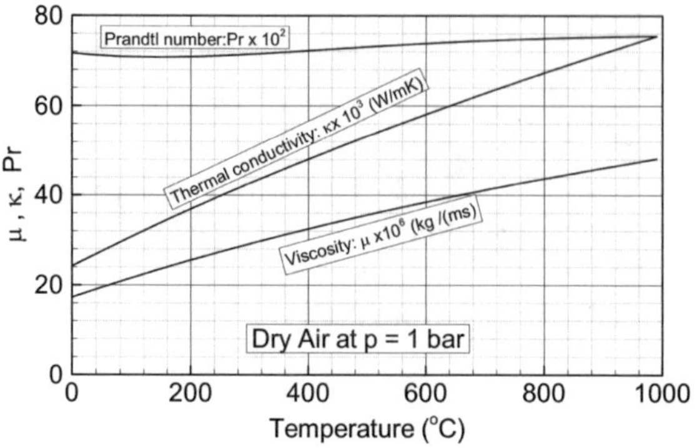

Fig. 7.4 Absolute viscosity, thermal conductivity and Pr-number as a function of temperature for dry air at $p = 1$ bar

$$\begin{aligned}
&\xi_2 = \xi_{2_{B1}} \equiv 0.1; \quad \mathbf{\Phi} = \mathbf{\Phi}_{B1} \equiv 0; \quad \mathbf{\Theta} = \mathbf{\Theta}_{B1} \equiv 0\\
&\xi_2 = \xi_{2_{B2}} \equiv 0.5; \quad \mathbf{\Phi} = \mathbf{\Phi}_{B2} \equiv 0; \quad \mathbf{\Theta} = \mathbf{\Theta}_{B2} \equiv 0.
\end{aligned} \tag{7.30}$$

To find the solution of Eq. (7.29) it must first be combined with the equation of motion (7.10). For the solution of the resulting system of two nonlinear, second-order differential equations, a numerical procedure based on the Predictor-Corrector method is applied. Starting from $\xi_2 = \xi_{2B1}$, and already known $\mathbf{\Phi}'_{B1}$ from Sect. 7.1.2, $\mathbf{\Theta}'_{B1}$ is first estimated that may lead to $\mathbf{\Theta}_{B2}$ .... The correct value can be obtained quickly with the iteration function:

$$\mathbf{\Theta}'_{Bl(i+1)} = \mathbf{\Theta}'_{B1(i)} - \mathbf{\Theta}_{B2(i-1)} \frac{\mathbf{\Theta}'_{B1(i)} - \mathbf{\Theta}'_{B1(i-1)}}{\mathbf{\Theta}_{B2(i)} - \mathbf{\Theta}_{B2(i-1)}}. \tag{7.31}$$

The iteration process is repeated until the accuracy ε is reached:

$$|\mathbf{\Theta}_{B2(i+1)} - \mathbf{\Theta}_{B2(i)}| \le \varepsilon = 10^{-6}. \tag{7.32}$$

7.2.2 *Case I: Curved Channel, Negative Pressure Gradient*

The effect of the different wall curvatures on temperature distributions is shown in Fig. 7.5 by asymmetrical temperature slopes at the convex and concave walls. For the accelerated flow with Re $= 500$, Fig. 7.5 shows the dimensionless temperature distribution for different Prandtl number as a parameter. As a consequence of energy dissipation, the temperature distribution near the channel walls experiences a steep gradient, with the maxima located close to the concave wall.

By approaching the channel middle, the temperature gradient gradually decreases for small Prandtl numbers and sharply for large ones. Increasing the Reynolds number to Re $= 3500$ causes pronounced temperature boundary layers, particularly for higher

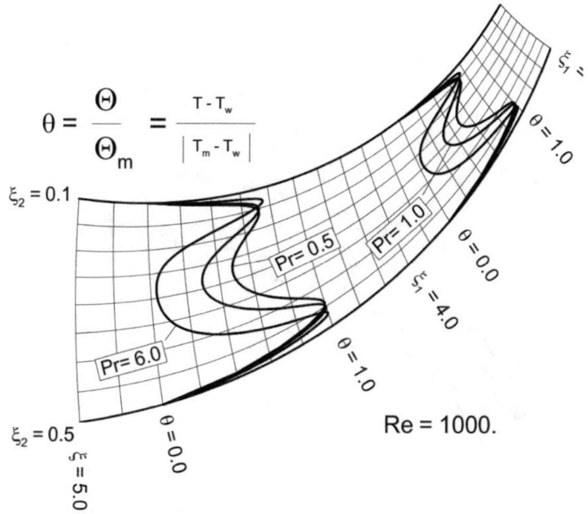

Fig. 7.5 Dimensionless temperature distribution for an accelerated laminar flow through a two-dimensional curved channel with Re- and Pr numbers as parameters, T_m = maximum temperature, T_W = wall temperature

Prandtl numbers. Moving towards the channel middle, the temperature distribution exhibits almost a constant value slightly above the wall temperature.

7.2.3 *Case I: Curved Channel, Positive Pressure Gradient*

The effect of different wall curvatures on temperature distributions is illustrated in Fig. 7.6 by asymmetric temperature slopes at the convex and concave walls. As we discussed in Sect. 7.1.5, the pressure gradient and the wall curvature were responsible for flow separation. For a positive pressure gradient and Re $= 500$, temperature distributions are shown in Fig. 7.6 with Prandtl number as the parameter. As with the accelerated flow, high energy dissipation occurs near the channel walls. When approaching the middle of the channel, the temperature gradient changes the sign. This effect might contribute to the instability of the laminar flow field under a positive pressure gradient.

7.2.4 *Case II: Radial Flow, Positive Pressure Gradient*

In this section we investigate the effect of pressure gradient in the absence of wall curvature. Similar to the case I in Sect. 7.1.5 we construct a straight walled channel by setting $a = 2$ and $b = 0$. With these new constants, Eq. (7.29) reduces to:

$$\boldsymbol{\Theta}'' + 4\boldsymbol{\Theta} + 2\,\mathrm{PrRe}\,\boldsymbol{\Phi}\boldsymbol{\Theta} + \mathrm{PrRe}\,\lfloor 4\boldsymbol{\Phi}^2 + \boldsymbol{\Phi}'^2 \rfloor = 0. \tag{7.33}$$

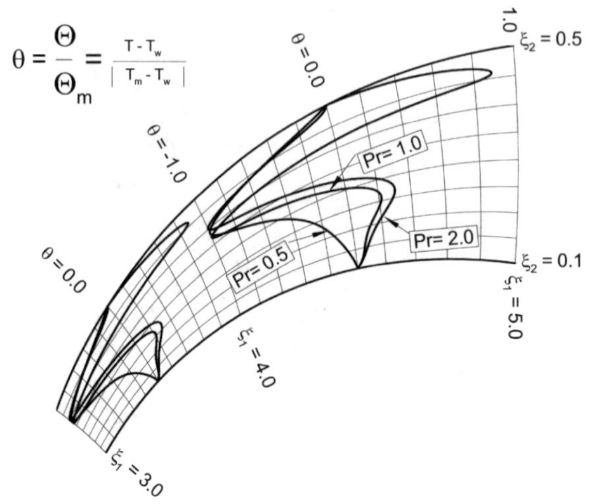

Fig. 7.6 Dimensionless temperature distribution for laminar decelerated flow through a two-dimensional curved channel with Re- and Pr- numbers as parameters, T_m = minimum temperature, T_W = wall temperature

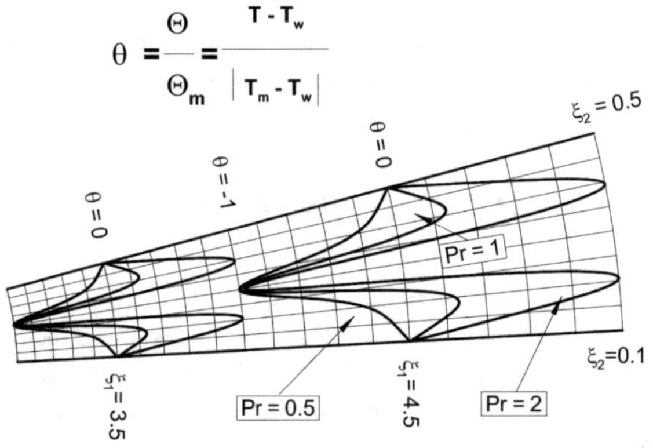

Fig. 7.7 Dimensionless temperature distribution within a straight walled channel with positive pressure gradient for Re = 1500 and different Pr-numbers, T_m = minimum temperature, T_W = wall temperature

To obtain the temperature distribution, Eq. (7.33) must be combined with Eq. (7.20), which is the exact solution of the Navier–Stokes equation. The solution is presented in Fig. 7.7.

As expected, the corresponding temperature distributions have symmetric profiles. A comparison with the results in Fig. 7.6 clearly indicates that the difference in temperature distributions is attributed to the wall curvature. Pronounced temperature boundary layer characteristics are exhibited for the flow at positive pressure gradient with higher Prandtl numbers. The separation tendency in the case of decelerated flow is apparent in the temperature distribution.

Another interesting case, namely the flow through concentric cylinders can be constructed by setting $a = 0$ and $b = 1$. In this case the Navier–Stokes and energy equation are:

$$\mathbf{\Phi}'' - 2\mathbf{\Phi}' + \mathbf{\Phi} + C_1 = 0 \tag{7.34}$$

$$\mathbf{\Theta}'' - 2\mathbf{\Theta} + \text{ Pr }(\mathbf{\Phi}^2 - 2\mathbf{\Phi}\mathbf{\Phi}' + \mathbf{\Phi}'^2) = 0. \tag{7.35}$$

As seen from the above equation, all terms with Re- number disappeared leading to the results that the temperature and velocity distribution do not dependent on Re-number.

References

1. Jeffery GB (1915) The two-dimensional steady motion of a viscous fluid. Phil. Mag. 6th Ser. 455
2. Hamel G (1916) Spiralförmige Bewegung zäher Flüssigkeiten. JahresBericht der deutschen Mathematikervereinigung 25:34–60
3. Schobeiri MT (1980) Geschwindigkeit- und Temperaturvereitlungen in Hamelscher Spiralströmung. Zeitschrift für angewandte Mathematik und Mechanik. ZAMM 60:195
4. Schobeiri MT (1990) The influence of curvature and pressure gradient on the flow and velocity distribution. Int J Mech Sci 32(10):851–861
5. Schobeiri MT (1976) Näherungslösung der Navier-StokesDifferentialgleichungen für eine zweidimensionale stationäre Laminarströmung konstanter Viskosität in konvexen und konkaven Diffusoren und Düsen. Zeitschrift für angewandte Mathematik und Physik. ZAMP 27:9
6. Schobeiri MT (2010) Fluid mechanics for engineers. Graduate Textbook. Springer, New York. 978-642-1193-6 published 2010
7. Schobeiri MT (2014) Engineering applied fluid mechanics. Graduate textbook. McGraw Hill, New York

Chapter 8
Curves, Curvature, Surfaces, Geodesics

Detailed treatment of this subject belongs to the domain of differential geometry which is an independent mathematical discipline and treated in numerous textbook, among others [1–5]. In what follows we barrow a few items from the differential geometry, which are directly related to the tensor analysis. We start with the differential geometry of two dimensional also called planar curves, define their different forms of representations, their curvatures and will move on to differential geometry of surfaces and their characteristics. A section about geodesics concludes the chapter.

8.1 Representation of the Plane Curves

There are three different forms of representing a plane curve: 1st *implicit*, 2nd *explicit* and 3rd *parametric*. An equation such as $f(x, y) = 0$ with x and y as the coordinates of the curve is an example of an implicit representation of a curve. The equation of a circle in implicit form is $f(x, y) = x^2 + y^2 - R^2 = 0$ with $R = const.$ as the radius of the circle. Considering the circle, its *explicit* form is $y = \pm\sqrt{R^2 - x^2}$. In this case, the variable y can directly be calculated for a given set of values for the variable x. The third form is the parametric form. Choosing t as a parameter, the coordinates of the plane curve can be expressed as a function of the parameter t. Again using the circle as an example, its parametric representation is: $x(t) = R\cos t$ and $y(t) = R\sin t$.

Generally a planar (2-D) curve for example S is a function of its two coordinates, in Cartesian coordinate system, it is represented by:

$$S = S(x, y) \tag{8.1}$$

M. T. Schobeiri, *Tensor Analysis for Engineers and Physicists - With Application to Continuum Mechanics, Turbulence, and Einstein's Special and General Theory of Relativity*, https://doi.org/10.1007/978-3-030-35736-8_8

where its coordinates can be represented in parametric form:

$$x = x(t), \qquad y = y(t) \tag{8.2}$$

with t as the curve parameter, and $t_1 \leq t \leq t_2$ as the lower and upper boundaries of the curve. It also can be represented as a vector valued function such as

$$\boldsymbol{S} = \boldsymbol{S}(t) = (x(t), y(t)) \equiv \begin{pmatrix} x(t) \\ y(t) \end{pmatrix}. \tag{8.3}$$

For the sake of completeness, the explicit representation in terms of x or y as the independent variable is:

$$x = g(y), \qquad y = f(x). \tag{8.4}$$

Considering an arbitrary plane curve with the parametric representation of its coordinates, the first and second derivatives with respect to the parameter t are:

$$\boldsymbol{S} = \begin{pmatrix} x(t) \\ y(t) \end{pmatrix}, \; \boldsymbol{S}' = \begin{pmatrix} x'(t) \\ y'(t) \end{pmatrix}, \; \boldsymbol{S}'' = \begin{pmatrix} x''(t) \\ y''(t) \end{pmatrix}. \tag{8.5}$$

8.2 Curvature of Plane Curves

To obtain the curvature of S, we first consider a segment of the curve as shown in Fig. 8.1 with the unit tangent vector $\boldsymbol{T}$ and normal vector $\boldsymbol{N}$ at a given point P. The normal vector $\boldsymbol{N}$ is perpendicular to the unit tangent vector $\boldsymbol{T}$ and points towards the center of curvature. It has the same direction as the derivative of the tangent vector $d\boldsymbol{T}$ that also points towards the center of curvature. Thus, the unit normal vector $\boldsymbol{N}$ of a curve is the derivative of its unit tangent vector, which is:

$$\boldsymbol{T} = \frac{\boldsymbol{S}'}{\| \boldsymbol{S}' \|} \tag{8.6}$$

with $\| S' \| = \sqrt{dx'^2 + dy'^2}$. To obtain the local radius R, we draw at neighboring points b and c, two tangent vectors $\boldsymbol{T_1}$ and $\boldsymbol{T_2}$. Furthermore, at the tangent points b and c, we draw two lines perpendicular to the tangent vectors. The lines intersect each other at the point a. For $b \Rightarrow c$ the length $\bar{ac}$ and $\bar{ab}$ approach the radius of the curvature R. In this case, the arc $bc = ds$ represents a segment of an osculation circle with the radius R that touches the curve S. Using the tangent vectors, a triangle $\boldsymbol{\Delta 2}$ is constructed, which is similar to the triangle $\boldsymbol{\Delta 1}$ in Fig. 8.1.

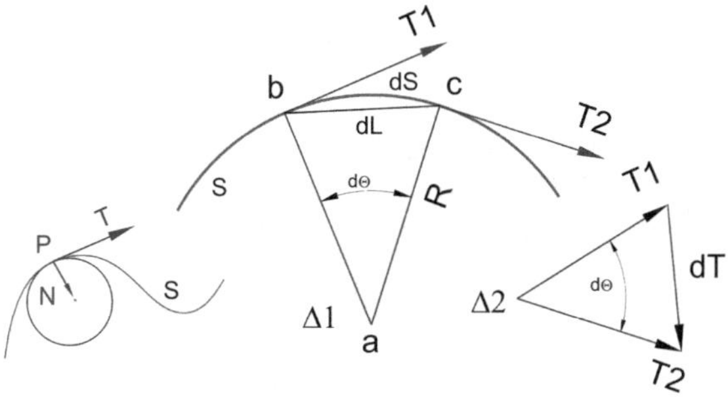

Fig. 8.1 Left, the curve S with the osculating circle and normal and tangent vectors at an arbitrary point P; Right, a segment of S with the tangent vectors

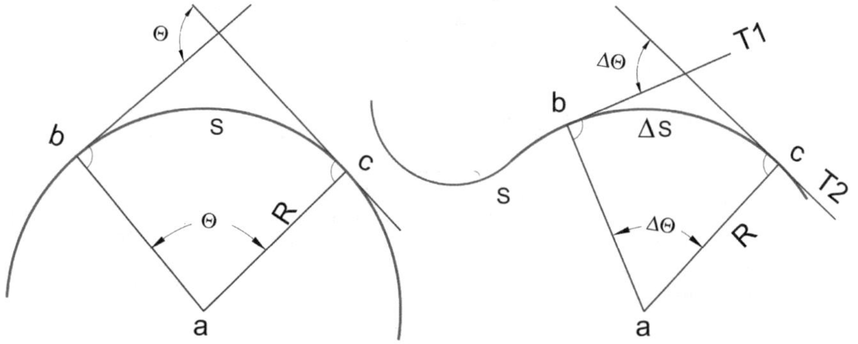

Fig. 8.2 (left) Explaining the curvature using the example of a circle with a constant curvature, (right) ΔS as a differential segments of a curve where the curvature is not constant

8.2.1 Curvature of Plane Curves, Derivation

To define the curvature of a planar curve $S = S(x, y)$, we consider S to be a segment of a circle with the radius R. At the intersection of the two radii with the circle, the tangents are perpendicular to the radii and build an angle Θ with each other as shown in Fig. 8.2.

For the circle the curvature is constant and is defined as the ratio of the angle Θ and the segment S, $\kappa = \frac{1}{R} = \frac{\Theta}{S} = \frac{\Theta}{R\Theta}$. For a curve with changing curvature, κ is defined as:

$$\kappa = \lim_{\Delta S \Rightarrow 0} \frac{\Delta\Theta}{\Delta S} = \frac{d\Theta}{dS} = \frac{d\Theta}{r d\Theta}. \tag{8.7}$$

Is the plane curve given as a function of its two coordinates in a Cartesian system, it is represented by $y = f(x)$ and its first derivative at the point under investigation is $\frac{dy}{dx} = \tan\Theta$. Taking the second derivative, we have:

$$\frac{d^2y}{dx^2} = (1 + \tan^2 \Theta)\frac{d\Theta}{dx}. \tag{8.8}$$

Inserting $\frac{dy}{dx} = \tan(\Theta)$ in Eqs. (8.8) and (8.9) and considering the arc length $ds^2 = dx^2 + dy^2$ with $\frac{ds}{dx} = \sqrt{1 + \left(\frac{dy}{dx}\right)^2}$ and the definition of the curvature, Eq. (8.7) is written as:

$$\kappa = \frac{d\Theta}{dS} = \frac{\frac{d\Theta}{dx}}{\frac{dS}{dx}}. \tag{8.9}$$

The numerator of Eq. (8.9) is

$$\frac{d\Theta}{dx} = \frac{\frac{d^2y}{dx^2}}{1 + \left(\frac{dy}{dx}\right)^2} \tag{8.10}$$

and the denominator of Eq. (8.9) reads:

$$\frac{ds}{dx} = \sqrt{1 + \left(\frac{dy}{dx}\right)^2} \tag{8.11}$$

thus, the curvature (8.9) becomes:

$$k = \frac{\frac{d^2y}{dx^2}}{\left(1 + \left(\frac{dy}{dx}\right)^2\right)^{3/2}}. \tag{8.12}$$

Equation (8.12) describes the curvature of a planar curve $S = S(x, y)$ represented by its coordinate $y = f(x)$. If the curve is smooth and at least twice differentiable, it can be parametrized. In this case the coordinates of the curve, (x, y), are represented as functions of a parameter. Taking t as the parameter, the coordinates are expressed in terms of t namely $x = x(t)$, $y = y(t)$. Starting with the first derivative $\frac{dy}{dx}$ and using the chain rule of differentiation, we have:

$$\frac{dy}{dx} = \frac{dy}{dt}\frac{dt}{dx} = \frac{y'}{x'} \tag{8.13}$$

where the prime "/" refers to differentiation with respect to the parameter t. It should be noted that and in some literature a dot "." for example $\dot{x}$, $\dot{y}$ is used to indicate the differentiation with respect to time and prime is used for differentiation with respect to the arc length. We keep the convention above and wherever confusion may arise, clarification will follow. The numerator of Eq. (8.12) reorganized in terms of the first and the second derivatives with respect to the parameter t:

$$\frac{d^2y}{dx^2} = \frac{d}{dx}\left(\frac{y'}{x'}\right) = \frac{d}{dt}\left(\frac{y'}{x'}\right)\frac{dt}{dx} = \frac{y''x' - x''y'}{x'^3}. \tag{8.14}$$

Applying the same differentiation procedure to the denominator of Eq. (8.12), we arrive at the curvature:

$$\kappa = \frac{y''x' - x''y'}{x'^3\left(1 + \frac{y'^2}{x'^2}\right)^{3/2}} = \frac{y''x' - x''y'}{(x'^2 + y'^2)^{3/2}}. \tag{8.15}$$

Equation (8.15) can be extended to 3D- space curves as follows:

$$\kappa = \frac{\sqrt{(z''y' - y''z')^2 + (x''z' - z''x')^2 + (y''x' - x''y')^2}}{(x'^2 + y'^2 + z'^2)^{3/2}}. \tag{8.16}$$

8.2.2 Derivation of Curvature and Torsion Using Tangent Vectors

Another alternative to obtain the curvature is using the tangent vectors. We consider a segment of the curve shown in Fig. 8.1 with the tangent vector $\boldsymbol{T}$ and normal vector $\boldsymbol{N}$ at a given point P, Fig. 8.2 (left). Expressing the curvature in terms of $d\boldsymbol{T}$, we set $k \propto \| d\boldsymbol{T} \|$ and consider the similarity between $\boldsymbol{\Delta 1}$ and $\boldsymbol{\Delta 2}$ we have:

$$\frac{dS}{R} = \frac{d\boldsymbol{T}}{1}, \text{ or } \frac{d\boldsymbol{T}}{dS} = \frac{1}{R} = \kappa \tag{8.17}$$

with 1 as the magnitude of the unit tangent vector. This relation can be re-written as:

$$\kappa = \| \frac{d\boldsymbol{T}}{dS} \| = \frac{\| \frac{d\boldsymbol{T}}{dt} \|}{\| \frac{dS}{dt} \|} \tag{8.18}$$

with κ as the curvature of S at the point, where the osculating circle touches the curve S. As shown in Fig. 8.3, based on the orientation of the curve S and the moving direction of the parameter t along S which indicates the changes of the second derivative of S, the curvature κ may assume positive, zero or negative sign.

Using Eq. (8.18), the curvature of space curves can be calculated, however because of the presence of the absolute values, it requires tedious work. A closer look at Eqs. (8.15) and (8.16), suggests that their numerators are the results of a cross product of two vectors. Therefor we may set:

$$\kappa = \frac{\| \boldsymbol{S}'(t) \times \boldsymbol{S}''(t) \|}{\| \boldsymbol{S}'(t) \|^3}. \tag{8.19}$$

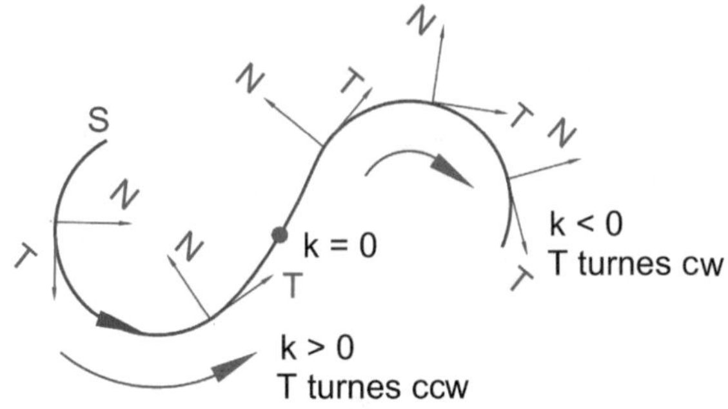

Fig. 8.3 Sign allocation to curvature κ based on the orientation of the curve S and its second derivative

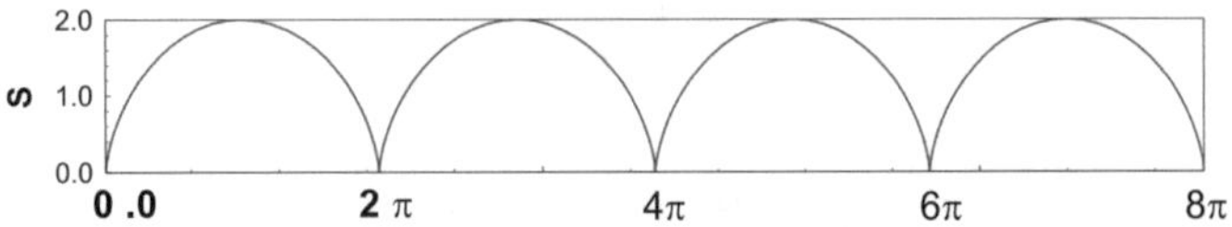

Fig. 8.4 The curve S with the osculating circle that touches S at an arbitrary point

The proof of the validity of Eq. (8.19) is presented in the following application section.

Application: Calculation of Curvature of a Cycloid

Applying the above procedure to a 2D-curve, we choose a cycloid shown in Fig. 8.4. A cycloid is a curve traced by a point on the rim of a circular disc with the radius R that rolls along a straight line without slipping. Without loss of generality the radius can be set $R = 1$. The coordinates of the cycloid in parametric form are given as:

$$\boldsymbol{S} = \begin{pmatrix} x(t) \\ y(t) \end{pmatrix} = \begin{pmatrix} t - \sin(t) \\ 1 - \cos(t) \end{pmatrix}. \tag{8.20}$$

In calculating the curvature, first we have to take the derivative of S with respect to the parameter t:

$$\boldsymbol{S}' = \begin{pmatrix} x'(t) \\ y'(t) \end{pmatrix} = \begin{pmatrix} 1 - \cos(t) \\ \sin(t) \end{pmatrix}. \tag{8.21}$$

The unit tangent vector is obtained by dividing $\boldsymbol{S}'$ by its magnitude:

$$\boldsymbol{T} = \frac{\boldsymbol{S}'}{\parallel \boldsymbol{S}' \parallel} = \begin{pmatrix} x'(t)/D \\ y'/D \end{pmatrix} = \begin{pmatrix} (1 - \cos(t)/D \\ \sin(t)/D \end{pmatrix} \tag{8.22}$$

with the denominator $D \equiv \parallel S' \parallel = \sqrt{(x'^2 + y'^2)}$ as the magnitude of $\boldsymbol{S}'$. To calculate the curvature as a function of t, we first establish the relationships for the elements of Eq. (8.18) namely $\parallel \frac{dT}{dt} \parallel$ and $\parallel \frac{dS}{dt} \parallel$. Inserting the results into Eq. (8.19), we find

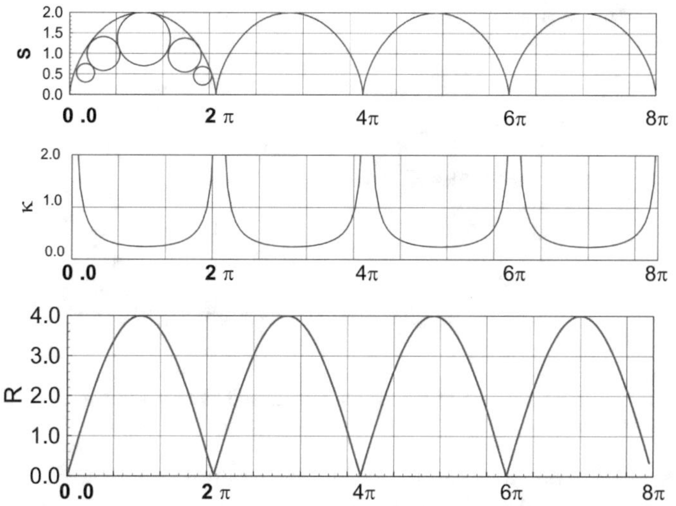

Fig. 8.5 A compound picture of S, κ and R

the curvature as a function of the parameter t:

$$\kappa = \frac{(1. - \cos(t))\cos(t) - \sin^2(t)}{[1. - \cos(t))^2 + \sin^2(t)]^{3/2}}. \tag{8.23}$$

Performing some trigonometric re-arrangements, we arrive at a simple result for the curvature:

$$\kappa = \frac{1}{4\left(\sin^2\left(\frac{t}{2}\right)\right)^{1/2}}. \tag{8.24}$$

The corresponding radius is:

$$R = \frac{1}{\kappa} = 4\left(\sin^2\left(\frac{t}{2}\right)\right)^{1/2}. \tag{8.25}$$

Figure 8.5 exhibits a compound picture of the cycloid, its curvature κ and its radius of curvature R. The radius of the curvature increases from 0 to π and decreases from π to 2π. The curvature κ has singularities at $0, 2\pi, 4\pi$ etc.

8.3 Space Curves, Torsion, Curvature

In addition to the curvature set out in previous section, a space curve may experience a torsion as seen in an example of a helix shown in Fig. 8.6. The torsion indicates the turning of the osculating circle of the actual space curve relative to the one of a plane curve. To quantitatively describe the torsion, in addition to the normal and tangential

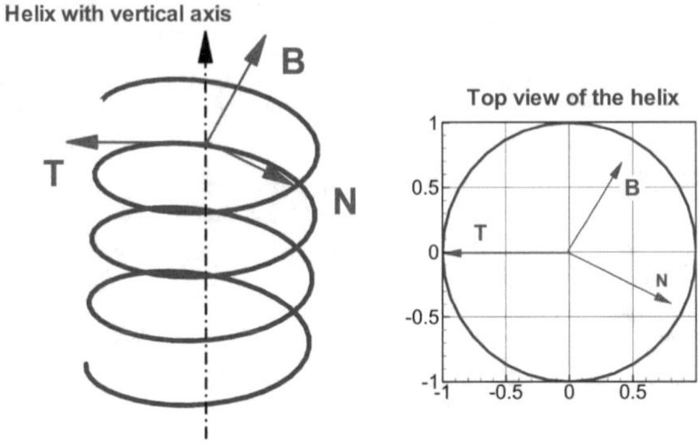

Fig. 8.6 (left) Plot of a helix with its tangential, normal and binormal vectors, (right) osculating circle with constant radius

vectors described in previous section, we introduce a third vector, the *binormal* vector $\boldsymbol{B}$, which is perpendicular to the plane described by the tangential vector $\boldsymbol{T}$ and the normal vector $\boldsymbol{N}$. These vectors along with the helix and the osculating plane are shown in Fig. 8.6. The angle between the vertical axis of the helix and the vector $\boldsymbol{B}$ indicates the turning of the osculating plane. Thus, the torsion vector is the vector product of the tangential and normal vectors.

$$B(t) = T(t) \times N(t) \tag{8.26}$$

Taking the first derivative of Eq. (8.26) and considering that $T' \times N = 0$, we find the changes of the binormal vector $\boldsymbol{B}' = \boldsymbol{T} \times \boldsymbol{N}'$. The normal vector N is calculated from:

$$\boldsymbol{N} = \frac{\boldsymbol{T}'(t)}{\| \boldsymbol{T}'(t) \|} \tag{8.27}$$

with $\boldsymbol{T}$ from Eq. (8.22) and $\boldsymbol{T}' = d\boldsymbol{T}/dt$. With $\boldsymbol{B}$, $\boldsymbol{N}$, B', $\boldsymbol{T}$, and $\boldsymbol{T}'$ we are now in a position to describe the curvature and torsional behavior of space curves. In the following we summarize the working equations for curvature and torsion expressed in terms of the normal vector $\boldsymbol{N}$, $\boldsymbol{N}'$, $\boldsymbol{B}$ and $\boldsymbol{B}'$. The curvature κ in Eq. (8.23) is an alternative version of Eq. (8.15) that can easily be transferred into (8.15). Equation (8.28) summarizes the equations for curvature and torsion.

$$\begin{aligned} \kappa &= \boldsymbol{N} \cdot \boldsymbol{T}' = -\boldsymbol{N}' \cdot \boldsymbol{T}, \\ \tau &= \boldsymbol{N}' \cdot \boldsymbol{B} = -\boldsymbol{N} \cdot \boldsymbol{B}'. \end{aligned} \tag{8.28}$$

8.3.1 Calculation of Curvature of Space Curves

To obtain the curvature of 3D space curves, the same procedure described in Eq. (8.2) is applied. This is demonstrated by using a helix shown in Fig. 8.6 as an example. The coordinates of the helix in parametric form are given as:

$$S(t) = \begin{pmatrix} a\cos t \\ a\sin t \\ bt \end{pmatrix}, \quad S'(t) = \begin{pmatrix} -a\sin t \\ a\cos t \\ b \end{pmatrix}, \quad S''(t) = \begin{pmatrix} -a\cos t \\ -a\sin t \\ 0 \end{pmatrix} \tag{8.29}$$

with a and b as constants. For calculation of the curvature κ, as an alternative, we utilize Eq. (8.19) and invoke Eq. (1.7). For the cross product in the numerator of (8.19) and its magnitude we find for the numerator:

$$\| \boldsymbol{S}'(t) \times \boldsymbol{S}''(t) \| = [a^2(a^2+b^2)]^{1/2} \tag{8.30}$$

and for the denominator of (8.19):

$$\| \boldsymbol{S}'(t) \|^3 = (a^2+b^2)^{3/2}. \tag{8.31}$$

Dividing Eq. (8.30) by Eq. (8.31) results in

$$\kappa = \frac{\| \boldsymbol{S}'(t) \times \boldsymbol{S}''(t) \|}{\| \boldsymbol{S}'(t) \|^3} = \frac{a(a^2+b^2)^{1/2}}{(a^2+b^2)^{3/2}} = \frac{a}{a^2+b^2}. \tag{8.32}$$

For $a = 1$, and $b = a/10$ of Fig. 8.6, we obtain a constant curvature of $\kappa = \frac{100}{101}$. This result reflects the constant curvature of the osculating circle shown in Fig. 8.6.

8.3.2 Calculation of Torsion τ

It is also of interest to find the torsion of the helix. An alternative to using Eq. (8.22) with the unit tangent vector and its derivative $\boldsymbol{T}$ and $\boldsymbol{T}' = d\boldsymbol{T}/dt$ to calculate $\boldsymbol{B}' = \boldsymbol{T} \times \boldsymbol{N}'$, an alternative is presented below for calculation of τ

$$\tau = \frac{\det . \lfloor \boldsymbol{S}'(t), \boldsymbol{S}''(t), \boldsymbol{S}'''(t) \rfloor}{\| \boldsymbol{S}'(t) \times S''(t) \|^2}. \tag{8.33}$$

The determinant of the numerator is expanded as:

$$\det . [\boldsymbol{S}'(t), \boldsymbol{S}''(t), \boldsymbol{S}'''(t)] = \det \begin{bmatrix} a\sin t & -a\cos t & a\sin t \\ a\cos t & -a\sin t & -a\cos t \\ b & 0 & 0 \end{bmatrix} = a^2 b \tag{8.34}$$

and the denominator:

$$\boldsymbol{S}'(t) \times S''(t) = \begin{pmatrix} -a\sin t \\ a\cos t \\ b \end{pmatrix} \times \begin{pmatrix} -a\cos t \\ -a\sin t \\ 0 \end{pmatrix} = \begin{pmatrix} ab\sin t \\ -ab\cos t \\ a^2 \end{pmatrix}$$

$$\| \boldsymbol{S}'(t) \times S''(t) \|^2 = a^2(a^2+b^2). \tag{8.35}$$

Dividing Eq. (8.30) by the second equation in (8.35) determines the torsion:

$$\tau = \frac{b}{a^2 + b^2}. \tag{8.36}$$

8.4 Surfaces

A surface in mathematical sense can be thought of as a geometrical shape which can be constructed from a plane through a simple or a complex change of its shape. A cylinder with a circular or elliptic cross section made from a sheet of paper is an example of a simple change of form, where no stretching is involved. Spheres, ellipsoids, hyperbolic torus and a spring, among many others, from which a few are shown in Fig. 8.7 exhibit a more complex form of surfaces. These surfaces cannot be converted to flat surfaces without deformation and distortion.

8.4.1 Description of a Surface

A surface is described by a set of points with given coordinates that can be expressed in terms of a continuous function. The defining equation can be described implicitly, explicitly or parametrically. In a Cartesian coordinate system, an implicit description of a surface is given by:

$$F = F(\boldsymbol{X}) = F(x, y, z) = F(x_i) = 0. \tag{8.37}$$

As an example, the implicit equations of a an ellipsoid and a unit sphere are:

$$\begin{aligned} F(x, y, z) &= \frac{x^2}{a^2} + \frac{y^2}{b^2} + \frac{z^2}{c^2} - 1 = 0 \text{ Ellipsoid} \\ F(x, y, z) &= x^2 + y^2 + z^2 - 1 = 0 \text{ Sphere.} \end{aligned} \tag{8.38}$$

The implicit equation of a unit sphere is obtained by setting the coefficients of Eq. (8.38) $a = b = c = 1$. The explicit form of the ellipsoid is:

$$z = c\sqrt{1 - \left(\frac{x^2}{a^2} + \frac{y^2}{b^2}\right)}. \tag{8.39}$$

The third form of defining a surface is its parametric representation. In this case the coordinates that constitute the surface are expressed in terms of the surface parameters u, v:

$$x = x(u, v), \ y = y(u, v), \ z = x(u, v). \tag{8.40}$$

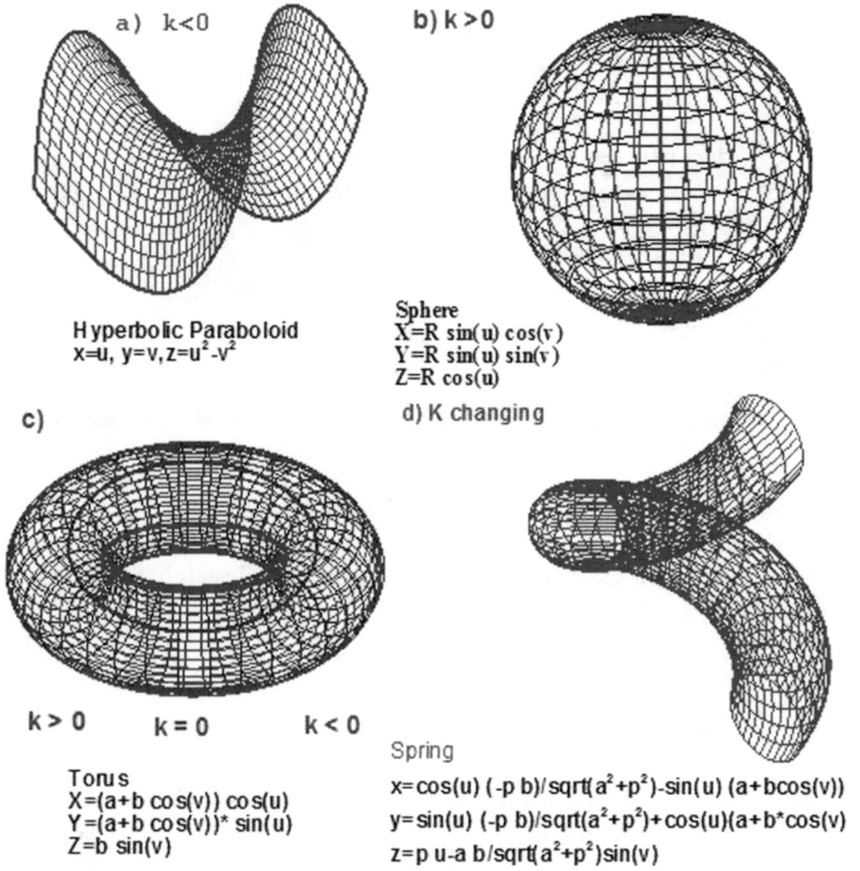

Fig. 8.7 Hyperboloid, **b** Sphere, **c** Torus, **d** Spring

Using the parameterization concept, it is easier and more straight forward to generate the desired surface. An example should clarify this: for the explicit equation of an ellipsoid to have a solution, the expression $(x^2/a^2 + y^2/b^2) \leq 0$ must apply. This requires a predefining the range of the variables involved. Using the parametric form, however, the function must be continuous and differentiable at least twice with respect to the parameters. Parametric version of an ellipsoid and a paraboloid coordinates reads:

$$\text{Ellipsoid: } x = a\cos u \sin v,\ y = b\sin u \sin v,\ z = c\cos v$$
$$\text{Paraboloid: } x = \sqrt{(u/h)}\cos v,\ y = a\sqrt{(u/h)}\sin v,\ z = u \tag{8.41}$$

with u and v as the parameters. Generating the above surfaces, the parameters u and v may be varied within the range of $[0, 2\pi]$. The coefficients a, b and c determine

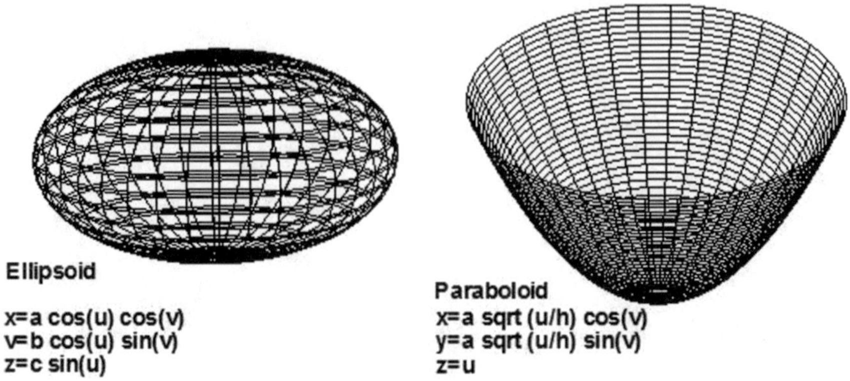

Fig. 8.8 Left, ellipsoid with $a = 5, b = 15$ and $c = 5$, right: Paraboloid with $a = 5$ and $h = 5$

the shape of the above surfaces. The plots of an ellipsoid and a paraboloid are shown in Fig. 8.8. It should be mentioned that for each configuration shown in Figs. 8.7 and 8.8 a family of figures can be created by changing the coefficients of that particular configuration.

8.4.2 Dimensions of a Surface

The surfaces presented above are embedded in a three-dimensional Euclidean space. The position of the points outside and inside these surfaces are defined by their three dimensions expressed in terms of curvilinear coordinates $\Theta^1, \Theta^2, \Theta^3$ or Cartesian coordinates x_1, x_2, x_3. The position of the points on the surface, however, are identified by two independent variables Θ^1, Θ^2 only. For the sake of simplicity we rename the coordinates $\Theta^1 \equiv u$, $\Theta^2 \equiv v$. Thus the surface is considered two-dimensional. The sphere exhibits a representative example for description of a two dimensional surface, Fig. 8.9. The points located outside and inside the surface are described by three coordinates.

8.4.3 Non-uniqueness of Parametric Representation

Given a surface equation in explicit or implicit form, infinite number of parametric representations can be produced. As an example, a hyperbolic paraboloid:

$$x^2 - y^2 = z \tag{8.42}$$

is parameterized as Surface (1), (2) and (3) as indicated below.

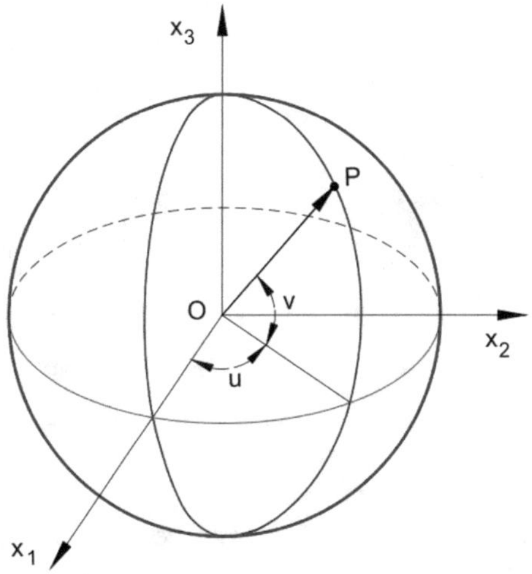

Fig. 8.9 A sphere as a two dimensional surface

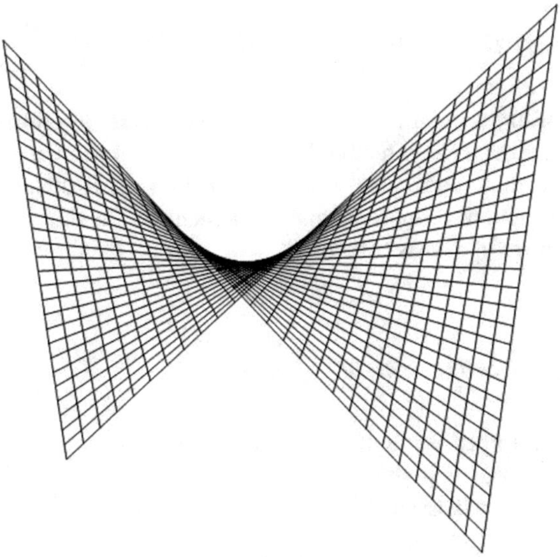

Fig. 8.10 Hyperbolic paraboloid as the solution of all three equations in (8.43)

$$\begin{aligned} &\text{Surface 1}: x = u, && y = v, && z = uv \\ &\text{Surface 2}: x = u + v, && y = u - v, && z = 4uv \\ &\text{Surface 3}: x = u \cos hv, && y = u \sin hv, && z = u^2. \end{aligned} \tag{8.43}$$

Any parametric configuration that satisfies Eq. (8.43) is the solution. The plot of these three functions is shown in Fig. 8.10.

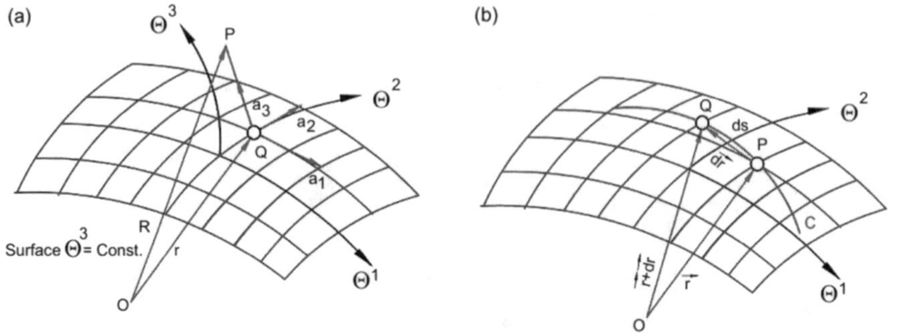

Fig. 8.11 A section of a 2-D surface with the surface coordinate $\Theta^3 = const.$

As seen all three equations in (8.43) deliver the same surface. This fact demonstrates that the parameter representation of surfaces is not unique.

8.5 Fundamental Forms of the Surface Theory

The fundamental forms of surface theory, its mathematical description and its treatment is found in almost every textbook about differential geometry. A straight forward treatment without unnecessary mathematical ballast is found in [4], which is reflected in this section. A more detailed treatment of the subject is found in the original publications by Gauss [6] and [7].

8.5.1 Surface Metric Tensor

Given is a surface embedded in the three dimensional space described by the coordinate system $\Theta^1, \Theta^2, \Theta^3$ as shown in Fig. 8.11a.

The surface is characterized by the constant coordinate $\Theta^3 = const.$ Although the surface is three dimensional, it is sufficient for its description to have only two curvilinear coordinates namely Θ^1 and Θ^2. The coordinate system that represents such a two dimensional surface is called surface coordinate system. The surface of a sphere shown in Fig. 8.9 exhibits an appropriate representative of two-dimensional surfaces. Figure 8.11 is considered a section of a 2-D surface shown in Fig. 8.7 or 8.8. Given the point P outside the surface with the position vector R, Fig. 8.11a, we draw a line perpendicular to the surface intersecting the latter at point Q with the position vector r. Thus, we have:

$$\boldsymbol{R}(\Theta^1, \Theta^2, \Theta^3) = \boldsymbol{r}(\Theta^1, \Theta^2) + S_c \mathbf{a_3}(\Theta^1, \Theta^2) \tag{8.44}$$

S_c as the surface constant and $\boldsymbol{a}_3$ as the unit vector perpendicular to the surface at point Q. The index notation for the general three dimensional coordinate systems uses Latin letters with the values 1, 2 and 3. For the 2-D Gauss surfaces, Greek letters are used with values 1, 2 only. So the Gauss coordinates is expressed in terms of Θ^α. The covariant base vectors of the position vector $\boldsymbol{R}$ and $\boldsymbol{r}$ are calculated the same way as shown in Eq. (6.6), namely

$$g_k = \boldsymbol{R}_{,k} = \frac{\partial \boldsymbol{R}}{\partial \Theta^k} = \frac{\partial (\boldsymbol{e}_i R_i)}{\partial \Theta^k} = \boldsymbol{e}_i \frac{\partial R_i}{\partial \Theta^k} \tag{8.45}$$

with $\boldsymbol{R} = e_i R_i$, similarly the covariant base vector for Gauss coordinates are:

$$\boldsymbol{a}_\alpha = \boldsymbol{r}_{,\alpha} = \frac{\partial \boldsymbol{r}}{\partial \Theta^\alpha} = \frac{\partial (\boldsymbol{e}_i r_i)}{\partial \Theta^\alpha} = \boldsymbol{e}_i \frac{\partial r_i}{\partial \Theta^\alpha}. \tag{8.46}$$

The differentiation of Eq. (8.44) results in:

$$g_\alpha = \boldsymbol{a}_\alpha + S_c \boldsymbol{a}_{3,\alpha}. \tag{8.47}$$

In this context it is obvious that:

$$g_3 = \boldsymbol{a}_3,\ \boldsymbol{a}_3 \cdot \boldsymbol{a}_3 = 1,\ \boldsymbol{a}_3 \cdot \boldsymbol{a}_\alpha = 0,\ \boldsymbol{a}_3 \cdot \boldsymbol{a}_{3,\alpha} = 0. \tag{8.48}$$

The question arises is, whether the base vector $\boldsymbol{a}_\alpha$ is a first order tensor or not. A simple rearrangement shows that it is indeed a first order tensor. Using Eq. (8.46), with

$$\boldsymbol{a}_\alpha = \frac{\partial \boldsymbol{r}}{\partial \Theta^\alpha} \tag{8.49}$$

and applying the differentiation rule to a new set of Gauss coordinates

$$\bar{\boldsymbol{a}}_\alpha = \frac{\partial \boldsymbol{r}}{\partial \bar{\Theta}_\alpha} \tag{8.50}$$

and applying the chain rule to Eq. (8.50) and implementing Eq. (8.49) leads to:

$$\bar{\boldsymbol{a}}_\alpha = \frac{\partial \boldsymbol{r}}{\partial \Theta^\beta} \frac{\partial \Theta^\beta}{\partial \bar{\Theta}^\alpha} = \boldsymbol{a}_\beta \frac{\partial \Theta^\beta}{\partial \bar{\Theta}^\alpha}. \tag{8.51}$$

Equation (8.51) fulfills the transformation requirements and proofs that the base vector $\boldsymbol{a}_\alpha$ of the Gauss surface is a first order covariant surface tensor. Now, using the base vectors, we create a matrix as the scalar product $a_{\alpha\beta} = \boldsymbol{a}_\alpha \cdot \boldsymbol{a}_\beta$. Here again the question associated with this operation is whether the matrix $a_{\alpha\beta}$ represents the matrix of a second order surface tensor. To show that, we construct $g_{\alpha\beta}$ from Eq. (8.47), considering Eq. (8.48) and after some rearrangements we arrive at:

$$\bar{a}_{\alpha\beta} = \frac{\partial\Theta^{\gamma}}{\partial\bar{\Theta}^{\alpha}} \frac{\partial\Theta^{\gamma}}{\partial\bar{\Theta}^{\beta}} a_{\gamma\lambda}. \tag{8.52}$$

With Eq. (8.52), it is shown that $a_{\alpha\beta} = \boldsymbol{a}_{\alpha} \cdot \boldsymbol{a}_{\beta}$ is indeed the matrix of a second order tensor that follows the transformation rule. It is also a symmetric tensor. To prove this claim, we decompose the second order tensor $a_{\alpha\beta}$ into a symmetric and an antisymmetric part. It is easy to show that the anti-symmetric part disappears identically. This shows that $a_{\alpha\beta} = a_{\beta\alpha}$ is the matrix of a covariant metric tensor of the surface. The metric tensor has in addition to the covariant components also the contravariant ones. The co- and contravariant components are connected with each others via the mixed Kronecker delta: $a_{\alpha\beta}a^{\beta\gamma} = \delta^{\gamma}_{\alpha}$. Likewise raising and lowering the indices follow exactly the same rule that explained in Chap. 6.

8.5.2 First Fundamental Form of Surface, Arc Length

The first fundamental form I is used to measure the curvature length, angles and areas on a two-dimensional surface, thus I defines the metric properties of a surface. Consider an arc length ds on the Gauss surface with Θ^1, Θ^2 as the coordinates, its length is obtained using scalar multiplication of:

$$ds^2 = d\boldsymbol{r} \cdot d\boldsymbol{r} = (\boldsymbol{a}_{\alpha} d\Theta^{\alpha}) \cdot (\boldsymbol{a}_{\beta} d\Theta^{\beta}) = a_{\alpha\beta} d\Theta^{\alpha} d\Theta^{\beta} \tag{8.53}$$

Figure 8.11b shows the differential element $d\boldsymbol{r}$ as the distance between P and Q on the two dimensional surface. Expanding $\boldsymbol{r} = f(\Theta^1, \Theta^2)$ in Eq. (8.53) we find:

$$d\boldsymbol{r} = \frac{\partial \boldsymbol{r}}{\partial \Theta^{\alpha}} d\Theta^{\alpha} = \frac{\partial \boldsymbol{r}}{\partial \Theta^{1}} d\Theta^{1} + \frac{\partial \boldsymbol{r}}{\partial \Theta^{2}} d\Theta^{2} \text{ or}$$

$$d\boldsymbol{r} = \boldsymbol{r}_{,\alpha} d\Theta^{\alpha} = \boldsymbol{r}_{,1} d\Theta^{1} + \boldsymbol{r}_{,2} d\Theta^{2}. \tag{8.54}$$

Introducing the parameter t and for conveniently writing the following equations, we denote the differentiation with respect to t with a dot "." and find:

$$d\Theta^{\alpha} = \frac{d\Theta^{\alpha}}{dt} dt = \dot{\Theta}^{\alpha} dt, \; d\Theta^{\beta} = \frac{d\Theta^{\beta}}{dt} = \dot{\Theta}^{\beta} dt, \tag{8.55}$$

as a result, we obtain the length of the arc as:

$$I = ds^2 = d\boldsymbol{r} \cdot d\boldsymbol{r} = (\boldsymbol{r}_{,1} d\Theta^{1} + \boldsymbol{r}_{,2} d\Theta^{2}) \cdot (\boldsymbol{r}_{,1} d\Theta^{1} + \boldsymbol{r}_{,2} d\Theta^{2}) \tag{8.56}$$

and with the introduction of Eq. (8.55) we find:

$$I = ds^2 = \boldsymbol{dr} \cdot \boldsymbol{dr} = \left(\frac{\partial \boldsymbol{r}}{\partial \Theta^1} d\dot{\Theta}^1 dt + \frac{\partial \boldsymbol{r}}{\partial \Theta^2} d\dot{\Theta}^2 dt \right) \cdot \left(\frac{\partial \boldsymbol{r}}{\partial \Theta^1} d\dot{\Theta}^1 dt + \frac{\partial \boldsymbol{r}}{\partial \Theta^2} d\dot{\Theta}^2 dt \right). \tag{8.57}$$

Equation (8.57) in a compact form is:

$$I = ds^2 = a_{\alpha\beta} \dot{\Theta}^\alpha \dot{\Theta}^\beta dt^2 = (a_{11} \dot{\Theta}^{1^2} + 2a_{12} \dot{\Theta}^1 \dot{\Theta}^2 + a_{22} \dot{\Theta}^{2^2}) dt^2. \tag{8.58}$$

Thus the surface tensor $a_{\alpha\beta}$ determines the *metric of the surface* and is called *metric tensor* of the surface. The counter part g_{ij} is the metric of the embedded three-dimensional space. Equation (8.57) and its compact version (8.58) is the *first fundamental form* of the surface theory. The symmetric property of $a_{\alpha\beta}$ enabled us to combine the two mixed elements to arrive at $2a_{12}\dot{\Theta}^1\dot{\Theta}^2$ in Eq. (8.58). For practical purposes we re-arrange the coefficients in (8.58):

$$E = a_{11} = \boldsymbol{r}_{,1} \cdot \boldsymbol{r}_{,1}, \; F = a_{12} = \boldsymbol{r}_{,1} \cdot \boldsymbol{r}_{,2}, \; G = \boldsymbol{r}_{,2} \cdot \boldsymbol{r}_{,2}. \tag{8.59}$$

With Eq. (8.59) we obtain the alternative of the first fundamental form:

$$I = ds^2 = \boldsymbol{dr} \cdot \boldsymbol{dr} = E d\theta^{1^2} + 2F d\theta^1 d\theta^2 + G d\theta^{2^2}. \tag{8.60}$$

Once ds is determined, its integration delivers the length of the curve. In this case, either one of the following relations can be used:

$$s = \int_{t_0}^{t_1} \sqrt{a_{\alpha\beta} \dot{\Theta}^\alpha \dot{\Theta}^\beta} dt,$$
$$s = \int_{t_0}^{t_1} \sqrt{E\dot{\theta}^{1^2} + 2F\dot{\theta}^1\dot{\theta}^2 + G\dot{\theta}^{1^2}} dt. \tag{8.61}$$

As an example of practical application, we apply the first fundamental form to a unit sphere, $R = 1$. Here again, as in Sect. 8.4.2, we rename the coordinates as Θ^1, Θ^2 as u, v and apply the parametric form for $\boldsymbol{r}(u, v)$:

$$\boldsymbol{r}(u, v) = \begin{pmatrix} \cos u \sin v \\ \sin u \sin v \\ \cos v \end{pmatrix} \tag{8.62}$$

and constructing the derivatives, first we find:

$$\boldsymbol{r}_{,u} = \begin{pmatrix} -\sin u \sin v \\ \cos u \sin v \\ 0 \end{pmatrix}, \; \boldsymbol{r}_{,v} = \begin{pmatrix} \cos u \cos v \\ \sin u \cos v \\ \sin v \end{pmatrix} \tag{8.63}$$

from which we calculate the coefficients:

$$\begin{aligned} E &= \boldsymbol{r}_{,1} \cdot \boldsymbol{r}_{,1} = \boldsymbol{r}_{,u} \cdot \boldsymbol{r}_{,u} = \sin^2 v \\ F &= \boldsymbol{r}_{,1} \cdot \boldsymbol{r}_{,2} = \boldsymbol{r}_{,u} \cdot \boldsymbol{r}_{,v} = 0 \\ G &= \boldsymbol{r}_{,2} \cdot \boldsymbol{r}_{,2} = \boldsymbol{r}_{,v} \cdot \boldsymbol{r}_{,v} = 1. \end{aligned} \tag{8.64}$$

Inserting the above coefficients into the modified Eq. (8.60), we find:

$$s = \int_0^{2\pi} \sqrt{E\left(\frac{d\Theta^1}{dt}\right)^2 + 2F\frac{d\Theta^1}{dt}\frac{d\Theta^2}{dt} + G\left(\frac{d\Theta^2}{dt}\right)^2}\, dt. \tag{8.65}$$

Having determined the coefficients E, F, G, the length of any given curve on the surface of the sphere or any other surface can be determined by defining the parametric form of the curve, whose arc length we want to measure. Using the sphere as the two-dimensional surface, we take
$u \equiv \Theta^1 = t,\ v \equiv \Theta^2 = const. = \pi/2$ with t ranging and $[0, 2\pi]$ and obtain:

$$s = \int_0^{2\pi} \sqrt{E\left(\frac{du}{dt}\right)^2 + 2F\left(\frac{du}{dt}\frac{dv}{dt}\right) + G\left(\frac{dv}{dt}\right)^2}\, dt. \tag{8.66}$$

Inserting the coefficient E, F, G and the derivatives of u and v with respect to t, we find the arc length which is the circumference of a *great circle* of a sphere with $R = 1$:

$$s = \int_0^{2\pi} |\sin v| dt = 2\pi. \tag{8.67}$$

For a sphere with $R \neq 1$ we have $s = 2\pi R$.

8.5.3 Surface Area

To calculate the area of a surface, we start from the infinitesimal arc length as discussed in Sect. 8.5.2 and use the differential form (8.46) $\boldsymbol{r}_{,1} d\Theta^1$ and $\boldsymbol{r}_{,2} d\Theta^2$. The infinitesimal surface element dF is found by a cross product of these two vectors: $dF = |\boldsymbol{r}_{,1} \times \boldsymbol{r}_{,2}| d\Theta^1 d\Theta^2$ and with Eq. (8.46) we have:

$$dF = \sqrt{(\boldsymbol{a}_1 \times \boldsymbol{a}_2) \cdot (\boldsymbol{a}_1 \times \boldsymbol{a}_2)} d\Theta^1 d\Theta^2. \tag{8.68}$$

Invoking the result of the multiplication of two permutation symbols (1.18), we obtain:

$$dF = \sqrt{a_{11}a_{22} - (a_{12})^2} d\Theta^1 d\Theta^2. \tag{8.69}$$

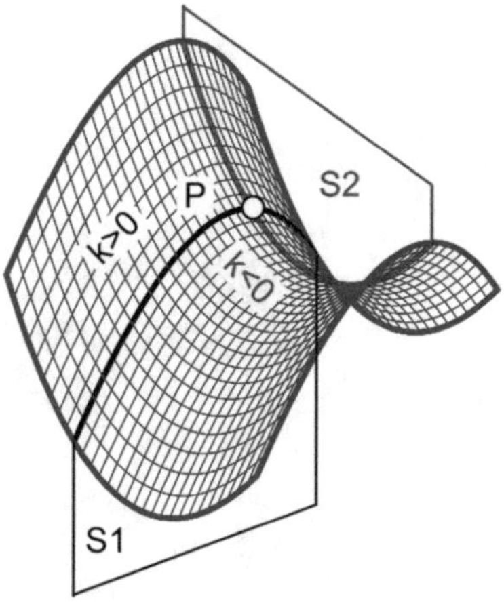

Fig. 8.12 Inspection of normal surfaces with $S1$ and $S2$

As a representative example we use the sphere with the coefficients E, F, G from Eq. (8.64) the surface area of a unit circle ($R = 1$) is calculated as:

$$S = \int_0^{\pi} \int_0^{2\pi} \sqrt{EG - F^2} du\, dv = \int_0^{\pi} \int_0^{2\pi} \sin v\, du\, dv = 4\pi \qquad (8.70)$$

and for $R \neq 1$, we obtain $S = 4R^2\pi$.

8.5.4 *Second Fundamental Form, Curvature Tensor*

The Sects. 8.2 and 8.3 were dedicated to understanding the curvatures of planar and space curves. This section treats the curvature of surfaces using the second fundamental form, which is the continuation of the first fundamental form. For calculating the arc length and the area on a Gauss surface we used the first fundamental which was based on the scalar product $\boldsymbol{dr} \cdot \boldsymbol{dr}$ in Eq. (8.53). The second fundamental form uses the scalar product $\boldsymbol{dr} \cdot \boldsymbol{da_3}$ with $\boldsymbol{da_3} = \boldsymbol{a}_{3,\beta} d\theta^3$. For calculating the surface curvatures, the second fundamental form includes: (1) the *principal curvatures* κ_1, κ_2 and their corresponding radii of curvature $R_1 = |\frac{1}{\kappa_1}|$ and $R_2 = |\frac{1}{\kappa_2}|$; (2) the mean curvature $H = (\kappa_1 + \kappa_2)/2$; and (3) the Gauss curvature $K = \kappa_1 \kappa_2$. To define the principal curvatures, we consider the surface of a hyperbolic paraboloid illustrated in Fig. 8.12. Two sets of curves define the contour of the surface. At an arbitrarily chosen point such as point P, Fig. 8.13, we introduce two normal planes. The intersection curves of these planes with the surface have two different curvature radii.

R_1 and R_2 with the corresponding curvatures $\kappa_1 = \frac{1}{R}$ and $\kappa_2 = \frac{1}{R_2}$. Among these Radii there is only one extreme R_1 and one extreme R_2 with reciprocal values $\kappa_1 = \frac{1}{R_1}$ and $\kappa_2 = \frac{1}{R_2}$ called principal curvature. The second fundamental form is utilized to determine the surface curvature.

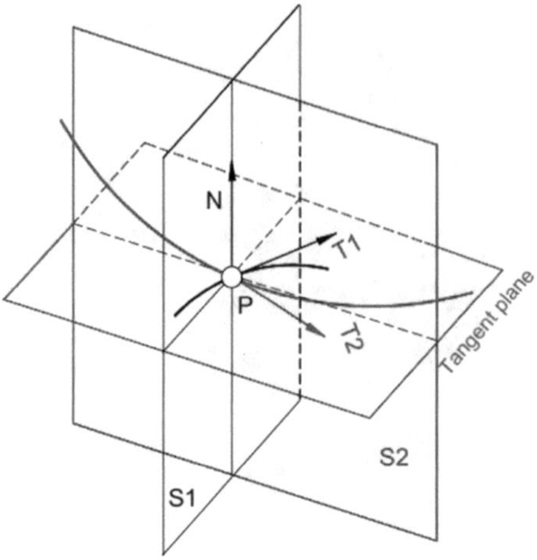

Fig. 8.13 Tangent plane at P

In the following, the derivation of the second fundamental form is presented. This subject is treated in almost all books about differential geometry using different approaches. A simple but detailed approach is found in [4], which we present in this section. Referring to Fig. 8.11 in conjunction with Fig. 8.12, the second fundamental form defined in [4] is $II = -d\boldsymbol{r} \cdot d\boldsymbol{a_3}$ with $d\boldsymbol{r}$ as:

$$d\boldsymbol{r} = \frac{\partial \boldsymbol{r}}{\partial \Theta^\alpha} d\boldsymbol{\theta}^\alpha = \frac{\partial \boldsymbol{r}}{\partial \Theta^1} d\boldsymbol{\theta}^1 + \frac{\partial \boldsymbol{r}}{\partial \Theta^2} d\boldsymbol{\theta}^2 \tag{8.71}$$

written in compact form

$$d\boldsymbol{r} = \frac{\partial \boldsymbol{r}}{\partial \Theta^\alpha} d\boldsymbol{\theta}^\alpha = \boldsymbol{r}_{,\alpha} d\Theta^\alpha = \boldsymbol{a}_\alpha d\Theta^\alpha. \tag{8.72}$$

Similar to Eq. (8.45) and under consideration of Eq. (8.46), we have with:

$$d\boldsymbol{a_3} = \boldsymbol{a}_{3,\beta} = \frac{\partial \boldsymbol{a_3}}{\partial \Theta^\beta} d\boldsymbol{\theta}^\beta = \frac{\partial \boldsymbol{a_3}}{\partial \Theta^1} d\boldsymbol{\theta}^1 + \frac{\partial \boldsymbol{a_3}}{\partial \Theta^2} d\boldsymbol{\theta}^2 \tag{8.73}$$

obtain the product

$$d\boldsymbol{r} \cdot d\boldsymbol{a_3} = \boldsymbol{a}_\alpha \cdot \boldsymbol{a}_{3,\beta} d\Theta^\alpha d\Theta^\beta \tag{8.74}$$

and introduce

$$b_{\alpha\beta} = -\boldsymbol{a}_\alpha \cdot \boldsymbol{a}_{3,\beta} \tag{8.75}$$

to arrive at the second fundamental form

$$II = b_{\alpha\beta} d\theta^\alpha d\theta^\beta. \tag{8.76}$$

Summarizing the above individual operations, we arrive at

$$II = -d\boldsymbol{r} \cdot d\boldsymbol{a_3}$$

$$II = -\left(\frac{\partial \boldsymbol{r}}{\partial \Theta^\alpha} d\boldsymbol{\theta}^\alpha\right) \cdot \left(\frac{\partial \boldsymbol{a_3}}{\partial \Theta^\alpha} d\boldsymbol{\theta}^\alpha\right)$$

$$II = -\left(\frac{\partial \boldsymbol{r}}{\partial \Theta^1} d\dot{\Theta}^1 dt + \frac{\partial \boldsymbol{r}}{\partial \Theta^2} d\dot{\Theta}^2 dt\right) \cdot \left(\frac{\partial \boldsymbol{a_3}}{\partial \Theta^1} d\dot{\Theta}^1 dt + \frac{\partial \boldsymbol{a_3}}{\partial \Theta^2} d\dot{\Theta}^2 dt\right)$$

$$II = L d\Theta^{1^2} + 2M d\Theta^1 d\Theta^2 + N d\Theta^{2^2} \tag{8.77}$$

with the coefficients L, M and N as:

$$L = -\frac{\partial \boldsymbol{r}}{\partial \boldsymbol{\Theta^1}} \cdot \frac{\partial \boldsymbol{a_3}}{\partial \boldsymbol{\Theta^1}}$$

$$M = -\frac{1}{2}\left(\frac{\partial \boldsymbol{r}}{\partial \boldsymbol{\Theta^1}} \cdot \frac{\partial \boldsymbol{a_3}}{\partial \boldsymbol{\Theta^2}} + \frac{\partial \boldsymbol{r}}{\partial \boldsymbol{\Theta^2}} \cdot \frac{\partial \boldsymbol{a_3}}{\partial \boldsymbol{\Theta^1}}\right)$$

$$N = -\frac{\partial \boldsymbol{r}}{\partial \boldsymbol{\Theta^2}} \cdot \frac{\partial \boldsymbol{a_3}}{\partial \boldsymbol{\Theta^2}}. \tag{8.78}$$

With the first and the second fundamental forms we are now able to calculate the radii of the surface curvature by invoking Eq. (8.17):

$$\frac{1}{R} = \frac{b_{\alpha\beta} d\theta^\alpha d\theta^\beta}{a_{\alpha\beta} d\theta^\alpha d\theta^\beta} = \frac{II}{I}. \tag{8.79}$$

For a given ratio $\lambda = \frac{d\theta^1}{d\theta^2}$ Eq. (8.79) can be rearranged as:

$$\frac{1}{R} = \frac{b_{11}\lambda^2 + 2b_{12}\lambda + b_{22}}{a_{11}\lambda^2 + 2a_{12}\lambda + a_{22}}. \tag{8.80}$$

Rearranging Eq. (8.80) with respect to the order of λ, we find

$$\left(\frac{1}{R}a_{11} - b_{11}\right)\lambda^2 + 2\left(\frac{1}{R}a_{12} - b_{12}\right)\lambda + \left(\frac{1}{R}a_{22} - b_{22}\right) = 0. \tag{8.81}$$

To find the principal curvature the discriminant of Eq. (8.81) must disappear

$$\begin{vmatrix} \frac{1}{R}a_{11} - b_{11} & -\frac{1}{R}a_{12} - b_{12} \\ \frac{1}{R}a_{12} - b_{12} & \frac{1}{R}a_{22} - b_{22} \end{vmatrix} = 0. \tag{8.82}$$

Sorting the results of Eq. (8.82) by the order of $\frac{1}{R}$, we find the following relation for the principal curvatures $\frac{1}{R_1}$ and $\frac{1}{R_2}$:

$$\frac{1}{R^2} - 2H\frac{1}{R} + K = 0 \tag{8.83}$$

with the coefficients:

$$2H = \frac{b_{11}a_{22} - 2b_{12}a_{12} + b_{22}a_{11}}{a_{11}a_{22} - a_{12}^2}$$

$$K = \frac{b_{11}b_{22} - b_{12}^2}{a_{11}a_{22} - a_{12}^2}. \tag{8.84}$$

Considering Eq. (8.84) the solution of Eq. (8.83) is:

$$H = \frac{1}{2}\left(\frac{1}{R_1} + \frac{1}{R_2}\right)$$

$$K = \frac{1}{R_1}\frac{1}{R_2}. \tag{8.85}$$

Thus the principal curvatures are:

$$\kappa_1 = \frac{1}{R_1}, \quad \kappa_2 = \frac{1}{R_2}. \tag{8.86}$$

8.6 Geodesics

8.6.1 Introduction

As indicated in Chap. 1, in a flat space, the shortest distance between two points is given by a straight line. In Euclidean geometry, this straight line exhibits a *geodesic*. In a curved space (Riemann space), we define the geodesic as a curve that connects two fixed points P_1 and P_2 that has a minimum arc length.

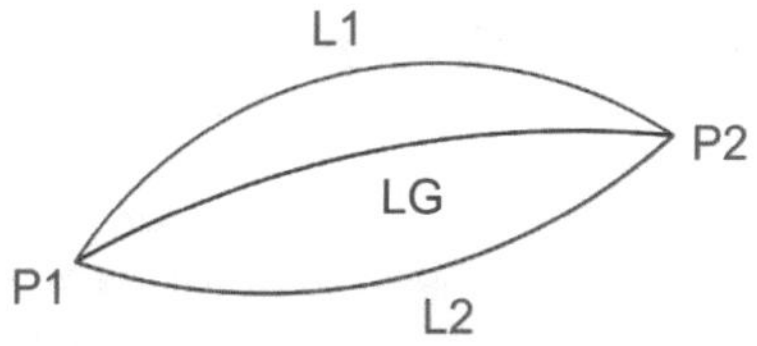

Fig. 8.14 Explaining the geodesic as shortest distance on a curved surface

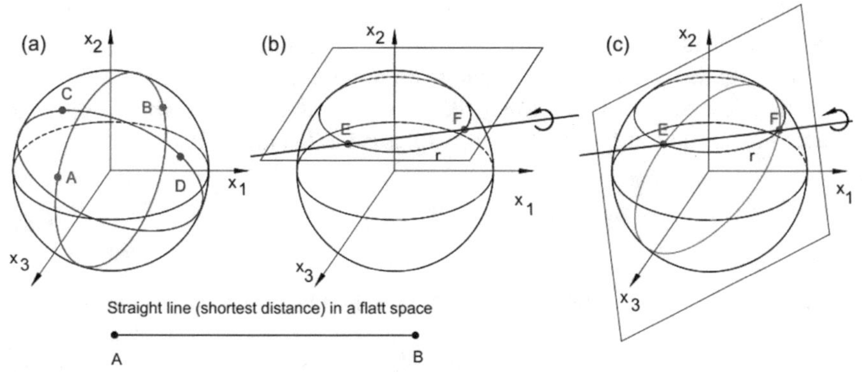

Fig. 8.15 Graphical construction of a geodesics on surface of a sphere, **a** two great circles with arc AB and CD, **b** a plane intersects the sphere and creates a small circle, **c** turning the plane about the axis EF, the small circle becomes a great circle

Figure 8.14 schematically shows the paths from P_1 to P_2. Between these points infinite number of arcs such as $L1$ and $L2$ can be constructed, each of which might have lengths with different no-zero *variations*. Among these arcs there is only one called geodesic which has zero variation and thus the shortest length.

The surface of a sphere shown in Fig. 8.15 is a descriptive example of constructing a geodesic. The shortest distance between two arbitrary points on a sphere is an arc of a *great circle*. Figure 8.15a shows the locations of points A and B and C and D on two different great circles. A great circle is constructed by intersecting the sphere with a plane where the center of the sphere is placed on the plane. If the two points under investigation for example E and F are not located on a great circle but rather on a small circle as shown in Fig. 8.15b, they can easily be transformed onto a great circle by turning the intersecting plane about the axis E-F that goes through those two points. Approaching the center of the sphere, the plane will touch its center at a certain angle. At this angle the intersection of the plane with the sphere surface constitutes a great circle.

8.6.2 Calculation of Geodesics for N-dimensional Riemann Space

In the following, we first present the calculation steps for an *N-dimensional Riemann* space where the condition $ds^2 = g_{ij}dx^i dx^j$ is satisfied and its metric is a continues function of the coordinates $g_{ij} = g_{ij}(x^1, x^2, ..., x^n)$. Then we specialize the steps for two-dimensional curved surfaces with $\Theta^\alpha = (\Theta^1, \Theta^2)$ and $\Theta^3 = const.$ Given any arbitrary two points on a Riemann surface, the particular curve with the shortest distance is found by using the Euler-Lagrange variation equation. It provides the condition that its integral becomes maximum or minimum.

Without loss of generality we may set $N = 3$ and assume that we have a curve on a three-dimensional surface that can be expressed as the integral:

$$I = \int_{P_1}^{P_2} ds. \tag{8.87}$$

Using the variational calculus, the integral (8.87) must disappear, this means that:

$$\delta I = 0 \tag{8.88}$$

with δ as the variation from the shortest distance. To evaluate Eq. (8.87), we first calculate the element ds:

$$ds = \sqrt{g_{ij}dx^i dx^j} \tag{8.89}$$

and introduce the parameter t such that

$$x^i = x^i(t) \tag{8.90}$$

then we take the first derivative of Eq. (8.90) with respect to t

$$\frac{dx^i}{dt} = \dot{x}^i \tag{8.91}$$

thus Eq. (8.89) becomes:

$$ds = \sqrt{g_{ij}\dot{x}^i \dot{x}^j} dt. \tag{8.92}$$

Now we introduce the *Lagrange* function:

$$L = \sqrt{g_{ij}\dot{x}^i \dot{x}^j} \tag{8.93}$$

with Eq. (8.93), the integral (8.87) becomes

$$I = \int_{P_1}^{P_2} L\,dt. \tag{8.94}$$

Since for all possible curves the variation must disappear at the ends of the fixed points P_1 and P_2, only those curves will be considered that go through them. Among these curves the one that fulfill the Euler-Lagrange equations:

$$\frac{d}{dt}\left(\frac{\partial L}{\partial \dot{x}^i}\right) - \frac{\partial L}{\partial x^i} = 0 \tag{8.95}$$

is the one with a minimal length. The solution of the N-differential equations (8.95) is:

$$\frac{d}{dt}\left(\frac{g_{ij}\dot{x}^j}{L}\right) - \frac{1}{2L} g_{jk,i}\dot{x}^j\dot{x}^k = 0. \tag{8.96}$$

Reintroduce the element ds along with the chain differentiation and considering Eqs. (8.92) and (8.93), we find:

$$\frac{dx^i}{dt} = \frac{dx^i}{ds}\frac{ds}{dt} = \frac{dx^i}{ds}L \tag{8.97}$$

thus, Eq. (8.96) is written as:

$$\frac{d}{ds}\left(\frac{g_{ij}dx^j}{ds}\right) - \frac{1}{2} g_{jk,i}\frac{dx^j}{ds}\frac{dx^k}{ds} = 0 \tag{8.98}$$

from which, we obtain:

$$g_{ij}\frac{d^2x^j}{ds^2} + g_{ij,k}\frac{dx^j}{ds}\frac{dx^k}{ds} - \frac{1}{2} g_{jk,i}\frac{dx^j}{ds}\frac{dx^k}{ds} = 0. \tag{8.99}$$

Considering the second term with summation over j and k, the symmetric part delivers a contribution:

$$\frac{1}{2}(g_{ij,k} + g_{ik,j})\frac{dx^j}{ds}\frac{dx^k}{ds} \tag{8.100}$$

and with Eqs. (8.100) (8.99) becomes:

$$g_{ij}\frac{d^2x^j}{ds^2} + \frac{1}{2}(g_{ij,k} + g_{ki,j} - g_{jk,i})\frac{dx^j}{ds}\frac{dx^k}{ds} = 0. \tag{8.101}$$

Now we eliminate the covariant metric g_{ij} by multiplying Eq. (8.101) with the contravariant metric tensor g^{ni} and obtain

$$\frac{d^2x^n}{ds^2} + \frac{1}{2} g^{ni}(g_{ij,k} + g_{ki,j} - g_{jk,i})\frac{dx^j}{ds}\frac{dx^k}{ds} = 0. \tag{8.102}$$

As seen the expression in the parentheses is the already known Christoffel symbol:

$$\frac{d^2x^n}{ds^2} + \Gamma^n_{jk}\frac{dx^j}{ds}\frac{dx^k}{ds} = 0. \quad (8.103)$$

Renaming the dummy indices, we have

$$\frac{d^2x^n}{ds^2} + \Gamma^n_{ij}\frac{dx^i}{ds}\frac{dx^j}{ds} = 0 \quad (8.104)$$

with s as a parameter. Equation (8.104) is the equation of geodesic for an N-dimensional Riemann space, which is characterized by a non-zero Christoffel symbol. For a flat space the Christoffel symbols $\Gamma^n_{ij} = 0$. As a result Eq. (8.104) reduces to:

$$\frac{d^2x^n}{ds^2} = 0. \quad (8.105)$$

Integrating Eq. (8.105) results in straight line solution as the shortest distance between two points:

$$x^n = a_n s + b_n \quad (8.106)$$

with the coefficients a_n, b_n that are independent constants.

8.6.3 Calculation of Geodesics for 2-Dimensional Surfaces

In what follows, we specialize Eq. (8.104) for two-dimensional surfaces with the coordinates $\Theta^\alpha = (\Theta^1, \Theta^2)$, thus Eq. (8.92) becomes:

$$ds = \sqrt{a_{\alpha\beta}\dot{\Theta}^\alpha\dot{\Theta}^\beta}dt. \quad (8.107)$$

From here on, the process of deriving the geodesic equation is similar to the one already discussed in Sect. 8.6.2. Replacing in Eq. (8.103) x^n with Θ^α we arrive at:

$$\frac{d^2\Theta^\lambda}{ds^2} + \Gamma^\lambda_{\alpha\beta}\frac{d\theta^\alpha}{ds}\frac{d\theta^\beta}{ds} = 0 \quad (8.108)$$

and in conjunction with Eq. (8.91)

$$\frac{d\Theta^\alpha}{dt} = \dot{\Theta}^\alpha \quad (8.109)$$

Equation (8.108) becomes:

$$\ddot{\Theta}^\lambda + \Gamma^\lambda_{\alpha\beta}\dot{\Theta}^\alpha\dot{\Theta}^\beta = 0. \quad (8.110)$$

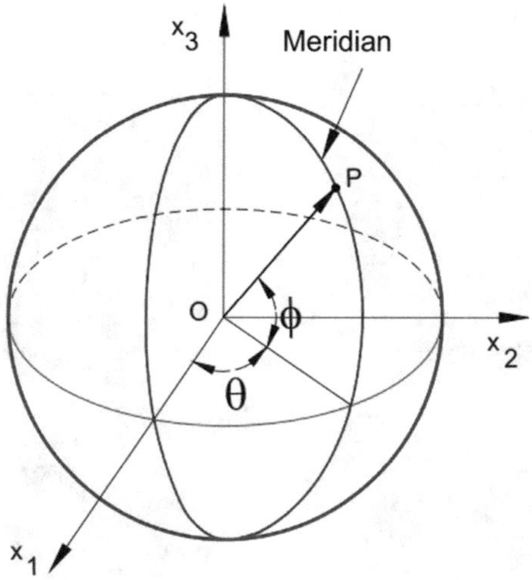

Fig. 8.16 On the definition of θ, Φ

As an example we apply Eq. (8.110) to the surface of a sphere decomposing it into its components. Considering the non-zero Christoffel symbols in Eq. (6.78), we have:

$$\ddot{\Theta}^1 + \Gamma^1_{22}\dot{\Theta}^2\dot{\Theta}^2 = 0$$
$$\ddot{\Theta}^2 + \Gamma^2_{12}\dot{\Theta}^1\dot{\Theta}^2 = 0. \tag{8.111}$$

To calculate the Christoffel symbols for a sphere with $R = 1$ Eq. (6.78) is re-written for the two-dimensional sphere as:

$$\Gamma^{\beta}_{\mu\nu} = \Gamma^{\beta}_{\nu\mu} = \frac{1}{2} g^{\beta\alpha}(g_{\alpha\mu,\nu} + g_{\alpha\nu,mu} - g_{\mu\nu,\alpha}). \tag{8.112}$$

Referring to Fig. 8.16, we set $\Theta^1 \equiv \theta$ and $\Theta^2 \equiv \Phi$, thus, the non-zero Christoffel symbols read:

$$\Gamma^1_{ij} = \begin{pmatrix} 0 & 0 \\ 0 & -sin\Phi\cos\Phi \end{pmatrix}, \Gamma^2_{ij} = \begin{pmatrix} 0 & \cot\Phi \\ \cot\Phi & 0 \end{pmatrix}, \Gamma^2_{ji} = \begin{pmatrix} 0 & \cot\Phi \\ \cot\Phi & 0 \end{pmatrix}. \tag{8.113}$$

Considering Eqs. (8.113), (8.111) and the Christoffel symbols above, we arrive at:

$$\ddot{\theta} - \dot{\Phi}^2 \sin\theta\cos\theta = 0$$
$$\ddot{\Phi} + 2\dot{\Phi}\dot{\theta}\cot\theta = 0. \tag{8.114}$$

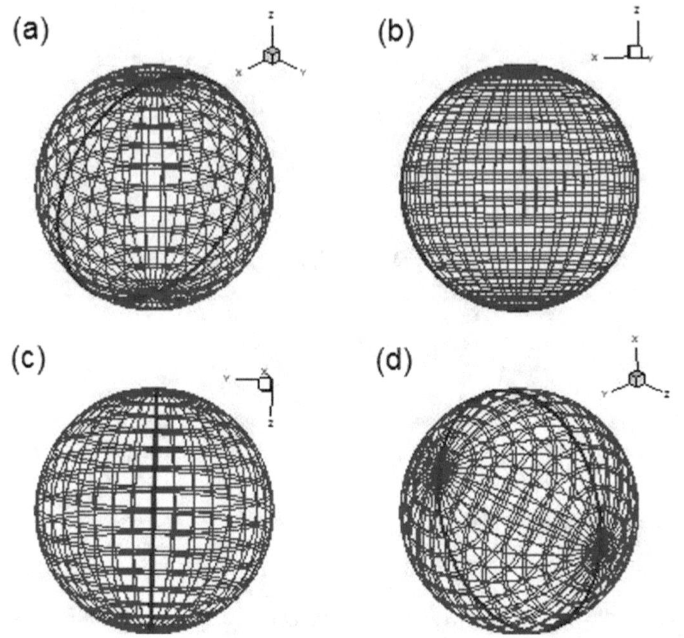

Fig. 8.17 Numerical solution of the geodesic differential equation

$$\theta(t)|_{t=0} = \theta_0 = \frac{\pi}{4}, \ \dot{\theta}(t)|_{t=0} = a = 3$$
$$\Phi(t)|_{t=0} = \Phi_0 = \frac{\pi}{4}, \ \dot{\Phi}(t)|_{t=0} = b = 1. \tag{8.115}$$

The initial conditions required to solve Eq. (8.114), Fig. 8.16 are.

With the initial conditions (8.115), the solution of Eq. (8.114), Fig. 8.16 which is a set of two second order ordinary differential equations is obtained using either *Runge-Kutta* or *Predictor-Corrector* method. Both are stable methods and deliver the same result. Figure 8.17 shows the result. Figure 8.17a reflects the geodesic solution of Eq. (8.114), Fig. 8.16 for the initial conditions (8.115). It shows the great circle rotated by 45°. Figure 8.17b shows the projection of the great circle on the plane zx as a full circle. On the yz-plane, Fig. 8.17c, the projection reduces to a straight line and Fig. 8.17d exhibits the great circle under the rotation of entire coordinate system.

References

1. Spivak M (1979) A comprehensive introduction to differential geometry, vols. 1 and 2. Publish or Perish
2. Blaschke W, Leichtweiss K (1975) Elementare Differentialgeometrie. Springer
3. Bishop RL, Crittenden RJ (1964) Geometry of manifolds. Academic. Reprinted later by Dover

4. Klingbeil E (197/197a) Tensorrechnung für Ingenieure, Mannheim: Bibliographisches Institut, 196
5. Kästner S (1964) Vektoren. Spinoren, Akademie-Verlag Berlin, Tensoren
6. Gauss CF (1903) Conforme Abbildung des Sphäroids in der Ebene (Projectionsmethode der Hannoverschen Landesvermessung) König. Ges. Wiss., Göttingen (1828), in Carl Friedrich Gauss Werke, Vol 9. (Ges. Wiss., Göttingen), pp 142–194
7. Gauss CF (1910) Untersuchungen über Gegenstände der höheren Geodäsie. Abhandl Math Cl Kön Ges Wiss zu Göttingen 2:3–45 (1843); 3:3–43 (1846). Reprint: Frischauf J (ed) Ostwald's Klass. ex. Wiss., No. 177. Engelmann, Leipzig

Chapter 9
Turbulent Flow, Modeling

9.1 Fundamentals of Turbulent Flows

Before starting with the application of tensors in turbulent flow, we need to know what a turbulent flow is. One of the parameters that defines the nature of the turbulent flow is the intermittency factor. We consider a flat plate with a smooth surface placed within a wind tunnel with statistically steady flow velocity and a low turbulence fluctuation velocity $\boldsymbol{V}'$[1] and measure the velocity with a high frequency probe such as a single hot sire probe, for detail we refer to [1, 2].

To explain the phenomenon of turbulent flow, we take a flat plate with a very smooth surface and insert it into a wind tunnel. Placing a high frequency senor, for example a single hot wire, [2], very close to the surface, for example 0.1 mm above the surface. For a given interval of T we measure the velocity as a function of time as shown in Fig. 9.1. Two distinguished patterns may transpire: (a) *laminar* pattern characterized by an orderly velocity distribution, without random fluctuation and (b) a *turbulent* pattern that occupies the intervals Δt_i. These two patterns are characterized by the intermittency factor γ defined in Fig. 9.1. Using the intermittency function as a parameter to describe the flow state under consideration, two distinct flow regimes are distinguished: (a) laminar flow regime characterized by the absence of irregular or random fluctuations with $\gamma = 0$ and, (b) turbulent flow state characterized by $\gamma = 1$ with irregularities expressed in terms of random variations in time and space. While the randomness is an inherent quality of a turbulent flow, it does not completely define the turbulent flow. In many engineering applications, however, turbulent flow can be described statistically by determining averaged values for flow quantities. Beside these two flow regimes, there is a third one that is characterized by $0 < \gamma < 1$.

[1] The superscript "′" pertains to stochastic fluctuations in contrast to "$\sim$" used in Sect. 9.4 that stands for deterministic disturbance.

M. T. Schobeiri, *Tensor Analysis for Engineers and Physicists - With Application to Continuum Mechanics, Turbulence, and Einstein's Special and General Theory of Relativity*, https://doi.org/10.1007/978-3-030-35736-8_9

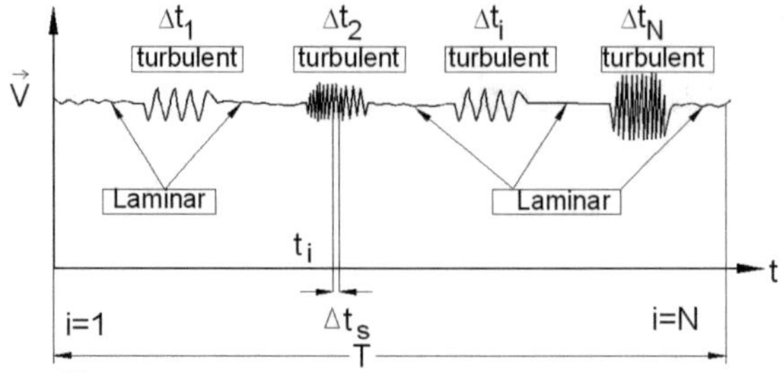

(b)

Intermittent velocity distrubution measured at point $P(x_1, x_2)$

$$\gamma = \frac{\Sigma \Delta t_i}{T}, \text{ Laminar flow} \gamma = 0, \text{ Turbulent flow } \gamma = 0$$

Fig. 9.1 Explaining the nature of laminar and turbulent flow using the intermittency factor γ

This is called a transitional flow. In fact the velocity distribution sketched in Fig. 9.1 is a transitional flow. This type of flow occurs in engineering and is prevalent in turbomachines, [3, 4].

Taking into consideration the complexity of turbulent flows encountered in physics and a multitude of definitions found in literature, among others, by G. I. Taylor [5], von Kármán [6], Hinze [7] and Rotta [8], we term a flow regime turbulent that has the following characteristics:

1. Turbulent flows are generally irregular and their properties continuously undergo stochastic spatial and temporal changes. As a result, no reproducible turbulence data with stochastic distribution can be obtained.
2. Despite its stochastic nature, using statistical tools, time or ensemble averaged values can be constructed that are perfectly reproducible.
3. Turbulent flows are rotational motions (vorticity $\nabla \times \boldsymbol{V} \neq 0$) with a wide variety of vortices with different sizes and vorticities.
4. Turbulent flows are generally unsteady and three-dimensional.

The above characteristics are implicitly indicative of the following features that are inherent in turbulence:

(a) There is no analytical *exact solution* for any type of turbulent flows, even for the simplest one.
(b) The inherent three-dimensional unsteady nature of turbulence associated with the velocity fluctuations is responsible for an intense mixing of fluid particles causing an enhanced momentum, and energy transfer between the fluid particles. This process is called *diffusion*. The enhanced *diffusivity* is due to the existence of Reynolds stress which is, in general, several order of magnitudes larger than the viscous stresses. Exceptions are flows very close to the wall, where the viscosity has a predominant effect.
(c) The high level of spatial and temporal fluctuations of velocity, pressure, and temperature causes fluctuating vortices, also called *eddies*, of different sizes. The eddies convect, rotate, stretch, decompose in smaller eddies, overlap and

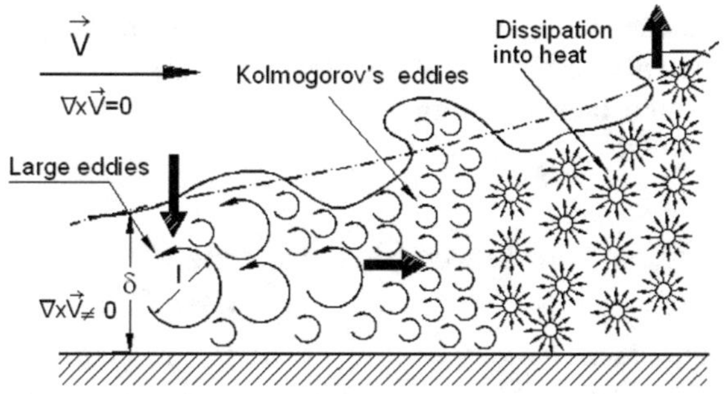

Fig. 9.2 Schematics of an instantaneous energy cascade in turbulent boundary layer. The arrows indicate energy extraction, transfer and dissipation

coalesce, as Fig. 9.2 schematically shows. Figure 9.2 schematically summarizes the energy cascade process taking place within a turbulent boundary layer. As seen, a flat plate is exposed to a constant, steady, non-turbulent flow that is separated from the rotational boundary layer ($\nabla \times \boldsymbol{V} \neq 0$) by a sharp interface. The averaged boundary layer thickness is shown as a dashed curve. Large eddies with a size l continuously extract energy from the main flow and transfer it to smaller eddies. The process of energy cascade leads to the smallest eddies whose energy dissipates as heat. The specific issues dealing with Kolmogorov's hypothesis, energy cascade, eddy structure, length, and time scale, are treated in more detail in the following sections.

(d) During the cascade process, the size of these eddies change from large to small. In a boundary layer flow, as shown in Fig. 9.2, the size of the largest eddy has the same order of magnitude as the local boundary layer thickness δ. It receives its energy from the mean flow. The larger eddies continuously transfer their kinetic energy to the smaller eddies. Once the eddy size is reduced to a minimum, its kinetic energy is dissipated by the viscous diffusion. A state of universal equilibrium is reached when the rate of energy received from larger eddies is nearly equal to the rate of energy
when the smallest eddies dissipate into heat. The process of transferring energy from the largest eddy to the smallest is called *energy cascade* process, introduced by Richardson [9]. While the statistics[2] of larger eddies change, Kolmogorov [10] introduced a hypothesis that enables quantifying the scale of the smallest eddy on the basis of *isotropy* of those eddies.

(e) Turbulent flow occurs at high Reynolds numbers. For engineering applications where a solid wall is present (boundary layer, wall turbulence), the order of magnitude for the Re-number to become fully turbulent depends on the pressure gradient along the surface, as well as the perturbation of the boundary layer by any incoming periodic unsteady disturbances such as wake impingements.

Note that in the course of the above introduction we utilized the rather vague term "eddy", which is used in the literature in context of turbulence research. In contrast

[2] The averages of a random quantity are called statistics. This includes mean and the rms (root-mean-square) of that quantity.

to the precisely defined term "vortex" with a descriptive circulation Γ and its direct relation to the vorticity vector $\nabla \times \boldsymbol{V}$, the term eddy lacks a precise definition and is loosely used for any individual turbulent structure with some length-scale.

9.2 Role of Tensors in Describing Turbulence

In Sect. 9.1, we introduced the very basics for understanding the nature if a turbulent flow without using any equations. In the following section we present the necessary tensor analytical tools that are inherent to turbulence equations. The turbulent flow is completely described by Navier–Stokes equation using Direct Navier–Stokes Simulation (DNS). This solution method, however requires significant amount of computational time and effort, making DNS impractical. Another, less accurate method of calculating the turbulent flow is to decompose the instantaneous velocity vector into an averaged velocity and a fluctuation velocity vector $\boldsymbol{V} = \overline{\boldsymbol{V}} + \boldsymbol{V}'$. The insertion of the decomposition into the Navier–Stokes equation and its subsequent averaging results in *Reynold Averaged Navier Stokes* equation known as *RANS*. Reynolds-averaging Navier Styokes equation has created an *apparent* stress tensor $\overline{\boldsymbol{V}'\boldsymbol{V}'} = e_i e_j \overline{V_i' V_j'}$ also called *Reynolds stress tensor* with nine components from which, for symmetric reasons, six are distinct. Thus, the creation of this tensor has added six more unknowns to Navier–Stokes equations. In order to find additional equations to close the equation set that consists of continuity, momentum, and energy balances, we need to construct additional equations. This is done by multiplying the ith component of the Navier–Stokes equation with the jth one. Thereby we expect to find turbulence models that establish relations between the new equations and the set of equations mentioned above. It should be pointed out that this purely mathematical manipulation does not represent any new conservation law with a physical background. However, it helps in providing additional tools that are necessary for turbulence modeling. In this context, *correlations* are indispensable tools for providing additional insight into turbulence.

9.2.1 Correlations, Length and Time Scales

The Reynold stress tensor $\overline{\boldsymbol{V}'\boldsymbol{V}'}$ is a second order tensor and is the mean product of the fluctuation components at a single point in space; it is called a *single point correlation*. It does not give any further information about the turbulence structure, such as the length and time scale of eddies. We obtain this information from a *two-point correlation*. It is a second order tensor of the mean product of fluctuation components at two different points in space and time, namely $(\boldsymbol{x}, t)$ and $(\boldsymbol{x} + \boldsymbol{r}, t + \tau)$. For a purely spatial correlation with $\tau = 0$, the same fluctuating quantity is measured at two different spatial positions $\boldsymbol{x}$ and $\boldsymbol{x} + \boldsymbol{r}$. Figure 9.3 shows the position

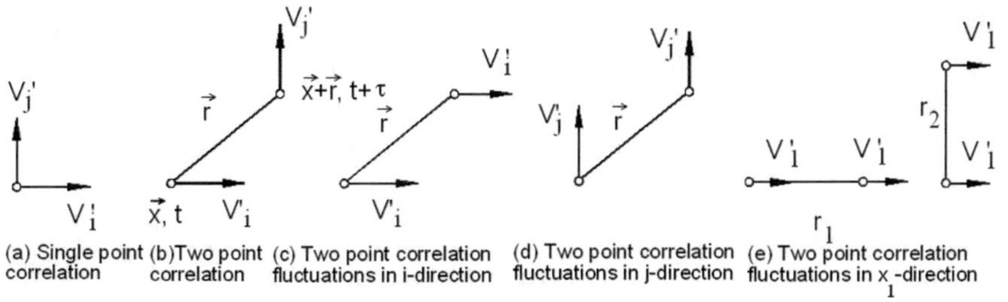

Fig. 9.3 Single- and two-point correlations

of the fluctuation components for (a) single point correlation and several *two-point correlations*. For a general two-point correlation, we construct the second order tensor

$$e_i e_j R_{ij}(\boldsymbol{x}, t, \boldsymbol{r}, \tau) = e_i e_j \overline{V_i'(\boldsymbol{x}, t) V_j'(\boldsymbol{x} + \boldsymbol{r}, t + \tau)}, \text{ with}$$
$$R_{ij}(\boldsymbol{x}, t, \boldsymbol{r}, \tau) = \overline{V_i'(\boldsymbol{x}, t) V_j'(\boldsymbol{x} + \boldsymbol{r}, t + \tau)} \tag{9.1}$$

with $r = e_i r_i$ and τ as the spatial and temporal distance between the two points. For $|r| \to \infty$ or $|\tau| \to \infty$, the fluctuation components V_i' and V_j' are independent from each other, leading to $R_{ij} = 0$. For a stationary or homogeneous process, the correlation tensor is symmetric and we may write:

$$R_{ij}(\boldsymbol{x}, t, \boldsymbol{r}, \tau) = R_{ji}(\boldsymbol{x} + \boldsymbol{r}, t + \tau, -\boldsymbol{r}, -\tau). \tag{9.2}$$

Normalizing the correlation (9.1), we obtain the dimensionless correlation, also called *correlation coefficient* as:

$$\rho_{ij}(\boldsymbol{x}, t, \boldsymbol{r}, \tau) = \frac{\overline{V_i'(\boldsymbol{x}, t) V_j'(\boldsymbol{x}\boldsymbol{r}, t + \tau)}}{[\overline{V_i'^2}(\boldsymbol{x}, t) \overline{V_j'^2}(\boldsymbol{x} + \boldsymbol{r}, t + \tau)]^{\frac{1}{2}}}. \tag{9.3}$$

For a stationary or homogeneous field, the tensor $R_{ij}(\boldsymbol{x}, t, \boldsymbol{r}, \tau)$ is independent of t and x. We can construct an auto-correlation when the fluctuation components are measured at the same position but at different times and have the same direction $(i = j)$. It is defined as

$$R_{ij}(t, \tau) = \overline{V_i'(t) V_j'(t + \tau)}. \tag{9.4}$$

Note that by setting $i = j$, we do not sum over the indices i and j. As an example, the auto-correlation for the fluctuation component V_1', is written as

$$R_{11}(t, \tau) = \overline{V_1'(t) V_1'(t + \tau)} \tag{9.5}$$

with the fluctuation V_1' at the same spatial position but at two different times t and $t+\tau$. On the other hand, the spatial correlation is obtained by setting in Eq. (9.1) $\tau = 0$.

$$R_{ij}(\boldsymbol{x}, \boldsymbol{r}) \equiv R_{ij}(\boldsymbol{r}) = \overline{V_i'(\boldsymbol{x})V_j'(\boldsymbol{x}+\boldsymbol{r})}. \tag{9.6}$$

The corresponding correlation coefficient is

$$\rho_{ij}(\boldsymbol{r}) = \frac{\overline{V_i'(\boldsymbol{x})V_j'(\boldsymbol{x}+\boldsymbol{r})}}{[\overline{V_i'^2(\boldsymbol{x})}\;\overline{V_j'^2(\boldsymbol{x}+\boldsymbol{r})}]^{\frac{1}{2}}}. \tag{9.7}$$

For the component in x_1-direction, Eq. (9.6) is simplified as

$$R_{11}(r_1, 0, 0) = \overline{V_1'(x_1)V_1'(x_1+r_1)}. \tag{9.8}$$

In Eq. (9.8), the reference position vector x_1 can be set $x_1 = 0$, resulting in

$$R_{11}(r_1, 0, 0) = \overline{V_1'(0)V_1'(r_1)}. \tag{9.9}$$

The right-hand side of Eq. (9.9) is called the *covariance* of the V_1'-component. The correlation coefficient then is obtained by setting in Eq. (9.7) $i = j$

$$\rho_{11}(r_1, 0, 0) = \frac{\overline{V_1'(0)V_1'(r_1)}}{[\overline{V_1'^2(0)}\;\overline{V_1'^2(r_1)}]^{\frac{1}{2}}}. \tag{9.10}$$

In most turbulence related literature, the term $\overline{V_1'^2(r_1)}$ is replaced by $\overline{V_1'^2(0)}$, thus, the modified coefficient is:

$$\rho_{11}(r_1, 0, 0) = \frac{\overline{V_1'(0)V_1'(r_1)}}{\overline{V_1'^2(0)}}. \tag{9.11}$$

In a similar manner, the coefficients in r_2 and r_3 may be constructed

$$\rho_{11}(0, r_2, 0) = \frac{\overline{V_1'(0)V_1'(r_2)}}{\overline{V_1'^2(0)}}. \tag{9.12}$$

The correlation functions are used to determine the length and time scales. The general definition of the integral length scale is

$$L_{ij}(\boldsymbol{x}, t) = \frac{\int_{-\infty}^{+\infty} R_{ij}(\boldsymbol{x}, t, t_k, 0)dr_k}{2\overline{V_i'(\boldsymbol{x}, t)V_j'(\boldsymbol{x}, t)}}. \tag{9.13}$$

Likewise, the time scale is defined as

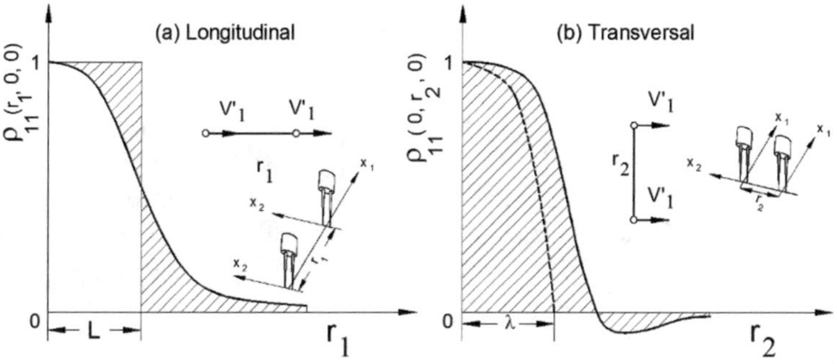

Fig. 9.4 Correlation coefficients with their osculating parabolas

$$T_{ij}(\boldsymbol{x}, t) = \frac{\int_{-\infty}^{+\infty} R_{ij}(\boldsymbol{x}, t, 0, \tau) d\tau}{2\overline{V_i'(\boldsymbol{x}, t) V_j'(\boldsymbol{x}, t)}}. \tag{9.14}$$

For the special cases discussed above, the length scale is schematically plotted in Fig. 9.4 for longitudinal $\rho_{11}(r_1, 0, 0)$, as well as lateral $\rho_{11}(0, r_2, 0)$ correlation coefficients. In both cases, the length scale is simply the area underneath the coefficient curves. Using the hot wire anemometry for measuring the velocity fluctuation, it is necessary to use two parallel wires separated either by r_1 in a longitudinal or by r_2 in a lateral direction. In the longitudinal case, Fig. 9.4a, the second wire is in velocity and thermal wakes of the first wire located upstream of the second wire. This configuration leads to an erroneous longitudinal length scale. The lateral length scale can be measured more accurately using the two hot wire probes as arranged in Fig. 9.4b. Re-arranging Eq. (9.13), we find the longitudinal length scale

$$L_{\text{long}} = \int_{-\infty}^{+\infty} \rho_{11}(r_1, 0, 0) dr_1 = \frac{1}{2\overline{V_1'^2(0)}} \int_{-\infty}^{+\infty} \overline{V_1'(0) V_1'(r_1)} dr_1 \tag{9.15}$$

and the lateral length scale

$$L_{\text{trans}} = \int_{-\infty}^{+\infty} \rho_{11}(0, r_1, 0) dr_1 = \frac{1}{2\overline{V_1'^2(0)}} \int_{-\infty}^{+\infty} \overline{V_1'(0) V_1'(r_2)} dr_2. \tag{9.16}$$

Although measuring the lateral length scale delivers a more accurate result than the measured longitudinal one, from an experimental point of view, both are not practical. This following hypothesis offers a practical alternative.

Taylor Hypothesis: An alternative method to estimate the length scale is the utilization of a *frozen turbulence hypothesis* proposed by G. I. Taylor [1]. Considering a large scale eddy with sufficiently high energy content, Taylor proposed an hypothesis that the energy transport contribution of small size eddies that are carried by a large scale eddy, as shown in Fig. 9.5, compared with the one produced by a larger

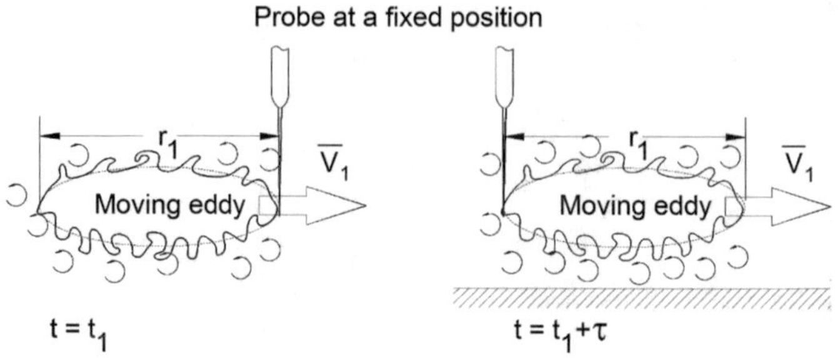

Fig. 9.5 Explaining Taylor "frozen" turbulence

eddy, is negligibly small. In such a situation, the transport of a turbulence field past a fixed point is due to the larger energy containing eddies. It states that "in certain circumstances, turbulence can be considered as 'frozen' as it passes by a sensor". This statement is illustrated in Fig. 9.5. It shows a large eddy moving with an averaged constant velocity of $\overline{V_1}$ in the x_1-direction, carrying a number of smaller eddies with fluctuating velocity V_1'. The hypothesis is considered valid only if the condition $V_1'^2 << \overline{V}_1^1$ holds. Despite this constraint, the Taylor hypothesis delivers a reasonable approximation for the variations of fluctuating eddies that are carried along by larger scale eddies. Taylor established his hypothesis using a spatial (Eulerian) description rather than a material (Lagrangian) one (see [2, Chap. 3]). The hypothesis relates the spatial variation to the temporal variation measured at a single point. From an experimental point of view, this approach exhibits a substantial reduction in efforts to determine the turbulence length and time scales. Mathematically speaking, the Taylor hypothesis implies that the substantial change of the velocity vector $\boldsymbol{V} = \overline{\boldsymbol{V}} + \boldsymbol{V}'$ must vanish. Utilizing the Taylor's assumption of constant mean velocity, we have

$$\frac{D(\overline{\boldsymbol{V}} + \boldsymbol{V}')}{Dt} = \frac{\partial \boldsymbol{V}'}{\partial t} + \overline{\boldsymbol{V}} \cdot \nabla \boldsymbol{V}' = 0. \tag{9.17}$$

Equation (9.17) is the mathematical formulation of the Taylor hypothesis. The component of (9.17) reads

$$\frac{\partial V_1'}{\partial x_1} = -\frac{1}{\overline{V_1}} \frac{\partial V_1'}{\partial t}. \tag{9.18}$$

Approximating the differentials by differences leads to:

$$\frac{V_1'(x_1 + r_1) - V_1'(x_1)}{r_1} = -\frac{V_1'(t + \tau) - V_1'(t)}{\overline{V}_1 \tau}. \tag{9.19}$$

Equation (9.19) implies that the spatial separation shown in Fig. 9.4a can be expressed in terms of a temporal separation. As seen, Eq. (9.17) is the left-hand side of the Navier–Stokes equation (5.62) with the right side $-1/\rho \nabla(\overline{p} + p') + \nu \Delta(\overline{\boldsymbol{V}} + \boldsymbol{V}') =$

0. This hypothesis is only valid if we assume that $V'/\overline{V} << 1$ and $\overline{p} = \text{const}$. As a consequence of this assumption, the pressure fluctuation p', which has the order of V'^2, can be neglected. With $r_1 = \overline{V}_1\tau$ in Eq. (9.19), we obtain

$$V_1(t) - V_1(t+\tau) = V_1(x_1 + r_1) - V_1(x_1). \tag{9.20}$$

Thus, the auto-correlation coefficient (9.11) becomes

$$\rho_{11}(r_1, 0, 0) \equiv \rho_{11}(\tau) = \frac{\overline{V_1'(t)V_1'(t+\tau)}}{\overline{V_1'^2}} \tag{9.21}$$

and the corresponding integral time scale follows from

$$T_1 = \frac{1}{2}\int_{-\infty}^{+\infty} \rho_{11}(\tau) d\tau \tag{9.22}$$

thus, the length scale results from

$$L = \overline{V}_1 T_1. \tag{9.23}$$

Shifting the time origin results in $\rho(\tau) = \rho(-\tau)$, meaning that Eq. (9.21) is an even function. Furthermore, Eq. (9.22) has the property that
$\rho(\tau) = 1$ at $\tau = 0$ and $\rho(\tau) = 0$ for $\tau \Rightarrow \infty$.

An alternative method to determine the time scale of small dissipating eddies uses Eq. (9.21). For this purpose we first expand the corresponding correlation coefficient (9.21) about $\tau = 0$ with respect to time, and truncate beyond the quadratic term; as a result, we arrive at

$$\rho_{11}(\tau) = 1 + \frac{\tau^2}{2}\left(\frac{\partial^2 \rho_{11}}{\partial \tau^2}\right)_{\tau=0} + \ldots.. \tag{9.24}$$

This crude approximation allows constructing an *osculating parabola* with the same vertex value and the derivative at $\tau = 0$ as the exact $\rho_{11}(\tau)$-curve. The parabola is described by

$$\rho_{11}(\tau) \approx 1 - \left(\frac{\tau}{\tau_1}\right)^2. \tag{9.25}$$

The intersection of this parabola with the τ-axis delivers the Taylor time scale τ_1, from which the Taylor micro length scale can be inferred . Equating Eqs. (9.25) and (9.24) gives the Taylor micro time scale

$$\tau_1 = \sqrt{\frac{-2}{\left(\frac{\partial^2 \rho_{11}}{\partial \tau^2}\right)_{\tau=0}}} \tag{9.26}$$

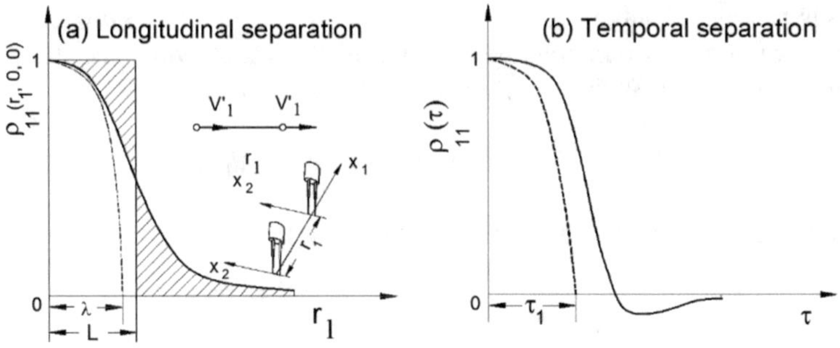

Fig. 9.6 Approximation of Taylor length and time scales by osculating parabolas

and the Taylor micro length scale

$$\lambda = \overline{V}_1 \tau. \tag{9.27}$$

The two scales are shown in Fig. 9.6 with the correlation coefficients for (a) spatial and (b) temporal separations.

It also includes the osculating parabola with the length scale and time scale intersects. The Taylor micro length scale can be found directly by using a similar procedure that leads to Eq. (9.26). In this case, we expand the correlation coefficient (9.11) about $r_1 = 0$ and truncate beyond the quadratic term. We may then approximate the result by the following parabola:

$$\rho_{11}(r_1, 0, 0) \approx 1 - \left(\frac{r_1}{\lambda}\right)^2 \tag{9.28}$$

and arrive directly at the Taylor length scale

$$\lambda = \sqrt{\frac{-\overline{V_1'^2}}{\left(\frac{\partial^2 \rho_{11}}{\partial r_1^2}\right)}}. \tag{9.29}$$

It should be pointed out that the Taylor micro length scale is only an approximate length scale. It does not represent the length scale of large energy containing eddies or the smallest dissipating eddies. However, for a homogeneous isotropic turbulence, λ provides a useful tool to estimate the turbulence dissipation. For this purpose, we first use the following length scale definition

$$\lambda = \sqrt{\frac{\overline{V_1'^2}}{\overline{\left(\frac{\partial V_1}{\partial x_1}\right)^2}}} \tag{9.30}$$

and then expand the dissipation equation defined in Sect. 9.4.1.4, Eq. (9.71) for isotropic turbulence and introduce Eq. (9.30). As a result, we find a relationship between the dissipation and the length scale[3]

$$\varepsilon = \nu \left(\overline{\frac{\partial V_i'}{\partial x_j} \frac{\partial V_i'}{\partial x_j}} \right) = \nu \overline{\left(\frac{\partial V_1'}{\partial x_1} \right)^2} (2\delta_{ii}\delta_{jj} - \delta_{ij}\delta_{ij}) = 15\nu \overline{\left(\frac{\partial V_1'}{\partial x_1} \right)^2} = 15\nu \left(\frac{\overline{V_1'}}{\lambda} \right)^2 . \tag{9.31}$$

Using dimensional analysis, Taylor established the following relationship for dissipation

$$\varepsilon \propto \frac{k^{3/2}}{l}. \tag{9.32}$$

Another important aspect is that length, time, and velocity scales describe the dissipative character of Kolmogorov's eddies as a result of energy cascading, as shown in Fig. 9.1. Using dimensional analysis, Kolmogorov arrived at his length scale (η), time scale (τ) and the velocity scale (υ) scales:

$$\eta = \left(\frac{\nu^3}{\varepsilon} \right)^{1/4}, \tau = \left(\frac{\nu}{\varepsilon} \right)^{1/2}, \upsilon = \left(\frac{\nu}{\varepsilon} \right)^{1/4} \tag{9.33}$$

which we will discuss in some details in the following section.

9.2.2 *Spectral Representation of Turbulent Flows*

As Fig. 9.1 shows, the scales of turbulence eddies distributed over a range of scales extend from the largest scales which interact with the mean flow, from which they extract their energy, to the smallest scales where their energy dissipates as heat. Utilizing a transformation from physical space into a wavenumber space, the energy of eddies can be expressed in terms of a spectral distribution represented by the function $E(k)dk$, which is the energy of eddies from k to $k + dk$ with k as the wavenumber. Since the wavenumber is expressed in terms of the wave length $\lambda = 2\pi/k$, the dimension of wavenumber is, L^{-1} in M, L, t dimension systems. If we assume that the eddy's length scale l is proportional to the wave length λ, then the wavenumber can be thought of as proportional to the inverse of an eddy's length l, i.e. $k \propto 1/l$. Figure 9.7 exhibits the energy spectral distribution $E(k)$ as a function of the wavenumber k. This energy spectrum corresponds to the formation and the scales of eddies within a boundary layer shown in Fig. 9.1. Kolmogorov introduced three distinct length scale/wavenumber regions that are marked in Fig. 9.7.

The first region is occupied by large eddies that contain most of the energy. These eddies interact with the mean flow from which they extract their energy (downward

[3] Detailed derivations of Eq. (9.31) is found in Hinze [3], p. 179 and Rotta [8], p. 80.

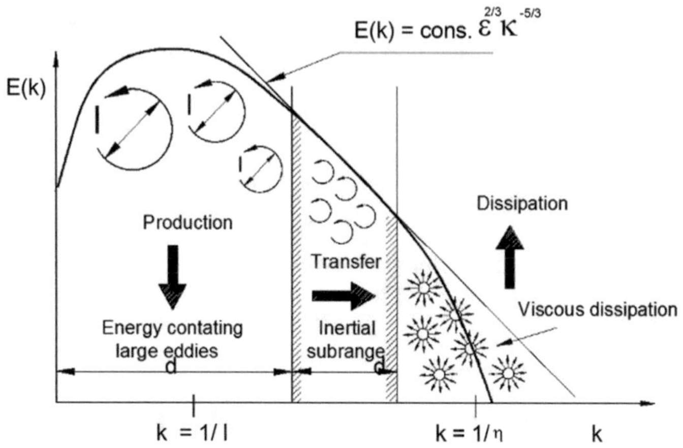

Fig. 9.7 Schematic of Kolmogorov energy spectrum as a function of wavenumber

arrow) and transfer it to smaller scale eddies. The large eddies affected by the flow boundary conditions are anisotropic. According to Kolmogorov, they loose their directional preference in the energy cascade process by which energy is transferred to successively smaller and smaller eddies. The second region is the *inertial subrange* (Kolmogorov).

In this region, a transport of energy takes place from the large eddies to the eddies that are in dissipation range (horizontal arrow). Since in this subrange, the energy transfer is accomplished by inertial forces, it is called the inertial subrange. As shown in Fig. 9.7, the existence of this region requires that the Reynolds number must be high to establish a fully turbulent flow. The third region is the dissipation range where the eddies are small and isotropic and their kinetic energy dissipates as heat. The scales of the eddies are described by the Kolmogorov scales, Eq. (9.33).

Kolmogorov Hypotheses: Utilizing the above scale decomposition, Kolmogorov established his universal equilibrium theory based on two similarity hypotheses for turbulent flows. The first hypothesis states that for a high Reynolds number turbulent flow, the small-scale turbulent motions are isotropic and independent of the detailed structure of large scale eddies. Furthermore, there is a range of high wavenumbers where the turbulence is statistically in equilibrium and uniquely determined by the energy dissipation $\varepsilon[L^2T^{-3}]$ and the kinematic viscosity $\nu[L^2T^{-1}]$. With this hypothesis and in conjunction with dimensional reasoning, Kolmogorov arrived at length (η), time (τ) and the velocity (υ) scales which have already been presented in Eq. (9.33). Considering the Kolmogorov's length and velocity scales, the corresponding Kolmogorov's equilibrium Reynolds number is

$$\mathrm{Re} = \frac{\upsilon\eta}{\nu} = 1. \tag{9.34}$$

To define the range of the equilibrium, we first introduce a dissipation wavenumber to emphasize the strong effects of viscosity

$$K_d = \frac{1}{\eta} \tag{9.35}$$

and the wavenumber of energy containing eddies with the length scale l that may be interpreted as the average size of the energy containing eddies

$$K_e = \frac{1}{l}. \tag{9.36}$$

The equilibrium range contains wavenumbers for which $k >> k_e$ with $k_e << k_d$. Thus, the equilibrium wavenumber must satisfy the condition

$$K_e << k << k_d. \tag{9.37}$$

The range defined by the above condition is exactly the Kolmogorov inertial subrange within which the turbulence is independent of the energy containing eddies and of the range of strong dissipation. Utilizing the dimensional analysis, we find for the energy spectrum $E(k)[L^3T^{-2}]$ the following relationship

$$E(k) = \nu^{5/4}\varepsilon^{1/4}F(k\eta) \tag{9.38}$$

with the function $F(k\eta)$ to be determined.

The second hypothesis states that when the Reynolds number is large enough for the energy containing eddies, there exists a subrange of wavenumbers in which the condition (9.37) is satisfied, then the energy spectrum is independent of ν and is determined by dissipation parameter ε only. In this hypothesis, within the inertial subrange and by virtue of dimensional analysis, where the Function $F(k\eta)$ becomes

$$F(k\eta) = \alpha(k\eta)^{-5/3} \tag{9.39}$$

with $\eta = \left(\frac{\nu^3}{\varepsilon}\right)^{1/4}$ from Eq. (9.33), Kolmogorov found the final equation for energy spectrum within the inertial subrange as

$$E(k) = \text{const. } \varepsilon^{2/3}k^{-5/3} \tag{9.40}$$

with const. $\equiv C_k$ as the universal Kolmogorov's constant shown in Fig. 9.7. Using tidal waves for measuring the spectrum, Grant et al. [11] found the values for $\alpha = 1.44 \pm 0.01$ and $C_k = 1.89 \pm 0.08$. Equation (9.40) is called the Kolmogorov spectrum, which is based on Kolmogorov's second hypothesis. Onsager [12] and Weizsäcker [13] arrived at the same equation independent of Kolmogorov and each other.

9.2.3 Spectral Tensor, Energy Spectral Function

This subject is treated in Hinze [7] and particularly by Rotta [8] from which we present the essentials for better understanding the mathematical structure for turbulence. As the energy spectrum schematically plotted in Fig. 9.7 reveals, in an energy cascade process, eddies with different length, time, and velocity scales interact with each other. Energy is continuously transferred from larger eddies to smaller and smaller ones reaching the dissipation as the final stage of the cascade process. To account for different scales in a more quantitative way, the Fourier analysis, as an appropriate tool, is utilized. To directly apply the Fourier analysis to the issues we discussed in the preceding section, we consider the two-point velocity correlation Eq. (9.1). To start with the simplest case, we assume that (a) the velocity field is spatially homogeneous, meaning that the two-point correlation is independent of the position vector $\boldsymbol{x}$ and, (b) there is no temporal separation between the two points measurement, then Eq. (9.1) reduces to

$$R_{ij}(t, \boldsymbol{r}) = \overline{V_i'(t)V_j'(\boldsymbol{r}, t)}. \tag{9.41}$$

We can now construct a second order velocity spectral tensor $\Phi(\boldsymbol{k}, t) = e_i e_j \Phi_{ij}(\boldsymbol{k}, t)$ in terms of *wavenumber spectrum* as the Fourier transform of the two point correlation (9.41)

$$\Phi_{ij}(\boldsymbol{k}, t) = \frac{1}{(2\pi)^3} \int_{V(\boldsymbol{r})} e^{-i\boldsymbol{k}\cdot\boldsymbol{r}} R_{ij}(t, \boldsymbol{r}) d\boldsymbol{r}. \tag{9.42}$$

$V(\boldsymbol{r})$ as the volume or space boundary for the integral. The inverse transform is found

$$R_{ij}(t, \boldsymbol{r}) = \int_{V(\boldsymbol{k})} e^{i\boldsymbol{k}\cdot\boldsymbol{r}} \Phi_{ij}(\boldsymbol{k}, t) d\boldsymbol{k} \tag{9.43}$$

with

$$e^{i\boldsymbol{k}\cdot\boldsymbol{r}} = \cos(\boldsymbol{k}\cdot\boldsymbol{r}) + i\sin(\boldsymbol{k}\cdot\boldsymbol{r}) \tag{9.44}$$

where $\boldsymbol{k} = e_i k_i$ is the wavenumber vector which is related to the wave length by $l = 2\pi/|\boldsymbol{k}|$. Since we transformed the physical space into the wavenumber space, the integral boundaries in Eqs. (9.42) and (9.43) constitute the volume in the wavenumber space. Furthermore, since the correlation tensor $R_{ij}(t, \boldsymbol{r})$ is real, the velocity spectrum tensor $\Phi_{ij}(\boldsymbol{k}, t)$ is, in general, of a complex nature. It also has the symmetry property

$$\Phi_{ij}(\boldsymbol{k}, t) = \Phi_{ji}(-\boldsymbol{k}, t) \tag{9.45}$$

and satisfies the orthogonality condition

$$k_i \Phi_{ij}(\boldsymbol{k}, t) = k_j \Phi_{ij}(\boldsymbol{k}, t) = 0. \tag{9.46}$$

One-Dimensional Spectral Function: Of practical interest is the one-dimensional version of Eq. (9.42), where the physical component in r_1 and the wavenumber component in k_1 are considered; the one-dimensional case of Eq. (9.42) reads

$$\Theta_{ij}(k_1, t) = \frac{1}{2\pi}\int_{-\infty}^{+\infty} R_{ij}(t, r_1)e^{-ik_1 r_1}dr_1. \tag{9.47}$$

The same result is obtained by integrating Eq. (9.42) over the other two wavenumber components

$$\Theta_{ij}(k_1, t) = \int\int_{-\infty}^{+\infty} \Theta_{ij}(\boldsymbol{k}, t)dk_2 dk_3. \tag{9.48}$$

Of particular practical interest is a spectral function which depends only on the magnitude of the wavenumber $k = |\boldsymbol{k}|$. It can be calculated as a surface integral over the surface of a sphere with the radius k in the wavenumber space as shown in Fig. 9.8.

$$\Psi_{ij}(k, t) = k^2 \oint \Phi_{ij}(\boldsymbol{k}, t)d\Omega \tag{9.49}$$

with $d\Omega$ as an infinitesimal solid angle. Equation (9.49) allows one to compute the energy spectral function as the half trace of (9.49)

$$E(k, t) = \frac{1}{2}\Psi_{ii}(k, t) = \frac{1}{2}k^2 \oint \Phi_{ii}(\boldsymbol{k}, t)d\Omega \tag{9.50}$$

with $E(k, t)$ as the specific kinetic energy of all wavenumbers with the magnitude $k \leq |\boldsymbol{dk}| \leq k + dk$. For isotropic turbulence, $\overline{V_1'^2} = \overline{V_2'^2} = \overline{V_3'^2} = \overline{V_i'V_i'}/3$, Eq. (9.50) is integrated and reduced to

$$E(k) = 4\pi k^2 \frac{1}{2}\Phi_{ii}(k) \tag{9.51}$$

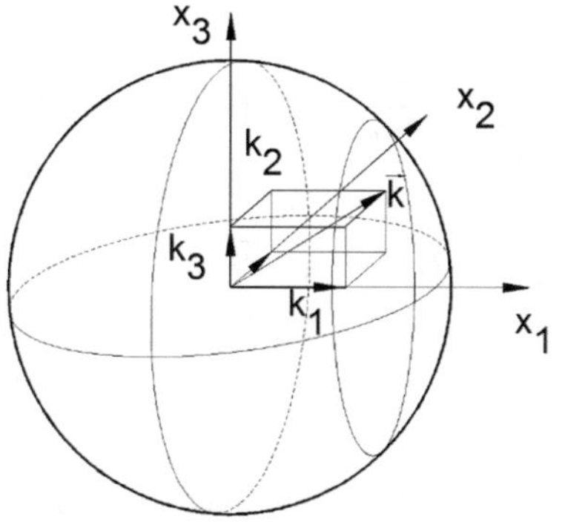

Fig. 9.8 Wavenumber vector and its components from Rotta [8]

with $\Phi_{ii}(k)$ as a scalar quantity. Without presenting the mathematical proof, the spectral tensor $\Phi(k) = e_i e_j \Phi_{ij}(k)$ can be reconstructed using Eq. (9.51)

$$\Phi(k) = \frac{E(k)}{4\pi k^4}(k^2 I - \boldsymbol{kk}) \tag{9.52}$$

with $I = e_i e_j \delta_{ij}$ as the unit tensor and $\boldsymbol{kk} = e_i e_j k_i k_j$ as the second order wavenumber tensor. Thus, Eq. (9.52) can be rewritten as

$$\Phi_{ij}(k) = \frac{E(k)}{4\pi k^4}(k^2 \delta_{ij} - k_i k_j). \tag{9.53}$$

The integration of the energy spectral function $E(k, t)$ over the entire wavenumber space leads to the total turbulent kinetic energy

$$\frac{1}{2} V_i' V_i' = \int_0^\infty E(k, t) dk. \tag{9.54}$$

9.3 Averaging Fundamental Equations of Turbulent Flow

In this section, we present the fundamental equations that describe the turbulent flows. For the sake of completeness, we also re-present some of those equations that were already presented in Chap. 5.

Turbulent flow is characterized by random fluctuations in velocity, pressure, temperature, density, etc. Any turbulent quantity can be decomposed in a mean and fluctuating part. Experimental observations revealed that average values with respect to time and space exist because distinct flow patterns are repeated regularly in time and space. Before preceding further, we apply the averaging formalism to two arbitrary quantities Q1 and Q2 with $Q_1 = \overline{Q}_1 + Q_1'$ and $Q_2 = \overline{Q}_2 + Q_2'$ where Q may be any quantity such as velocity, density, pressure or any other thermodynamic property. If we deal with a statistically steady flow, re-applying the averaging procedure results in:

$$\begin{aligned}
&\overline{Q}_1 = \overline{\overline{Q_1} + Q_1'} = \overline{\overline{Q_1}} + \overline{Q_1'}, \text{ and } \overline{Q_1'}, = 0\\
&\overline{\overline{Q_1}}\,\overline{\overline{Q_2}} = \overline{\overline{Q_1}\,\overline{Q_2}} = \overline{Q_1}\,\overline{Q_2}\\
&\overline{\overline{Q_1} Q_2'} = \overline{\overline{Q_1} Q_2'} = \overline{Q_1}\,\overline{Q_2'} = 0\\
&\overline{\overline{Q_2} Q_1'} = \overline{\overline{Q_2} Q_1'} = \overline{Q_2}\,\overline{Q_1'} = 0\\
&\overline{Q_1 Q_2} = \overline{Q_1}\,\overline{Q_2} + \overline{Q_1' Q_2'}.
\end{aligned} \tag{9.55}$$

Applying operators to a tensor valued function results in:

$$\overline{\nabla(\overline{Q}+Q')} = \nabla\overline{Q} + \nabla\overline{Q'} = \nabla\overline{Q}$$

$$\overline{\frac{D}{Dt}(\overline{Q}+Q')} = \frac{D\overline{Q}}{Dt} + \overline{V' \cdot \nabla Q'}.$$

9.3.1 Averaging Conservation Equations

In this section, we apply the averaging procedure (9.55) to the conservation equations of continuity, motion, mechanical energy, and thermal energy presented in Chap. 5.

9.3.1.1 Averaging the Continuity Equation

Averaging the continuity equation reads:

$$\frac{\partial\overline{\rho}}{\partial t} + \nabla.(\overline{\rho}\,\overline{\boldsymbol{V}} + \overline{\rho'\boldsymbol{V}'}) = 0 \tag{9.56}$$

with the index notation

$$\frac{\partial\overline{\rho}}{\partial t} + \frac{\partial}{\partial x_i}(\overline{\rho}\,\overline{V} + \overline{\rho' V'}) = 0. \tag{9.57}$$

9.3.1.2 Averaging the Navier–Stokes Equation

First, we decompose the velocity vector in the Navier–Stokes equation and find:

$$\frac{\partial\overline{\boldsymbol{V}}}{\partial t} + \frac{\partial\boldsymbol{V}'}{\partial t} + \overline{\boldsymbol{V}}\cdot\nabla\overline{\boldsymbol{V}} + \overline{\boldsymbol{V}}\cdot\nabla\boldsymbol{V}' + \boldsymbol{V}'\cdot\nabla\overline{\boldsymbol{V}} + \boldsymbol{V}'\cdot\nabla\boldsymbol{V}' =$$
$$-\frac{1}{\rho}\nabla\overline{p} - \frac{1}{\rho}\nabla p' + \nu\Delta\overline{\boldsymbol{V}} + \nu\Delta\boldsymbol{V}' + \boldsymbol{g}. \tag{9.58}$$

then we apply the averaging rules Eqs. (9.55)–(9.58) and arrive at the Reynolds averaged Navier Stokes equation:

$$\frac{\partial\overline{\boldsymbol{V}}}{\partial t} + \overline{\boldsymbol{V}}\cdot\nabla\overline{\boldsymbol{V}} + \nabla\cdot\overline{(\boldsymbol{V}'\boldsymbol{V}')} = -\frac{1}{\rho}\nabla\overline{p} + \nu\Delta\overline{\boldsymbol{V}} + \boldsymbol{g}. \tag{9.59}$$

The index notation gives:

$$\frac{\partial\overline{V}_i}{\partial t} + \overline{V}_j\frac{\partial\overline{V}_i}{\partial x_j} = -\frac{1}{\rho}\frac{\partial\overline{p}}{\partial x_i} + \nu\frac{\partial^2\overline{V}_i}{\partial x_j\partial x_j} - \frac{\partial(\overline{V_i'V_j'})}{\partial x_j} + g_i. \tag{9.60}$$

9.3.1.3 Averaging the Mechanical Energy Equation

The mechanical energy equation for a turbulent flow is obtained using Eq. (4.71), which includes the compressibility term ($\nabla \cdot \boldsymbol{V} \neq 0$). Dividing the involved flow quantities into the mean and the fluctuating parts and applying the averaging procedure outlined in Sect. 9.2, results in a complex equation. To reduce the degree of complexity, we consider an incompressible flow with the mechanical energy equation given by Eq. (5.101) and also below:

$$\frac{D}{Dt}\left(\frac{V^2}{2}\right) = \nabla \cdot \left(-\frac{p}{\rho}\boldsymbol{V} + 2\nu \boldsymbol{V} \cdot \boldsymbol{D}\right) - 2\nu \boldsymbol{D} : \boldsymbol{D} + \boldsymbol{V} \cdot \boldsymbol{g}. \tag{9.61}$$

Using the identity $\boldsymbol{V} \cdot \nabla(V^2) = \nabla \cdot (\boldsymbol{V} V^2)$ for an incompressible flow and its index notation $V_i \frac{\partial}{\partial x_i}(V_j V_j) = \frac{\partial}{\partial x_i}(V_i V_j V_j)$, we find the index notation of Eq. (9.61):

$$\begin{aligned}\frac{\partial}{\partial t}\left(\frac{V_j V_j}{2}\right) &= -\frac{\partial}{\partial x_i} V_i \left(\frac{p}{\rho} + \frac{V_j V_j}{2}\right)\\ &+ \nu \frac{\partial}{\partial x_i} V_j \left(\frac{\partial V_i}{\partial x_j} + \frac{\partial V_j}{\partial x_i}\right) - \nu \left(\frac{\partial V_i}{\partial x_j} + \frac{\partial V_j}{\partial x_i}\right)\frac{\partial V_j}{\partial x_i} + V_i g_i.\end{aligned} \tag{9.62}$$

We introduce the following decompositions:

$$\begin{aligned}&\boldsymbol{V} = \overline{\boldsymbol{V}} + \boldsymbol{V}' \text{ with } V_i = \overline{V}_i + V_i',\\ &V^2 = V_i V_i = \overline{V}_i \overline{V}_i + 2\overline{V}_i V_i' + V_i' V_i'\\ &p = \overline{p} + p'\end{aligned} \tag{9.63}$$

and substitute the quantities in Eq. (9.61) by (9.63) and average the results, we find:

$$\begin{aligned}\frac{\partial}{\partial t}\left(\frac{\overline{V}_i \overline{V}_i}{2}\right) + \frac{\partial}{\partial t}\left(\frac{\overline{V_i' V_i'}}{2}\right) &= -\frac{\partial}{\partial x_i}\overline{V}_i\left(\frac{\overline{p}}{\rho} + \frac{\overline{V_j V_j}}{2}\right) + \nu \frac{\partial}{\partial x_i}\overline{V}_j\left(\frac{\partial \overline{V}_i}{\partial x_j} + \frac{\partial \overline{V}_j}{\partial x_i}\right)\\ &- \nu\left(\frac{\partial \overline{V}_i}{\partial x_j} + \frac{\partial \overline{V}_j}{\partial x_i}\right)\frac{\partial \overline{V}_j}{\partial x_i} - \frac{\partial}{\partial x_i}\overline{V_i'\left(\frac{p'}{\rho} + \frac{V_j V_j}{2}\right)}\\ &- \frac{\partial}{\partial x_i}(\overline{V}_j \overline{V_i' V_j'}) - \frac{\partial}{\partial x_i}\left(\overline{V}_i \frac{\overline{V_j' V_j'}}{2}\right)\\ &+ \nu \frac{\partial}{\partial x_i}\overline{V_j'\left(\frac{\partial V_i'}{\partial x_j} + \frac{\partial V_j'}{\partial x_i}\right)} - \nu\overline{\left(\frac{\partial V_i'}{\partial x_j} + \frac{\partial V_j'}{\partial x_i}\right)\frac{\partial V_j'}{\partial x_i}} + \overline{V}_i g_i.\end{aligned} \tag{9.64}$$

Equation (9.64) is the mechanical energy equation for turbulent flow, where the energy of the Reynolds stress tensor appears on the right-hand side. In deriving (9.64), we considered the gravitation force as the only field force with the component in the x_3-direction. If other field forces such as electromagnetic or electrostatic forces are present, they are added to (9.64) in the same way as the above gravitational work.

9.3.1.4 Averaging the Thermal Energy Equation

A thermal energy equation can be expressed in terms of specific internal energy u or specific static enthalpy h. In both cases, the specific internal energy and specific static enthalpy can be expressed in terms of temperature $\boldsymbol{u} = c_\nu T$ and $\boldsymbol{h} = c_p T$. Both forms are fully equivalent and one can be converted into the other by $\boldsymbol{h} = \boldsymbol{u} + p\nu$, which is the defining equation for the specific static enthalpy. For averaging the thermal energy equation in terms of specific static enthalpy which we replace by the static temperature, we resort to Eq. (5.125)

$$c_p \frac{DT}{Dt} = \frac{k}{\varrho}\nabla^2 T + \frac{1}{\varrho}\frac{Dp}{Dt} + \frac{1}{\rho}\boldsymbol{T} : \boldsymbol{D} \tag{9.65}$$

with the friction stress tensor $\boldsymbol{T}$ from Eq. (5.54):

$$\begin{aligned} &\boldsymbol{T} = \lambda(\nabla \cdot \boldsymbol{V})\boldsymbol{I} + 2\mu\boldsymbol{D} \text{ with} \\ &\lambda = \overline{\mu} - \frac{2}{3}\mu. \end{aligned} \tag{9.66}$$

Decomposing, in Eq. (9.65), the temperature and pressure T and p as well as the friction and deformation tensors $\boldsymbol{T}$ and $\boldsymbol{D}$ while neglecting the density fluctuation, we find:

$$\begin{aligned} c_p \frac{D(\overline{T} + T')}{Dt} &= \frac{k}{\varrho}\nabla^2(\overline{T} + T') + \frac{1}{\varrho}\frac{D(p + p')}{Dt} \\ &+ \frac{1}{\varrho}(\overline{\boldsymbol{T}} + \boldsymbol{T}') : (\overline{\boldsymbol{D}} + \boldsymbol{D}'). \end{aligned} \tag{9.67}$$

Averaging the entire Eq. (9.67) and considering the rule for averaging the spatial and substantial derivatives of tensor valued functions listed in Eq. (9.55), we arrive at:

$$c_p \left(\frac{D\overline{T}}{Dt} + \overline{\boldsymbol{V}' \cdot \nabla T'} \right) = \frac{k}{\varrho}\nabla^2 \overline{T} + \frac{1}{\varrho}\frac{D\overline{p}}{Dt} + \frac{1}{\varrho}(\overline{\boldsymbol{T}} : \overline{\boldsymbol{D}} + \overline{\boldsymbol{T}' : \boldsymbol{D}'}). \tag{9.68}$$

Further expansion of Eq. (9.68) and considering Eq. (9.66) leads to:

$$c_p\left(\frac{\partial \overline{T}}{\partial T}+\overline{V}\cdot\nabla\overline{T}+\overline{V'\cdot\nabla T'}\right)=\frac{k}{\varrho}\nabla^2\overline{T}+\frac{1}{\varrho}\frac{D\overline{p}}{Dt}+$$
$$+\frac{\lambda}{\varrho}(\nabla\cdot\overline{V})^2+2\nu\overline{D}:\overline{D}+\frac{\lambda}{\varrho}\overline{(\nabla\cdot V')^2}+2\nu\overline{D':D'}. \tag{9.69}$$

Following Eq. (5.103), the viscous dissipation of the mean flow :

$$\overline{\Phi}=\varrho\left(\frac{\lambda}{\varrho}(\nabla\cdot\overline{V})^2+2\nu\overline{D}:\overline{D}\right)\equiv\varrho\overline{\varepsilon}\text{ with }\overline{\varepsilon}=\frac{\lambda}{\varrho}(\nabla\cdot\overline{V})^2+2\nu\overline{D}:\overline{D}. \tag{9.70}$$

Correspondingly, the turbulent dissipation reads

$$\Phi_{\text{tur}}=\lambda\overline{(\nabla\cdot V')^2}+2\mu\overline{D'}:\overline{D'}\equiv\varrho\varepsilon$$
$$\text{with }\varepsilon=\frac{\lambda}{\varrho}\overline{(\nabla\cdot V')^2}+2\nu\overline{D':D'}. \tag{9.71}$$

Thus, the total dissipation reads

$$\Phi_{\text{total}}=\overline{\Phi}+\Phi_{\text{tur}}=\varrho(\overline{\varepsilon}+\varepsilon). \tag{9.72}$$

In Equations (9.70)–(9.72) $\overline{\varepsilon}$ and ε are the specific dissipation (dissipation per unit of mass) of the mean flow and the turbulent flow, respectively. The latter is also called the turbulent dissipation. The index notation of Eq. (9.71) reads:

$$\Phi_{\text{tur}}=\lambda\overline{\left(\frac{\partial V_i'}{\partial x_i}\right)^2}+\frac{2}{4}\mu\overline{\left(\frac{\partial V_i'}{\partial x_j}+\frac{\partial V_j'}{\partial x_i}\right)\left(\frac{\partial V_i'}{\partial x_j}+\frac{\partial V_j'}{\partial x_i}\right)} \tag{9.73}$$

and its expansion is:

$$\Phi_{\text{tur}}=\lambda\overline{\left(\frac{\partial V_1'}{\partial x_1}+\frac{\partial V_2'}{\partial x_2}+\frac{\partial V_3'}{\partial x_3}\right)^2}+2\mu\left[\overline{\left(\frac{\partial V_1'}{\partial x_1}\right)^2}+\overline{\left(\frac{\partial V_2'}{\partial x_2}\right)^2}+\overline{\left(\frac{\partial V_3'}{\partial x_3}\right)^2}\right]$$
$$+\mu\left[\overline{\left(\frac{\partial V_1'}{\partial x_2}+\frac{\partial V_2'}{\partial x_1}\right)^2}+\overline{\left(\frac{\partial V_1'}{\partial x_3}+\frac{\partial V_3'}{\partial x_1}\right)^2}+\overline{\left(\frac{\partial V_2'}{\partial x_3}+\frac{\partial V_3'}{\partial x_2}\right)^2}\right]. \tag{9.74}$$

Generally, the total dissipation expresses the conversion of mechanical energy into heat and causes the system to heat up. Comparing the specific dissipation of the mean flow $\overline{\varepsilon}$ with the turbulent flow ε shows that the order of magnitude of ε by far surmounts the one of $\overline{\varepsilon}$. The reason is that, in spite of the fact that $|V'| << \overline{V}$, the changes of the fluctuating velocity $|\partial V_i'/\partial x_j|$ is much larger than changes of the mean flow velocity $|\partial V_i'/\partial x_j| >> \partial\overline{V}_i/\partial x_j$. This circumstance is an inherent characteristic of all turbulent flows and allows the mean flow dissipation to drop in all turbulence equations. Thus, with Eqs. (9.69) and (9.70), Eq. (9.68) becomes:

$$c_p\left(\frac{\partial \overline{T}}{\partial t}+\overline{V}\cdot\nabla\overline{T}+\overline{V'\cdot\nabla T'}\right)=\frac{k}{\varrho}\nabla^2\overline{T}+\frac{1}{\varrho}\frac{D\overline{p}}{Dt}+\overline{\varepsilon}+\varepsilon. \tag{9.75}$$

9.3.1.5 Averaging the Total Enthalpy Equation

The quantities in total enthalpy Eq. (5.134) are given below,

$$\rho\left(\frac{\partial H}{\partial t}+\boldsymbol{V}\cdot\nabla H\right)=\kappa\nabla^2 T+\frac{\partial p}{\partial t}+\nabla\cdot(\boldsymbol{T}\cdot\boldsymbol{V})+\rho\boldsymbol{V}\cdot\boldsymbol{g} \tag{9.76}$$

with $\boldsymbol{T}$ as the temperature and $\boldsymbol{T}=\lambda(\nabla\cdot\boldsymbol{V})\boldsymbol{I}+2\mu\boldsymbol{D}$ as the friction stress tensor. Before further treating the total enthalpy equation, we re-arrange the term $\nabla\cdot(\boldsymbol{T}\cdot\boldsymbol{V})$ as:

$$\nabla\cdot(\boldsymbol{T}\cdot\boldsymbol{V})=(\nabla\cdot\boldsymbol{T})\cdot\boldsymbol{V}+\boldsymbol{T}:\nabla\boldsymbol{V}=(\nabla\cdot\boldsymbol{T})\cdot\boldsymbol{V}+\boldsymbol{T}:(\boldsymbol{\Omega}+\boldsymbol{D}). \tag{9.77}$$

Since the friction tensor $\boldsymbol{T}$ is symmetric and the rotation tensor Ω is antisymmetric, their double dot product identically disappears leading to

$$\nabla\cdot(\boldsymbol{T}\cdot\boldsymbol{V})=(\nabla\cdot\boldsymbol{T})\cdot\boldsymbol{V}+\boldsymbol{T}:\nabla\boldsymbol{V}=(\nabla\cdot\boldsymbol{T})\cdot\boldsymbol{V}+\boldsymbol{T}:\boldsymbol{D}. \tag{9.78}$$

The friction tensor $\boldsymbol{T}$ in Eqs. (9.76)–(9.78) includes $\nabla\cdot\boldsymbol{V}$, which is nonzero for compressible flows. In the context of turbulent flow treatment, its contribution is insignificant and brings only additional complexity to a topic which is complex anyway. Setting $\nabla\cdot\boldsymbol{V}=0$ reduces Eq. (9.78) to

$$\nabla\cdot(\boldsymbol{T}\cdot\boldsymbol{V})=2\mu\left((\nabla\cdot\boldsymbol{D})\cdot\boldsymbol{V}+\boldsymbol{D}:\boldsymbol{D}\right). \tag{9.79}$$

Implementing Eq. (9.79) into Eq. (9.76), we find:

$$\rho\frac{DH}{Dt}=\kappa\nabla^2 T+\frac{\partial p}{\partial t}+2\mu((\nabla\cdot\boldsymbol{D})\cdot\boldsymbol{V}+\boldsymbol{D}:\boldsymbol{D})+\rho\boldsymbol{V}\cdot\boldsymbol{g}. \tag{9.80}$$

Thus, Eq. (9.80) is identical with

$$\rho\left(\frac{\partial H}{\partial t}+\boldsymbol{V}\cdot\nabla H\right)=\kappa\nabla^2 T+\frac{\partial p}{\partial t}+2\mu\nabla\cdot(\boldsymbol{D}\cdot\boldsymbol{V})+\rho\boldsymbol{V}\cdot\boldsymbol{g}. \tag{9.81}$$

Decomposing the quantities in Eq. (9.81) gives:

$$\varrho\frac{\partial}{\partial t}\left(\overline{h}+h'+\frac{1}{2}(\overline{V}+V')^2\right)+\varrho(\overline{\boldsymbol{V}}+\boldsymbol{V}')\cdot\nabla\left(\overline{h}+h'+\frac{1}{2}(\overline{V}+V')^2\right)=$$
$$\kappa\nabla^2(\overline{T}+T')+\frac{\partial(\overline{p}+p')}{\partial t}+2\mu\nabla\cdot\left((\overline{\boldsymbol{D}}+\boldsymbol{D}')\cdot(\overline{\boldsymbol{V}}+\boldsymbol{V}')\right)+\rho(\overline{\boldsymbol{V}}+V')\cdot\boldsymbol{g}. \tag{9.82}$$

Comparing the order of magnitude of the fluctuation kinetic energy with the one of the mean flow shows that $V'^2/2 << V^2/2$. This is true even for flow situations with relatively high turbulence intensities of 10% and above. This order of magnitude comparison can directly be related to the square of turbulence intensity defined as $Tu = \sqrt{\overline{V'^2}}/V$. For a large turbulence intensity of $Tu = 10\%$, we obtain a ratio of $V'^2/V^2 = 0.01$. This comparison allows neglecting the fluctuation kinetic energy. After averaging Eq. (9.82), we find:

$$\varrho\left(\frac{\partial\overline{H}}{\partial t}+\overline{\boldsymbol{V}}\cdot\nabla\overline{H}\right)=\kappa\nabla^2\overline{T}-\varrho\overline{\boldsymbol{V}'\cdot\nabla h'}-\varrho\overline{\boldsymbol{V}'\cdot\nabla(\overline{V}\cdot V')}+$$
$$\frac{\partial\overline{p}}{\partial t}+2\mu(\nabla\cdot(\overline{\boldsymbol{D}}\cdot\overline{V})+\nabla\cdot\overline{(\boldsymbol{D}'\cdot V')}+\varrho\overline{\boldsymbol{V}}\cdot\boldsymbol{g}. \tag{9.83}$$

For steady case and neglecting the gravitational work, we obtain:

$$\varrho(\overline{\boldsymbol{V}}\cdot\nabla\overline{H})=\kappa\nabla^2\overline{T}-\varrho\overline{\boldsymbol{V}'\cdot\nabla h'}-\varrho\overline{\boldsymbol{V}'\cdot\nabla(\overline{V}\cdot V')}+$$
$$2\mu(\nabla\cdot(\overline{\boldsymbol{D}}\cdot\overline{V})+\nabla\cdot\overline{(\boldsymbol{D}'\cdot V')}). \tag{9.84}$$

In Eq. (9.84) we replace the averaged static temperature with static enthalpy, add and subtract the kinetic energy to introduce the total enthalpy, and the Prandtl number $Pr = \mu c_p/\kappa$. Furthermore, for the sake of practicability, we modify the third term in (9.84) by adding a zero $\nabla\cdot\boldsymbol{V}' = 0$:

$$\overline{\boldsymbol{V}'\cdot\nabla(\overline{V}\cdot V')}+\overline{(\nabla\cdot V')\overline{\boldsymbol{V}}\cdot\boldsymbol{V}'}=\nabla\cdot\lfloor\overline{(\overline{\boldsymbol{V}}\cdot\boldsymbol{V}')\boldsymbol{V}'}\rfloor \tag{9.85}$$

as a result, we find:

$$\varrho(\overline{\boldsymbol{V}}\cdot\nabla\overline{H})=\frac{\mu}{Pr}\nabla^2\overline{H}-\frac{\mu}{Pr}\nabla^2\left(\frac{1}{2}\overline{V}^2\right)-\varrho\overline{\boldsymbol{V}'\cdot\nabla h'}-\varrho\nabla\cdot[\overline{(\overline{\boldsymbol{V}}\cdot\boldsymbol{V}')\boldsymbol{V}'}]$$
$$+2\mu[\nabla\cdot(\overline{\boldsymbol{D}}\cdot\overline{V})+\nabla\cdot\overline{(\boldsymbol{D}'\cdot V')}]. \tag{9.86}$$

Equation (9.86) written in Cartesian index notation is:

$$
\begin{aligned}
\varrho\left(\overline{V}_i\frac{\partial\overline{H}}{\partial x_i}\right) = & \frac{\mu}{Pr}\frac{\partial^2\overline{H}}{\partial x_i\partial x_i} - \frac{\mu}{Pr}\left[\frac{\partial}{\partial x_i}\left(\overline{V}_m\frac{\partial\overline{V}_m}{\partial x_i}\right)\right] - \varrho\overline{V_i'\frac{\partial h'}{\partial x_i}} - \varrho\frac{\partial(\overline{V_m V_m' V_i'})}{\partial x_i}\\
& + \mu\left(\frac{\partial\overline{V}_i}{\partial x_j}\frac{\partial\overline{V}_j}{\partial x_i}\right) + \mu\left[\frac{\partial}{\partial x_i}\left(\overline{V}_m\frac{\partial\overline{V}_m}{\partial x_i}\right)\right]\\
& + \mu\left(\overline{\frac{\partial V_i'}{\partial x_j}\frac{\partial V_j'}{\partial x_i} + \frac{\partial^2 V_j'}{\partial x_i\partial x_i}V_j' + \frac{\partial V_j'}{\partial x_i}\frac{\partial V_j'}{\partial x_i}}\right).
\end{aligned}
\tag{9.87}
$$

Combining the second and the sixth term on the right hand side of Eq. (9.87) results in a more compact version:

$$
\begin{aligned}
\varrho\left(\overline{V}_i\frac{\partial\overline{H}}{\partial x_i}\right) = & \frac{\mu}{Pr}\frac{\partial^2\overline{H}}{\partial x_i\partial x_i} + \frac{\partial}{\partial x_i}\left[\mu\left(1-\frac{1}{Pr}\right)\left(\overline{V}_m\frac{\partial\overline{V}_m}{\partial x_i}\right)\right]\\
& - \varrho\overline{V_i'\frac{\partial h'}{\partial x_i}} - \varrho\frac{\partial(\overline{V_m V_m' V_i'})}{\partial x_i} + \mu\left(\frac{\partial\overline{V}_i}{\partial x_j}\frac{\partial\overline{V}_j}{\partial x_i}\right)\\
& + \mu\left(\overline{\frac{\partial V_i'}{\partial x_j}\frac{\partial V_j'}{\partial x_i} + \frac{\partial^2 V_j'}{\partial x_i\partial x_i}V_j' + \frac{\partial V_j'}{\partial x_i}\frac{\partial V_j'}{\partial x_i}}\right).
\end{aligned}
\tag{9.88}
$$

To apply Eq. (9.88) to boundary layer problems it is more appropriate to deal with the correlation $\overline{\partial(V_i'h')/\partial x_i}$ rather than $\overline{(V_i'\partial h')/\partial x_i}$. This requires a modification of (9.88) by introducing the following identity for incompressible flows

$$
\begin{aligned}
\varrho\overline{\boldsymbol{V}'\cdot\nabla h'} &= \varrho\nabla\cdot(\overline{\boldsymbol{V}'h'}) - \varrho\overline{\nabla\cdot\boldsymbol{V}'\boldsymbol{h}'}\\
\varrho\overline{V_i'\frac{\partial h'}{\partial x_i}} &= \varrho\frac{\partial(\overline{V_i'h'})}{\partial x_i} + 0.
\end{aligned}
\tag{9.89}
$$

With Eqs. (9.89), (9.88) becomes

$$
\begin{aligned}
\varrho\left(\overline{V}_i\frac{\partial\overline{H}}{\partial x_i}\right) = & \frac{\mu}{Pr}\frac{\partial^2\overline{H}}{\partial x_i\partial x_i} + \frac{\partial}{\partial x_i}\left[\mu\left(1-\frac{1}{Pr}\right)\left(\overline{V}_m\frac{\partial\overline{V}_m}{\partial x_i}\right)\right]\\
& - \varrho\frac{\partial(\overline{V_i'h'})}{\partial x_i} - \varrho\frac{\partial(\overline{V_m V_m' V_i'})}{\partial x_i} + \mu\left(\frac{\partial\overline{V}_i}{\partial x_j}\frac{\partial\overline{V}_j}{\partial x_i}\right)\\
& + \mu\left(\overline{\frac{\partial V_i'}{\partial x_j}\frac{\partial V_j'}{\partial x_i} + \frac{\partial^2 V_j'}{\partial x_i\partial x_i}V_j' + \frac{\partial V_j'}{\partial x_i}\frac{\partial V_j'}{\partial x_i}}\right).
\end{aligned}
\tag{9.90}
$$

Equation (9.88) (or Eq. (9.90)) is the complete equation of the total enthalpy for steady incompressible three-dimensional flows. Summing over the range of indices, (9.88) can easily be expanded. The expanded version contains several terms that are insignificant for a two-dimensional flow and may be deleted altogether as shown in [2, Chap. 11], when dealing with the boundary layer theory.

9.3.1.6 Quantities Resulting from Averaging to Be Modeled

In addition to the viscous and turbulent dissipation terms, Eq. (9.86) includes a new correlation $-\overline{V_i'\frac{\partial h'}{\partial x_i}}$ and a transport term $\frac{\partial(\overline{V_m V_m' V_i'})}{\partial x_i}$ as a result of averaging the enthalpy equation. As a result of the averaging procedure, the Reynolds stress tensor $\overline{\boldsymbol{V}'\boldsymbol{V}'}$ was created in Eq. (9.59) with nine components from which six are distinct:

$$-\rho\overline{V_i'V_j'} = \begin{pmatrix} \overline{V_1'V_1'} & \overline{V_1'V_2'} & \overline{V_1'V_3'} \\ \overline{V_2'V_1'} & \overline{V_2'V_2'} & \overline{V_2'V_3'} \\ \overline{V_3'V_1'} & \overline{V_3'V_2'} & \overline{V_3'V_3'} \end{pmatrix} = \boldsymbol{T}' = \begin{pmatrix} \tau_{11}' & \tau_{12}' & \tau_{13}' \\ \tau_{21}' & \tau_{22}' & \tau_{23}' \\ \tau_{31}' & \tau_{32}' & \tau_{33}' \end{pmatrix} \quad (9.91)$$

with $\tau_{12}' = \tau_{21}'$, $\tau_{13}' = \tau_{31}'$ and $\tau_{23}' = \tau_{32}'$. Considering the molecular friction tensor for an incompressible fluid, the total friction tensor of a turbulent flow $\boldsymbol{T}_{\text{flow}}$ consists of the molecular friction stress tensor $\overline{\boldsymbol{T}}$ and the turbulent stress tensor $\boldsymbol{T}'$:

$$\boldsymbol{T}_{\text{total}} = \overline{\boldsymbol{T}} + \boldsymbol{T}' = 2\mu\overline{\boldsymbol{D}} - \rho\overline{\boldsymbol{V}'\boldsymbol{V}'}. \quad (9.92)$$

Experimental results show that close to a solid wall, the order of magnitude of the Reynolds stress is comparable with the molecular stress. In free turbulent flow cases such as wake flow, jet flow and jet boundary, where the flow is not affected by a solid wall, the order of magnitude of $\boldsymbol{T}'$ can be much higher than $\overline{\boldsymbol{T}}$ such that the latter can be neglected.

The elements of the tensor $\overline{\boldsymbol{V}'\boldsymbol{V}'}$ in Eqs. (9.59) or (9.60) have added six more unknowns to the Navier–Stokes equation (8.62). With three velocity components, the pressure and six Reynolds stress terms, we have totally ten unknowns with only four differential equations resulting from (9.59) together with the continuity equation. Additional unknowns such as $\overline{\boldsymbol{V}' \cdot \nabla H'}$ and static temperature $\overline{\boldsymbol{T}' \cdot \boldsymbol{V}'}$, are added to the system of differential equations for solving the energy equation. In order to find a solution, one has to provide additional equations that relate the Reynolds stress tensor (9.60) to the quantities of the main flow. Likewise, empirical correlations need to be found that relate $\overline{\boldsymbol{V}' \cdot \nabla H'}$ and $\overline{\boldsymbol{T}' \cdot \boldsymbol{V}'}$ to the quantities of the main flow. Such correlations can be constructed by mathematically manipulating the equations of motion and by establishing empirical models.

These additional equations are called closure equations. To obtain these equations, in the following we perform certain time consuming, yet mathematically simple operations to drive new equations from the already existing ones. As we will see, these new equations contain additional unknowns that need to be determined. It should be pointed out that these new equations do not have any new physical background and are just simple mathematical manipulations. The purpose of these manipulations is to find some empirical correlations to close our new system of equations. To easily follow the sequence of operations that generates the new equation, we introduce a new operator $\boldsymbol{N}(\boldsymbol{V})$, which we call the Navier–Stokes operator, where the velocity is assumed to be a function of time and space. This assumption is valid for statistically stationary/non-stationary, with a constant time dependent mean and stochastic fluctuations. Resorting to Eq. (9.58), we define

$$N(\boldsymbol{V}) = \frac{\partial \boldsymbol{V}}{\partial t} + \boldsymbol{V} \cdot \nabla \boldsymbol{V} + \frac{1}{\rho} \nabla p - \nu \Delta \boldsymbol{V} - \boldsymbol{g} = 0. \tag{9.93}$$

With $\boldsymbol{N}$ as the operator and $\boldsymbol{V}$ the tensor valued argument, upon which the operator acts and builds the Navier–Stokes equation. The argument may be a vector such as $\boldsymbol{V} = \overline{\boldsymbol{V}} + \boldsymbol{V}'$ or a component of a vector such as V_i. If the argument is the component V_i, then $\boldsymbol{N}(V_i)$ describes the ith component of the Navier–Stokes equation. In case the vector is decomposed into a mean and a fluctuation part, then the argument of the operator is replaced by $\boldsymbol{V} = \overline{\boldsymbol{V}} + \boldsymbol{V}'$ leading to $\boldsymbol{N}(\overline{\boldsymbol{V}} + \boldsymbol{V}')$. If the entire Navier–Stokes equation is averaged, we replace the operator argument by $\overline{(\overline{\boldsymbol{V}} + \boldsymbol{V}')}$. Before discussing different turbulence models, we present equations of turbulent kinetic energy and its dissipation as the two major closure equations. Similarly, we may write Eq. (9.93) in index form

$$\boldsymbol{N}(V_i) = \frac{\partial V_i}{\partial t} + V_j \frac{\partial V_i}{\partial x_j} + \frac{1}{\rho} \frac{\partial p}{\partial x_i} - \nu \frac{\partial^2 V_i}{\partial x_j \partial x_j} - g_i = 0. \tag{9.94}$$

Equation (9.94) describes the ith component of the Navier–Stokes equation. We may also obtain $\boldsymbol{N}(\overline{\boldsymbol{V}} + \boldsymbol{V}')$ and $\overline{\boldsymbol{N}(\overline{\boldsymbol{V}} + \boldsymbol{V}')}$. In the course of the following derivations, we encounter cases where second order tensors such as, $\overline{V_j \boldsymbol{N}(\overline{V}_i + V_i)}$, the jth derivative of the ith component such as, $\overline{\frac{\partial}{\partial x_j}(\boldsymbol{N}(\overline{V}_i + V_i))}$, or a second order tensor product such as, $\overline{\frac{\partial V_i'}{\partial x_j} \frac{\partial}{\partial x_j}(\boldsymbol{N}(\overline{V}_i + V_i))}$, are necessary to close the equation system.

9.3.2 Equation of Turbulence Kinetic Energy

To arrive at the equation of turbulence kinetic energy for an incompressible turbulent flow, we first subtract Eq. (9.59) from Eq. (9.58):

$$\frac{\partial \boldsymbol{V}'}{\partial t} + \overline{\boldsymbol{V}} \cdot \nabla \boldsymbol{V}' + \boldsymbol{V}' \cdot \nabla \overline{\boldsymbol{V}} + \boldsymbol{V}' \cdot \nabla \boldsymbol{V}' = -\frac{1}{\rho} \nabla p' + \nu \Delta \boldsymbol{V}' - \nabla \cdot \overline{(\boldsymbol{V}'\boldsymbol{V}')} \tag{9.95}$$

and scalarly multiply Eq. (9.95) with $\boldsymbol{V}'$:

$$V' \cdot \frac{\partial V'}{\partial t} + V' \cdot (\overline{V} \cdot \nabla V') + V' \cdot (V' \cdot \nabla \overline{V}) + V' \cdot (V' \cdot \nabla V') =$$
$$-\frac{1}{\rho} V' \cdot \nabla p' + \nu V' \cdot \Delta V' - V' \cdot \nabla \cdot \overline{(V'V')} \tag{9.96}$$

and rearrange the Reynolds stress tensor $\nabla \cdot \overline{(V'V')}$ in Eq. (9.96) by subtracting the continuity equation:

$$\nabla \cdot \overline{(V'V')} = \overline{V' \cdot \nabla V'} + \overline{V' \nabla \cdot V'} = \overline{V' \cdot \nabla V'}. \tag{9.97}$$

Inserting Eq. (9.97) into Eq. (9.96) results in:

$$V' \cdot \frac{\partial V'}{\partial t} + V' \cdot (\overline{V} \cdot \nabla V') + V' \cdot (V' \cdot \nabla \overline{V}) + V' \cdot (V' \cdot \nabla V') + V' \cdot \overline{V' \cdot \nabla V'} =$$
$$-\frac{1}{\rho} V' \cdot \nabla p' + \nu V' \cdot \Delta V'. \tag{9.98}$$

Now we average Eq. (9.98) by considering the following identities for the second term on the left-hand-side:

$$V' \cdot (\overline{V} \cdot \nabla V') = \overline{V} \cdot \nabla \left(\frac{V' \cdot V'}{2} \right). \tag{9.99}$$

Using the index notation, it can be shown that the third term on the left-hand-side of (9.98) is:

$$V' \cdot (V' \cdot \nabla \overline{V}) = (V'V') : \nabla \overline{V}. \tag{9.100}$$

Since the gradient of the mean velocity $\nabla \overline{V}$ is the sum of the deformation and rotation tensor $\nabla \overline{V} = \overline{D} + \overline{\Omega}$, Eq. (9.100) can be modified as:

$$V' \cdot (V' \cdot \nabla \overline{V}) = (V'V') : (\overline{D} + \overline{\Omega}). \tag{9.101}$$

Since the product $(V'V') : \overline{\Omega} = 0$, Eq. (9.101) can be modified as:

$$V' \cdot (V' \cdot \nabla \overline{V}) = (V'V') : \overline{D}. \tag{9.102}$$

Now we define the *turbulent kinetic energy* as:

$$k = \frac{1}{2} \overline{V' \cdot V'} = \frac{1}{2} \overline{V_i' V_i'} = \frac{1}{2} (\overline{V_1'^2} + \overline{V_2'^2} + \overline{V_3'^2}) = \frac{1}{2} \overline{V'^2} \tag{9.103}$$

and insert into Eq. (9.98) and average:

$$\frac{\partial k}{\partial t} + \overline{\overline{\boldsymbol{V}} \cdot \nabla k} + \overline{(\boldsymbol{V}'\boldsymbol{V}') : \overline{\boldsymbol{D}}} + \overline{\boldsymbol{V}' \cdot (\boldsymbol{V}' \cdot \nabla \boldsymbol{V}')} = -\frac{1}{\rho}\overline{\boldsymbol{V}' \cdot \nabla p'} + \nu\overline{\boldsymbol{V}' \cdot \Delta \boldsymbol{V}'}. \tag{9.104}$$

The forth term on the left-hand-side can be written as

$$\overline{\boldsymbol{V}' \cdot (\boldsymbol{V}' \cdot \nabla \boldsymbol{V}')} = \overline{\boldsymbol{V}' \cdot \nabla(\boldsymbol{V}' \cdot \boldsymbol{V}'/2)}. \tag{9.105}$$

Considering Eq. (9.105), the equation of turbulence kinetic energy (9.104) becomes:

$$\frac{\partial k}{\partial t} + \overline{\overline{\boldsymbol{V}} \cdot \nabla k} + \overline{(\boldsymbol{V}'\boldsymbol{V}') : \overline{\boldsymbol{D}}} + \overline{\boldsymbol{V}' \cdot \nabla k} = -\frac{1}{\rho}\overline{\boldsymbol{V}' \cdot \nabla p'} + \nu\overline{\boldsymbol{V}' \cdot \Delta \boldsymbol{V}'}. \tag{9.106}$$

A simple rearrangement of Eq. (9.106) yields:

$$\frac{\partial k}{\partial t} + \overline{\overline{\boldsymbol{V}} \cdot \nabla k} + \overline{(\boldsymbol{V}'\boldsymbol{V}') : \overline{\boldsymbol{D}}} = -\overline{\boldsymbol{V}' \cdot \left(\nabla k + \frac{\nabla p'}{\rho}\right)} + \nu\overline{\boldsymbol{V}' \cdot \Delta \boldsymbol{V}'}. \tag{9.107}$$

We add to the argument in the parenthesis on the right-hand side of Eq. (9.107) the following zeros:

$$\overline{k\nabla \cdot V'} = 0, \qquad \frac{\overline{p'\nabla \cdot V'}}{\rho} = 0 \tag{9.108}$$

and obtain:

$$\frac{\partial k}{\partial t} + \overline{\overline{\boldsymbol{V}} \cdot \nabla k} + \overline{(\boldsymbol{V}'\boldsymbol{V}') : \overline{\boldsymbol{D}}} = -\overline{V' \cdot \nabla k} + \overline{k\nabla \cdot \boldsymbol{V}'} + \frac{\overline{\boldsymbol{V}' \cdot \nabla p'}}{\rho} + \frac{\overline{p'\nabla \cdot V^i}}{\rho}$$
$$\nu\overline{\boldsymbol{V}' \cdot \Delta \boldsymbol{V}'}. \tag{9.109}$$

Rearranging the terms in the parentheses on the right-hand side of Eq. (9.107) results in the final equation of turbulence kinetic energy for incompressible flow in a coordinate invariant form:

$$\frac{\partial k}{\partial t} + \overline{\overline{\boldsymbol{V}} \cdot \nabla k} + \overline{(\boldsymbol{V}'\boldsymbol{V}') : \overline{\boldsymbol{D}}} = -\nabla \cdot \left(\overline{V'k} + \frac{\overline{\boldsymbol{V}'p'}}{\rho}\right) + \nu\overline{\boldsymbol{V}' \cdot \Delta \boldsymbol{V}'}. \tag{9.110}$$

The Cartesian index notation is:

$$\frac{\partial k}{\partial t} + \overline{V}_i\frac{\partial k}{\partial x_i} = -\overline{V_i'V_j'}\frac{\partial \overline{V}_i}{\partial x_j} - \frac{\partial}{\partial x_i}\left(\overline{V_i'\left(k + \frac{p'}{\rho}\right)}\right) + \nu\overline{V_i'\frac{\partial^2 V_i'}{\partial x_j \partial x_j}}. \tag{9.111}$$

Equation (9.110) with its index notation (9.111) is the balance equation of the turbulence kinetic energy per unit of mass. Before interpreting the individual terms in Eq. (9.111), we first modify the last term on the right-hand side. The modification is aimed at providing a detailed mathematical explanation that describes the dissipative nature of this term. We use the following identity

$$\nu\overline{\boldsymbol{V}' \cdot \Delta \boldsymbol{V}'} = 2\nu\{\overline{\nabla \cdot (\boldsymbol{V}' \cdot \boldsymbol{D}')} - \overline{\boldsymbol{D}' : \nabla \boldsymbol{V}'}\} \tag{9.112}$$

with $\boldsymbol{D}'$ as the deformation tensor of the turbulence fluctuation and its components $D'_{ij} = \frac{1}{2}\left(\frac{\partial V'_j}{\partial x_i} + \frac{\partial V'_i}{\partial x_j}\right)$. The first term on the right-hand side of Eq. (9.112) written in index notation:

$$2\nu\{\overline{\nabla \cdot (\boldsymbol{V}' \cdot \boldsymbol{D}'}\} = \nu \frac{\partial}{\partial x_i}\left\{\overline{V'_j\left(\frac{\partial V'_i}{\partial x_j} + \frac{\partial V'_j}{\partial x_i}\right)}\right\} \text{ differentiating gives:}$$

$$2\nu\{\overline{\nabla \cdot (\boldsymbol{V}' \cdot \boldsymbol{D}'}\} = \nu\overline{\frac{\partial V'_j}{\partial x_i}\left(\frac{\partial V'_i}{\partial x_j} + \frac{\partial V'_j}{\partial x_i}\right)} + \nu\overline{V'_j\frac{\partial^2 V'_j}{\partial x_i \partial x_i}} \tag{9.113}$$

with $\nu\overline{V'_j\frac{\partial^2 V'_i}{\partial x_j \partial x_j}} = \nu\overline{V'_j\frac{\partial}{\partial x_j}\left(\frac{\partial V'_i}{\partial x_i}\right)} = 0$ in Eq. (9.113) as a consequence of the incompressibility requirement. The second term on the right-hand side of Eq. (9.112) written in index notation is:

$$-2\nu\overline{\boldsymbol{D}' : \nabla \boldsymbol{V}'} = -\nu\overline{\left(\frac{\partial V'_i}{\partial x'_j} + \frac{\partial V'_j}{\partial x_i}\right)\frac{\partial V'_j}{\partial x_i}} \tag{9.114}$$

with the velocity gradient in that can be set as $\nabla \boldsymbol{V}' = \boldsymbol{D}' + \boldsymbol{\Omega}'$ and since $\boldsymbol{D}' : \boldsymbol{\Omega}' = 0$, Eq. (9.114) is rearranged as:

$$\varepsilon_c \equiv 2\nu\overline{\boldsymbol{D}' : \boldsymbol{D}'} = \nu\overline{\frac{\partial V'_i}{\partial x_j}\frac{\partial V'_j}{\partial x_i}} + \nu\overline{\frac{\partial V'_j}{\partial x_i}\frac{\partial V'_j}{\partial x_i}}. \tag{9.115}$$

Equation (9.115) exhibits the complete turbulence dissipation as found, among others, in Hinze [7] and Rotta [8]. The above definition of dissipation differs from the definition of the dissipation we will use in conjunction with the modeling, which is defined as

$$\varepsilon \equiv \nu\overline{\frac{\partial V'_j}{\partial x_i}\frac{\partial V'_j}{\partial x_i}}. \tag{9.116}$$

Thus, Eq. (9.116) is expressed in terms of Eq. (9.115) through

$$\varepsilon = \varepsilon_c - \nu\overline{\frac{\partial V'_i}{\partial x_j}\frac{\partial V'_j}{\partial x_i}}. \tag{9.117}$$

Bradshaw and Pitt [14] have shown that for cases with strong velocity gradients such as shock waves, the maximum difference $\Delta\varepsilon = \varepsilon_c - \varepsilon$ is about 2% and is for other flow situations negligibly small. Returning to Eq. (9.112), the sum of Eqs. (9.113) and (9.114) yields:

$$\nu \overline{\boldsymbol{V}' \cdot \Delta \boldsymbol{V}'} = \nu \frac{\partial}{\partial x_i} \overline{\left\{ V'_j \left(\frac{\partial V'_i}{\partial x_j} + \frac{\partial V'_j}{\partial x_i} \right) \right\}} - \nu \overline{\left(\frac{\partial V'_i}{\partial x_j} + \frac{\partial V'_j}{\partial x_i} \right) \frac{\partial V'_j}{\partial x_i}}. \tag{9.118}$$

Expressing the left-hand side of Eq. (9.118) in index notation, we get

$$\nu \overline{V'_i \frac{\partial^2 V'_i}{\partial x_j \partial x_j}} = \nu \frac{\partial}{\partial x_i} \overline{\left\{ V'_j \left(\frac{\partial V'_i}{\partial x_j} + \frac{\partial V'_j}{\partial x_i} \right) \right\}} - \nu \overline{\left(\frac{\partial V'_i}{\partial x_j} + \frac{\partial V'_j}{\partial x_i} \right) \frac{\partial V'_j}{\partial x_i}}. \tag{9.119}$$

We replace in Eq. (9.110) the last term on the right-hand side by Eq. (9.112) and obtain:

$$\frac{\partial k}{\partial t} + \overline{\boldsymbol{V}} \cdot \nabla k = -\overline{(\boldsymbol{V}'\boldsymbol{V}')} : \overline{\boldsymbol{D}} - \nabla \left(\overline{\boldsymbol{V}'k} + \frac{\overline{\boldsymbol{V}'p'}}{\rho} \right)$$
$$+ 2\nu \{ \overline{\nabla \cdot (\boldsymbol{V}' \cdot \boldsymbol{D}')} - \overline{\boldsymbol{D}' : \nabla \boldsymbol{V}'} \}. \tag{9.120}$$

The index notation of Eq. (9.120) is:

$$\frac{\partial k}{\partial t} + \overline{V}_i \frac{\partial k}{\partial x_i} = -\overline{V'_i V'_j} \frac{\partial \overline{V}_j}{\partial x_i} - \frac{\partial}{\partial x_i} \left(\overline{V'_i \left(k + \frac{p'}{\rho} \right)} \right) +$$
$$+ \nu \frac{\partial}{\partial x_i} \overline{\left\{ V'_j \left(\frac{\partial V'_i}{\partial x_j} + \frac{\partial V'_j}{\partial x_i} \right) \right\}} - \nu \overline{\left(\frac{\partial V'_i}{\partial x_j} + \frac{\partial V'_j}{\partial x_i} \right) \frac{\partial V'_j}{\partial x_i}}. \tag{9.121}$$

Equation (9.121) expresses the same physical content as Eq. (9.111), thus, it does not represent a new physical relationship that can be used to reduce the number of unknowns. Using some mathematical manipulations, we merely decomposed the last term of Eq. (9.111) to explicitly introduce the dissipation process into the turbulence kinetic energy balance.

Interpretation of individual terms in Eq. (9.121): The two terms on the lefthand side of Eqs. (9.120) and (9.121) describe the substantial change of the turbulence kinetic energy per unit of mass consisting of the local and convective changes of the kinetic energy. The first term on the right-hand side is the energy transferred from the mean flow through the turbulent shear stress. This term is also called the *production of turbulence energy*. This is explicitly expressed in terms of the double scalar product of the mean flow deformation tensor $\overline{\boldsymbol{D}}$ and the second order Reynolds stress tensor $\overline{\boldsymbol{V}'\boldsymbol{V}'}$. The second term is the spatial change of the work by the total pressure of the fluctuating motion. It exhibits the *convective diffusion by turbulence of the total turbulence energy*. The third term on the right-hand side is the spatial change of the specific work (work per unit mass) by the viscous shear stress of the turbulent motion. The last term expresses the *viscous dissipation by the turbulent motion*.

Introducing the dissipation: There are a variety of alternative forms for turbulence kinetic energy. The following alternative form is used for the purpose of turbulence modeling. We further re-arrange the first term on the right-hand side of Eq. (9.119):

$$\nu \overline{\boldsymbol{V}' \cdot \Delta \boldsymbol{V}'} = \nu \frac{\partial}{\partial x_i}\left(\overline{V'_j \frac{\partial V'_i}{\partial x_j}} + \overline{V'_j \frac{\partial V'_j}{\partial x_i}}\right) - \nu \overline{\left(\frac{\partial V'_i}{\partial x_j} + \frac{\partial V'_j}{\partial x_i}\right) \frac{\partial V'_j}{\partial x_i}}. \tag{9.122}$$

The second term within the first parentheses of Eq. (9.122) is the spatial change of the kinetic energy

$$\nu \overline{\boldsymbol{V}' \cdot \Delta \boldsymbol{V}'} = \nu \frac{\partial}{\partial x_i}\left(\overline{V'_j \frac{\partial V'_i}{\partial x_j}} + \frac{\partial k}{\partial x_i}\right) - \nu \overline{\left(\frac{\partial V'_i}{\partial x_j} + \frac{\partial V'_j}{\partial x_i}\right) \frac{\partial V'_j}{\partial x_i}}. \tag{9.123}$$

We differentiate the expression in the first parentheses of Eq. (9.123) with respect to x_i and obtain:

$$\begin{aligned}\nu \overline{\boldsymbol{V}' \cdot \Delta \boldsymbol{V}'} = {} & \nu \left\{ \overline{\frac{\partial V'_j}{\partial x_i} \frac{\partial V'_i}{\partial x_j}} + \overline{V'_j \frac{\partial}{\partial x_i}\left(\frac{\partial V'_i}{\partial x_j}\right)} + \frac{\partial^2 k}{\partial x_i \partial x_i} \right\} \\ & - \nu \left(\overline{\frac{\partial V'_i}{\partial x_j} \frac{\partial V'_j}{\partial x_i}} + \overline{\frac{\partial V'_j}{\partial x_i} \frac{\partial V'_j}{\partial x_i}}\right).\end{aligned} \tag{9.124}$$

Because of the continuity requirement for an incompressible flow, the second term in the first parenthesis of Eq. (9.124) identically vanishes. Moreover, the first terms within the first parenthesis and the second parenthesis cancel each other out reducing (9.124) to:

$$\nu \overline{\boldsymbol{V}' \cdot \Delta \boldsymbol{V}'} = \nu \overline{V'_i \frac{\partial^2 V'_i}{\partial x_j \partial x_j}} + \nu \frac{\partial^2 k}{\partial x_i \partial x_i} - \nu \left(\overline{\frac{\partial V'_i}{\partial x_j} \frac{\partial V'_i}{\partial x_j}}\right). \tag{9.125}$$

With Eqs. (9.125) and (9.116), Eq. (9.111) reads:

$$\frac{\partial k}{\partial t} + \overline{V}_i \frac{\partial k}{\partial x_i} + \frac{1}{2}\overline{V'_i V'_j}\left(\frac{\partial \overline{V}_i}{\partial x_j} + \frac{\partial \overline{V}_j}{\partial x_i}\right) = -\frac{\partial}{\partial x_i}\left(\overline{V'_i\left(k + \frac{p'}{\rho}\right)}\right) + \nu \frac{\partial^2 k}{\partial x_j \partial x_j} - \varepsilon. \tag{9.126}$$

Equation (9.126) establishes a relationship between the substantial change of the turbulence kinetic energy and its dissipation. In (9.126), ε can be replaced by the complete dissipation ε_c, leading to:

$$\frac{\partial k}{\partial t} + \overline{V}_i \frac{\partial k}{\partial x_i} + \frac{1}{2}\overline{V_i' V_j'}\left(\frac{\partial \overline{V}_i}{\partial x_j} + \frac{\partial \overline{V}_j}{\partial x_i}\right) = -\frac{\partial}{\partial x_i}\left(\overline{V_i'\left(k + \frac{p'}{\rho}\right)}\right)$$
$$+ \nu \frac{\partial^2 k}{\partial x_j \partial x_j} + \nu \overline{\frac{\partial V_i'}{\partial x_j}\frac{\partial V_j'}{\partial x_i}} - \varepsilon_c. \tag{9.127}$$

9.3.3 *Equation of Dissipation of Kinetic Energy*

As we will see later in turbulence modeling, besides the equations of continuity, motion, and energy, the equation of dissipation is also used. To arrive at this equation, we write Eq. (9.95) in index notation

$$\frac{\partial V_i'}{\partial t} + V_k' \frac{\partial \overline{V}_i}{\partial x_k} + \overline{V}_k \frac{\partial V_i'}{\partial x_k} + V_k' \frac{\partial V_i'}{\partial x_k} = -\frac{1}{\rho}\frac{\partial p'}{\partial x_i} + \nu \frac{\partial^2 V_i'}{\partial x_k \partial x_k} + \frac{\partial (\overline{V_i' V_k'})}{\partial x_k}. \tag{9.128}$$

We differentiate Eq. (9.128) with respect to x_j and scalarly multiply the result with $2\nu \frac{\partial V_i'}{\partial x_j}$. After averaging, we arrive at the following exact dissipation equation by Launder et al. [15]

$$\frac{\partial \varepsilon}{\partial t} + \overline{V}_k \frac{\partial \varepsilon}{\partial x_k} = -2\nu \left\{ \frac{\partial^2 \overline{V}_j}{\partial x_l \partial x_k}\left(\overline{\frac{V_k' \partial V_j'}{\partial x_l}}\right) + \frac{\partial \overline{V}_j}{\partial x_k}\left(\overline{\frac{\partial V_j'}{\partial x_l}\frac{\partial V_k'}{\partial x_l}} + \overline{\frac{\partial V_i'}{\partial x_j}\frac{\partial V_i'}{\partial x_k}}\right)\right.$$
$$+ \overline{\frac{\partial V_j'}{\partial x_k}\frac{\partial V_j'}{\partial x_l}\frac{\partial V_k'}{\partial x_l}} + \frac{1}{2}\frac{\partial}{\partial x_k}\left(\overline{V_k' \frac{\partial V_j'}{\partial x_l}\frac{\partial V_j'}{\partial x_l}}\right)$$
$$\left. + \frac{1}{\rho}\frac{\partial}{\partial x_k}\left(\overline{\frac{\partial V_k' \partial p'}{\partial x_l \partial x_l}}\right) + \nu \overline{\left(\frac{\partial^2 V_j}{\partial x_k \partial x_l}\right)^2}\right\} + \nu \frac{\partial^2 \varepsilon}{\partial x_k \partial x_k}. \tag{9.129}$$

Equation (9.129) exhibits an exact derivation of the dissipation equation and is more complicated than Eq. (9.126) for the turbulence kinetic energy. For modeling purposes, an empirical relation proposed by Launder and Spalding [16] is used as the standard model equation, which we present in the following section.

9.4 Turbulence Modeling

Equation (9.59) indicates that in order to obtain solutions for the Reynolds averaged Navier–Stokes equations (RANS), it is necessary to provide further information about the Reynolds stress Tensor $T' = \rho \overline{V'V'}$ which has generally nine components from which six are distinct. Many studies investigated the possibilities to establish a relationship between $T' = \rho \overline{V'V'}$ and the mean velocity field. This approach is called turbulence modeling. Tremendous amount of papers published in the last three decades show that none of the existing turbulence models can be universally applied to arbitrary types of turbulent flows. Very recent direct Navier–Stokes simulations (DNS) performed successfully for different flow situations exhibit a major breakthrough, making the turbulence modeling and its use superfluous. However, the computational effort to perform DNS makes it, at least for the time being, not attractive. This situation certainly will change in the near future. Until then, one has to work with several turbulence models, each of which is appropriate for certain types of flows. In the context of this Chapter, we intend to make readers familiar with the most representative models that are being used. To obtain a more detailed insight into the turbulence and its modeling, we refer to a list of almost all turbulence models that have been used in the past three decades presented in [2], Chap. 10. Of this multitude of turbulence models, only few have survived and these are the twoequation models, which we treat in the following section.

9.4.1 Two-Equation Models

Among the many two-equation models, three are the most established ones: (1) the standard $k - \varepsilon$ model, first introduced by Chou [17] and enhanced to its current form by Jones and Launder [18], (2) $k - \omega$ model first developed by Kolmogorov and enhanced to its current version by Wilcox [14] and (3) the shear stress transport (SST) model developed by Menter [19], who combined $\kappa - \varepsilon$ and $\kappa - \omega$ models by introducing a blending function with the objective to get the best out of these two models. All three models are built-in models of commercial codes that are used widely. In the following, we present these models and discuss their applicability.

9.4.1.1 Two-Equation *k*-ε Model

The two equations utilized by this model are the transport equations of kinetic energy k and the transport equation for dissipation ε. These equations are used to determine the turbulent kinematic viscosity ν_t. For fully developed high Reynolds number turbulence, the exact transport equations for k (9.126) can be used. The transport equation for ε (9.129) includes triple correlations that are almost impossible to measure. Therefore, relative to ε, we have to replace it with a relationship that approximately resembles the terms in (9.129). To establish such a purely empirical relationship, dimensional analysis is heavily used. Launder and Spalding [16] used the following equations for kinetic energy

$$\frac{\mathrm{D}k}{\mathrm{D}t} = \frac{1}{\rho}\frac{\partial}{\partial x_j}\left(\frac{\mu_t}{\sigma_k}\frac{\partial k}{\partial x_j}\right) + \frac{\mu_t}{\rho}\left(\frac{\partial \overline{V}_i}{\partial x_j} + \frac{\partial \overline{V}_j}{\partial x_i}\right)\frac{\partial \overline{V}_i}{\partial x_j} - \varepsilon \tag{9.130}$$

and for dissipation

$$\frac{\mathrm{D}\varepsilon}{\mathrm{D}t} = \frac{1}{\rho}\frac{\partial}{\partial x_j}\left(\frac{\mu_t}{\sigma_\varepsilon}\frac{\partial \varepsilon}{\partial x_j}\right) + C_{\varepsilon 1}\frac{\mu_t}{\rho}\frac{\varepsilon}{k}\left(\frac{\partial \overline{V}_i}{\partial x_j} + \frac{\partial \overline{V}_j}{\partial x_i}\right)\frac{\partial \overline{V}_i}{\partial x_j} - \frac{C_{\varepsilon 2}\varepsilon^2}{k}, \tag{9.131}$$

and the turbulent viscosity, μ_t, can be expressed as

$$\mu_t = \nu_t \rho = \frac{C_\mu \rho k^2}{\varepsilon}. \tag{9.132}$$

The constants $\sigma_k, \sigma_\varepsilon, C_{\varepsilon 1}, C_{\varepsilon 2}, c_\mu$ listed in the following table are calibration coefficients that are obtained from simple flow configurations such as grid turbulence. The models are applied to such flows and the coefficients are determined to make the model simulate the experimental behavior. The values of the above constants recommended by Launder and Spalding [11] are given in the following Table 9.1. As seen, the simplified Eqs. (9.131) and (9.132) do not contain the molecular viscosity. They may be applied to free turbulence cases where the molecular viscosity is negligibly small compared to the turbulence viscosity. However, one cannot expect to obtain reasonable results by simulation of the wall turbulence using these equations. This deficiency is corrected by introducing the *standard* $k - \varepsilon$ *model*. This model uses the wall functions where the velocity at the wall is related to the wall shear stress by the logarithmic law of the wall. Jones and Launder [26] extended the original $k - \varepsilon$ model to the low Reynolds number form, which allows calculations right up to a solid wall. In the recent three decades, there have been many two-equation models, some of which Wilcox has listed in his book [14]. In general, the modified k- and ε-equations, setting $\nu = \mu/\rho$ and $\nu_t = \mu_t/\rho$, are expressed as

$$\frac{\mathrm{D}k}{\mathrm{D}t} = \frac{\partial}{\partial x_j}\left\{\left(\nu + \frac{\nu_t}{\sigma_k}\right)\frac{\partial k}{\partial x_j}\right\} - \overline{V_i' V_j'}\frac{\partial \overline{V}_i}{\partial x_j} - \varepsilon \tag{9.133}$$

$$\frac{\mathrm{D}\varepsilon}{\mathrm{D}t} = \frac{\partial}{\partial x_j}\left\{\left(\nu + \frac{\nu_t}{\sigma_\varepsilon}\right)\frac{\partial \varepsilon}{\partial x_j}\right\} - C_{\varepsilon 1}\frac{\varepsilon}{k}\overline{V_i' V_j'}\frac{\partial \overline{V}_i}{\partial x_j} - C_{\varepsilon 2}\frac{\varepsilon^2}{k}. \tag{9.134}$$

The closure coefficients are listed in Table 9.1. The Reynolds stress, $-\rho\overline{V_i' V_j'}$, can be expressed as

$$\nu_t = C_\mu f_\mu k^{1/2}\ell_t = \frac{C_\mu f_\mu k^2}{\varepsilon}, \tag{9.135}$$

where $\ell_t = k^{3/2}/\varepsilon$ is the eddy length scale.

Using the $k - \varepsilon$ model, successful simulations of a large variety of flow situations have been reported in a number of papers that deal with internal and external aerodynamics, where no or minor separation occurs. However, no satisfactory results are achieved whenever major separation is involved, indicating the lack of sensitivity to adverse pressure gradient. The model tends to significantly overpredict the shear-

Table 9.1 Closure Coefficients of Eq. 9.131

C_μ	σ_k	σ_g	$C_{\varepsilon 1}$	$C_{\varepsilon 2}$
0.09	1	1.3	1.44	1.92

stress levels and thereby delays (or completely prevents) separation. This exhibits a major shortcoming, which Rodi [20] attributes to the overprediction of the turbulent length-scale in the near wall region. Menter [21] pointed to another shortcoming of the $k - \varepsilon$ model which is associated with the numerical stiffness of the equations when integrated through the viscous sublayer.

9.4.1.2 Two-Equation k-ω Model

This model replaces the ε-equation with the ω-transport equation, first introduced by Kolmogorov. It combines the physical reasoning with dimensional analysis. Following the Kolmogorov hypotheses, two quantities, namely ε and κ, seem to play a central role in his turbulence research. Therefore, it seemed appropriate to establish a transport equation in terms of a variable that is associated with the smallest eddy and includes ε and κ. This might be a ratio such as $\omega = \varepsilon/\kappa$ or $\omega = \kappa/\varepsilon$. Kolmogorov postulated the following transport equation

$$\frac{\partial \omega}{\partial t} + \overline{V}_i \frac{\partial \omega}{\partial x_i} = -\beta \omega^2 + \frac{\partial}{\partial x_i}\left(\sigma \nu_t \frac{\partial \omega}{\partial x_i}\right) \tag{9.136}$$

with β and σ as the two new closure coefficients. As seen, unlike the k-equation, the right-hand-side of Eq. (9.136) does not include the production term. This equation has undergone several changes where different researchers tried to add additional terms. The most current form developed by Wilcox [22], reads

$$\frac{\partial \omega}{\partial t} + \overline{V}_i \frac{\partial \omega}{\partial x_i} = \alpha \frac{\omega}{\kappa} \tau_{ij} \frac{\partial \overline{V}_j}{\partial x_i} - \beta \omega^2 + \frac{\partial}{\partial x_i}\left(\left(\nu + \sigma \frac{k}{\omega}\right) \frac{\partial \omega}{\partial x_i}\right) + \frac{\sigma_d}{\omega} \frac{\partial k}{\partial x_i} \frac{\partial \omega}{\partial x_i} \tag{9.137}$$

with $\tau_{ij} = -\overline{V_i' V_j'}$ as the specific Reynolds stress tensor. Wilcox also modified the k-equation as

$$\frac{\mathrm{D}k}{\mathrm{D}t} = \frac{\partial}{\partial x_i}\left\{\left(\nu + \sigma^* \frac{\kappa}{\omega}\right) \frac{\partial k}{\partial x_i}\right\} - \overline{V_i' V_j'} \frac{\partial \overline{V}_i}{\partial x_j} - \beta^* k \omega. \tag{9.138}$$

He also introduced the kinematic turbulent viscosity

$$\nu_t = \frac{k}{\omega}, \text{ with } \tilde{\omega} = \max\left(\omega,\ C_{\text{lim}} \frac{\sqrt{2 D_{ij} D_{ij}}}{\beta^*}\right), \text{ and } C_{\text{lim}} = \frac{7}{8}. \tag{9.139}$$

With $\overline{D_{ij}} = \frac{1}{2}\left(\frac{\partial \overline{V}_i}{\partial x_j} + \frac{\partial \overline{V}_j}{\partial x_i}\right)$ as the matrix of the mean deformation tensor. Wilcox defined the following closure coefficients and auxiliary relations

$$\begin{aligned}
\alpha &= \frac{13}{25}, \beta = \beta_0 f_\beta, \\
\beta^* &= \frac{9}{100}, \ \sigma = \frac{1}{2}, \\
\sigma^* &= \frac{3}{5}, \ \sigma_{do} = \frac{1}{8} \\
\sigma_d &= 0, \ \text{if} \ \frac{\partial k}{\partial x_i}\frac{\partial \omega}{\partial x_i} \le 0 \\
\sigma_d &= \sigma_{do}, \ \text{if} \ \frac{\partial k}{\partial x_i}\frac{\partial \omega}{\partial x_i} \prec 0.
\end{aligned} \tag{9.140}$$

Furthermore,

$$\begin{aligned}
\beta_0 = 0.0708, \ f_\beta &= \frac{1 + 85\chi_\omega}{1 + 100\chi_\omega}, \ \chi_\omega \equiv \left| \frac{\Omega_{ij}\Omega_{jk}Ski}{(\beta^*\omega)^3} \right|, \\
\varepsilon &= \beta^*\omega\kappa, \ l_m = k^{1/2}.
\end{aligned} \tag{9.141}$$

The $k-\omega$ model performs significantly better under adverse pressure gradient conditions than the $k-\varepsilon$ model. Another strong point of the model is the simplicity of its formulation in the viscous sublayer. The model does not employ damping functions, and has straightforward Dirichlet boundary conditions. This leads to significant advantages in numerical stability, Menter [23]. A major shortcoming of the $k-\omega$ model is its strong dependency on freestream values. Menter investigated this problem in detail, and showed that the magnitude of the eddy-viscosity can be changed by more than 100% just by using different values for $\tilde{\omega}$.

9.4.1.3 Two-Equation *k*-ω-SST-Model

Considering the strength and the shortcomings of $\kappa-\varepsilon$ and $\kappa-\omega$ models briefly discussed in the previous two sections, Menter [23–25] introduced a *blending function* that combines the best of the two models. He modified the Wilcox $k-\omega$ model to account for the transport effects of the principal turbulent shear-stress. The resulting *SST*-model (Sear Stress Transport model) uses a $\kappa-\omega$ formulation in the inner parts of the boundary layer down to the wall through the viscous sublayer. Thus, the *SST*-$k-\omega$ model can be used as a low-Re turbulence model without any extra damping functions. The SST formulation also switches to a $k-\varepsilon$ mode in the free-stream and thereby avoids the common $k-\omega$ problem that the model is too sensitive to the turbulence free-stream boundary conditions and inlet free-stream turbulence properties. For the sake of completeness, we present the Menter's SST-model in terms of $\boldsymbol{T}$-equation with the blending function F_1

$$\frac{\partial(\rho\omega)}{\partial t} + \frac{\partial(\rho V_i \omega)}{\partial x_i} = \frac{\alpha}{\nu_t}\tilde{P}_k - \beta\rho k\omega^2 + \frac{\partial}{\partial x_i}\left[(\mu + \sigma_\omega \mu_t)\frac{\partial \omega}{\partial x_i}\right] + 2(1 - F_1)\rho\sigma_{\omega 2}\frac{1}{\omega}\frac{\partial k}{\partial x_i}\frac{\partial \omega}{\partial x_i} \tag{9.142}$$

and the turbulence Kinetic Energy

$$\frac{\partial(\rho k)}{\partial t} + \frac{\partial \rho V_i k)}{\partial x_i} = \tilde{P}_k - \beta^* \rho k\omega + \frac{\partial}{\partial x_i}\left[(\mu + \sigma_k \mu_t)\frac{\partial k}{\partial x_i}\right]. \tag{9.143}$$

The term $\tilde{P}_k$ in Eqs. (9.142) and (9.143) is a production limiter and is defined in Eq. (9.149). The blending function F_1 is determined from

$$F_1 = \tanh(arg_1^4) \tag{9.144}$$

with the argument arg_1

$$arg_1 = \min\left(\max\left(\frac{\sqrt{k}}{\beta^*\omega y}, \frac{500\nu}{y^2\omega}\right), \frac{4\rho\sigma_{\omega 2}k}{CD_{k\omega}y^2}\right) \tag{9.145}$$

$$CD_{k\omega} = \max\left(2\rho\sigma_{\omega 2}\frac{1}{\omega}\frac{\partial k}{\partial x_i}\frac{\partial \omega}{\partial x_i}, 10^{-10}\right) \tag{9.146}$$

and F_1 is equal to zero away from the surface (k-model), and switches over to one inside the boundary layer (k-model). The turbulent eddy viscosity is defined as follows:

$$\nu_t = \frac{a_1 k}{\max(a_1\omega, SF_2)} \tag{9.147}$$

with $S = \sqrt{D_{ij}D_{ij}}$ and F_2 as a second blending function defined by:

$$F_2 = \tanh\left\{\left[\max\left(\frac{2\sqrt{k}}{\beta^*\omega y}, \frac{500\nu}{y^2\omega}\right)\right]^2\right\}. \tag{9.148}$$

A production limiter in Eq. (9.149) is used in the SST model to prevent the build-up of turbulence in stagnation regions. It is defined as

$$\tilde{P}_k = \min(P_k, 10\beta^*\rho k\omega) \tag{9.149}$$

with

$$P_k = \mu_t \frac{\partial V_i}{\partial x_j}\left(\frac{\partial V_i}{\partial x_j} + \frac{\partial V_j}{\partial x_i}\right). \tag{9.150}$$

Fig. 9.9 **a** Simulation with $k - \varepsilon$ model

All constants are computed by a blend from the corresponding constants of the $k - \boldsymbol{g}$ and the $k - \boldsymbol{T}$ model via $\alpha = \alpha_1 F + \alpha_2(1 - F)$, etc. The constants for this model are

$$\beta^* = 0.09,\ \alpha_1 = 5/9,\ \beta_1 = 3/40,$$
$$\sigma_{k1} = 0.85,\ \sigma_{\omega 1} = 0.5,\ \alpha_2 = 0.44,$$
$$\beta_2 = 0.0828,\ \sigma_{k2} = 1,\ \sigma_{\omega 2} = 0.856.$$

According to [26], the above version of the $k - \varepsilon$ and $k - \omega$ equations, including constants listed above, is the most updated version.

9.4.1.4 Two Examples of Two-Equation Models

Internal Flow, Sudden Expansion: The following representative examples should illustrate the substantial differences between the two-equation models we presented above. The flow through a sudden expansion is appropriate for comparison purposes for two reasons: (1) It has a flow separation associated with a circulation zone and (2) it is very easy to obtain experimental data from this channel.

Standard $k - \varepsilon$ verses $k - \omega$: Figures 9.9 and 9.10 show flow simulations through a channel with a sudden expansion ratio of 2/1 using $k - \omega$ and standard $k - \varepsilon$ models. The purpose was to simulate the flow separation. The $k - \varepsilon$ simulation,

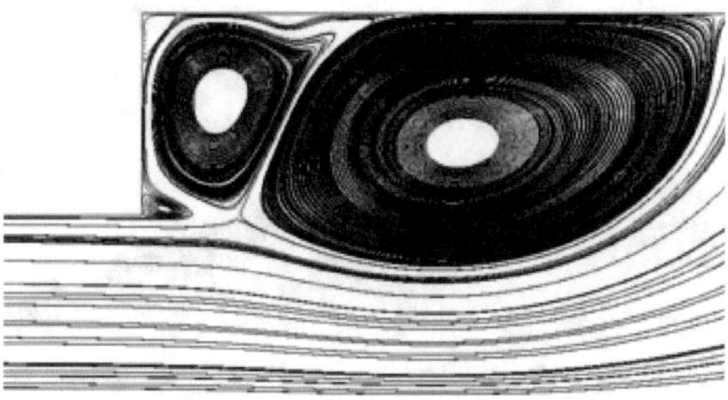

Fig. 9.10 Simulation with $k - \omega$ model

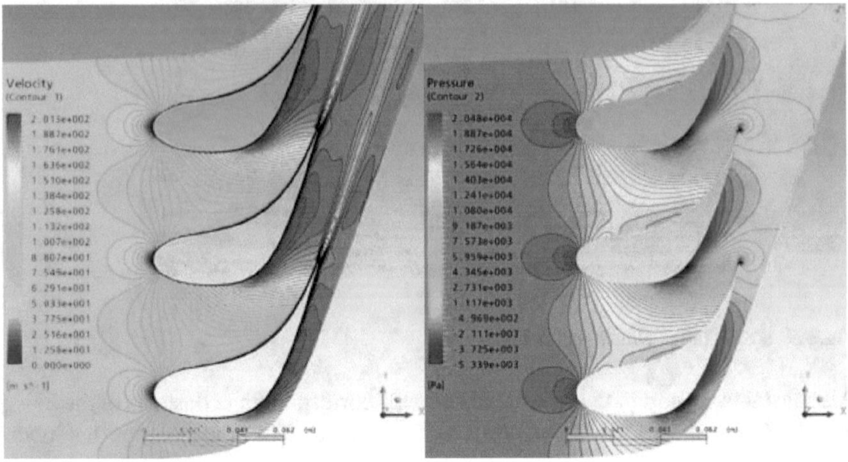

Fig. 9.11 Flow simulation through a turbine cascade, TPFL-Design

Fig. 9.9 delivers a single large corner vortex. However, experiments show that for this type of flow generally a system of two or more vortices, are present, Fig. 9.10.

Internal Flow, Turbine Cascade: Flow simulation with CFD has a wide application in engineering in general and in aerodynamic design of turbines, compressors, gas turbine inlet nozzles and exit diffusers in particular. As an example, Fig. 9.11 shows contour plots of velocity and pressure distributions in a high efficiency turbine blade using SST-turbulence model. On the convex surface (suction surface), the flow is initially accelerated at a slower rate from the leading edge and exits the channel at a higher velocity close to the trailing edge. The acceleration process is reflected in pressure contour.

External Flow, Lift-Drag Polar Diagram: This example presents two test cases to predict the lift-drag polar diagram of an aircraft without and with engine integration, Fig. 9.12.

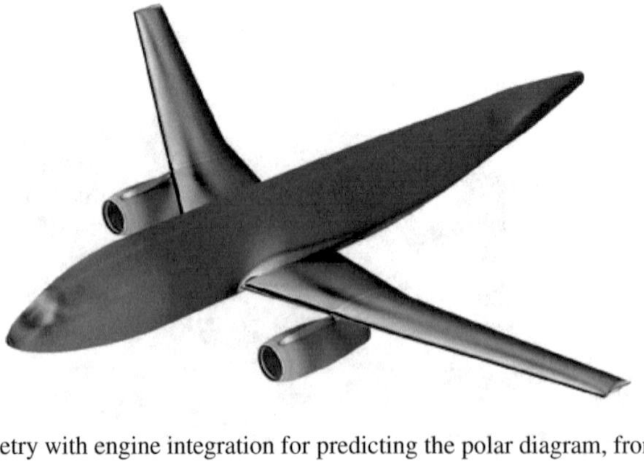

Fig. 9.12 Geometry with engine integration for predicting the polar diagram, from [32]

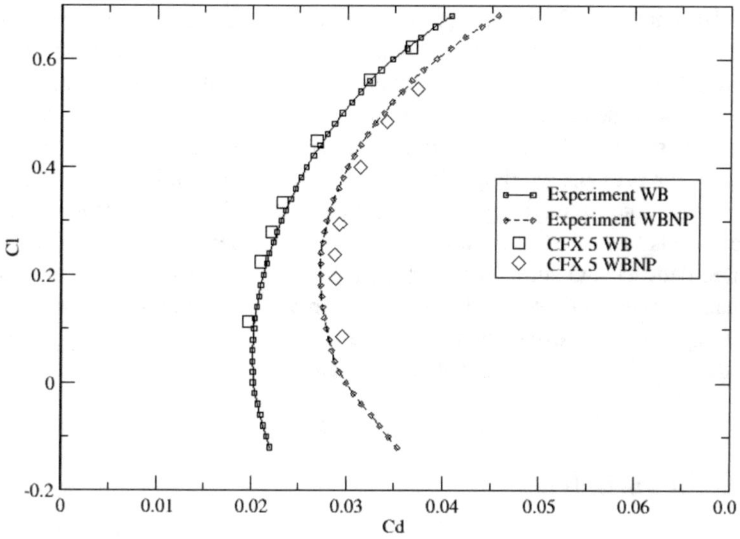

Fig. 9.13 Lift-Drag polar diagram for an aircraft model without engine (WB) and with engines (WVBN), from [31]

Figure 9.13 shows the predicted lift-drag *polar diagram* for the geometries presented in Fig. 9.12. The computation was performed using the SST-turbulence model and the results compared with the experiments. The lift and drag coefficients plotted in Fig. 9.13 are integral quantities that represents the lift and drag forces acting on the entire aircraft. Thus, they represent the lift and drag distribution integrated over the entire surface. The polar diagram is obtained by varying the angle of attack and measuring or computing the lift and drag forces. These forces are then non-dimensionalized with respect to a constant reference force, which is a product of a constant dynamic pressure and a characteristic area of the aircraft. Once a complete set of data for a given range of angle of attack is generated, then for each angle the lift coefficient is plotted against the drag coefficient as shown in Fig. 9.13.

Closing Remarks: The multitude of the closure constants in the above discussed turbulence models have been calibrated using different experimental data. Since the geometry, Re-number, Mach number, pressure gradient, boundary layer transition and many more flow parameters differ from case to case, the constants may require new calibrations. The question that arises is this: can any of the models discussed above a priori predict an arbitrary flow situation? The answer is a clear no. Because all turbulence models are of purely empirical nature with closure constants that are not universal and require adjustments whenever one deals with a completely new case. As we saw, in implementing the exact equations for k and ε that constitute the basis for $k - \varepsilon$ as well as $k - \omega$ model, major modifications had to be performed. Actually, in the case of ε-equations, the exact equation is surgically modified beyond recognition. Under this circumstance, none of the existing turbulence models can be regarded as universal. Considering this situation, however, satisfactory results can be obtained if the closure constants are calibrated for certain groups of flow situations. Following this procedure, numerous papers show quantitatively excellent results for groups of flow cases that have certain commonalities. Examples are flow cases at moderate pressure gradients and simple geometries. More complicated cases where the sign of the pressure gradient changes, flow separation and re-attachment occur and boundary layer transition plays a significant role still not adequately predicted.

The models presented above are just a few among many models published in the past three decades and summarized in [14]. In selecting these models, efforts have been made to present those that have been improved over the last three decades and have a longer lasting prospect of survival before the full implementation of DNS that makes the use of turbulence models unnecessary.

9.5 Grid Turbulence

Calibration of closure coefficients and a proper model assessment require accurate definition of boundary conditions for experiments as well as computation. These include, among other things, information about inlet turbulence such as the turbulence intensity, length, and time scales. This information can be provided by using *turbulence grids*, Fig. 9.14.

The grids may consist of an array of bars with cylindrical or quadratic cross sections. The thickness of the grid bars and the grid openings determine the turbulence intensity, length, and time scale of the flow downstream of the grid. Immediately downstream of the grid, a system of discrete wakes with vortex streets are generated that interact with each other. Their turbulence energy undergoes a continuous decay process leading to an almost homogeneous turbulence. The grid is positioned at a certain distance upstream of the test section in such a way that it generates homogeneous turbulence. The examples show.

how to achieve a defined inlet turbulence condition. Table 9.2 shows the data of three different turbulence grids for producing inlet turbulence intensities $Tu = 3.0\%, 8.0\%$, and 13.0%. The grids consist of square shaped aluminum rods with

Table 9.2 Turbulence grids: geometry, turbulence intensity and length scale

Turbulence grid	Grid opening gO (%)	Rod thickness RT (mm)	Turbulence intensity Tu (%)	Length scale 7 (mm)
No Grid	100	0	1.9	41.3
TG1	77	6.35	3.0	32.5
TG2	55	9.52	8.0	30.1
TG3	18	12.7	13.0	23.4

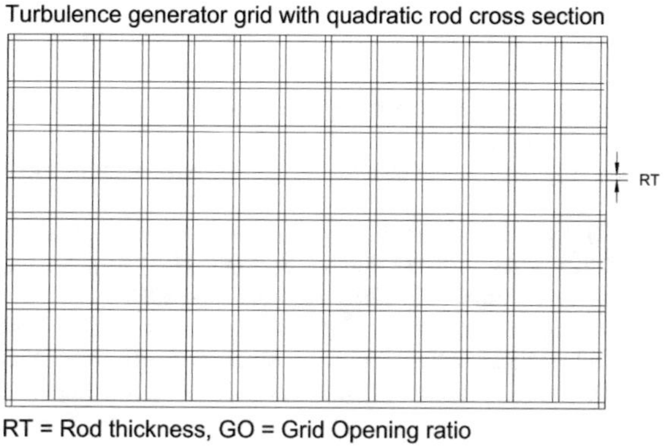

Fig. 9.14 Turbulence grid

the thickness RT and opening GO. The turbulence quantities were measured at the test section inlet with a distance of 130 mm from the grid. Figure 9.15a shows the power spectral density of the velocity signals from a hot wire sensor as a function of signal frequency. The length scale is calculated from $\Lambda = \overline{V} E_{(f=0)}/\nu_{rms}^2$ [mm], Fig. 9.15b.

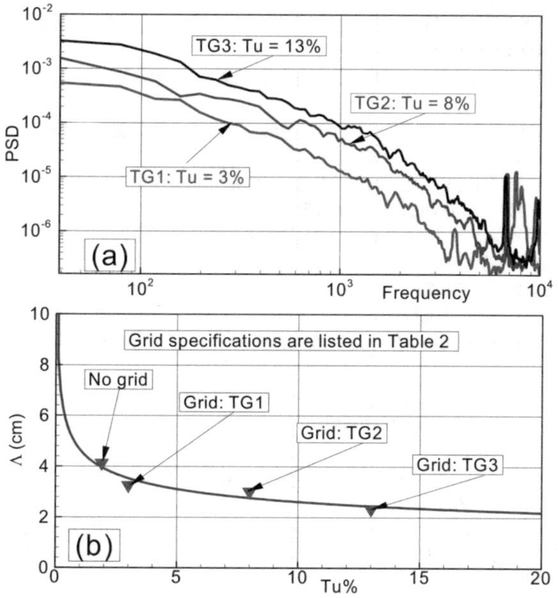

Fig. 9.15 **a** Power spectral distribution PSD as a function of frequency for three different grids described in Table 9.2. The results from **a** is used to generate the turbulence length scales as a function of turbulence intensity (**b**)

References

1. Meinhard S (2010) Fluid mechanics for engineers. Graduate textbook. Springer, New York, Berlin, Heidelberg. 978-6421193-6
2. Meinhard S (2014) Applied fluid mechanics. Graduate textbook. McGraw Hill. ISBN 978-0-07-180004-4
3. Meinhard S (2010) Turbomachinery flow physics and dynamic performance, Second and Enhanced Edition, 725 pages with 433 Figures. Springer, New York, Berlin, Heidelberg. 978-3-642-24675-3. library of congress 2012935425 published 2012
4. Meinhard S (2017) Gas turbine design. Components, system and off-design operation. 13: 978-3319583761, 509 PAGES with 327 Figures, Springer, Berlin, New York, Heidelberg
5. Taylor GI (1921) Diffusion by continuous movements. Proc. Lond. Math. Soc. Ser 2(20):196–211
6. Kármán Th, von (1937) Aeronaut. Sci. 4:137
7. JO H (1975) Turbulence, 2nd edn. McGraw Hill Book company, New York
8. Rotta JC (1972) Turbulente Strömungen. Teubner-Verlag Stuttgart, B.C
9. Richardson LF (1922) Weather prediction by numerical process. Cambridge University Press
10. Kolmogorov AN (1941) Local structure of turbulence in incompressible viscous fluid for very large reynolds number. Doklady Akademia Nauk, SSSR 30:299–303
11. Grant H, Stewart H, R. W. and Moilliet A (1962) Turbulence spectra from a tidal channel. J Fluid Mech 12:241
12. Onsager L (1945) Phys Rev 68:286
13. Weizsäcker CF (1948) 1948, Zeitschrift Physik, 124, 628, also proc. Roy Soc Lond 195A:402
14. Bradshaw P, Perot JB (1993) A note on turbulent energy dissipation in the viscous wall region. Phys Fluids A %:3305
15. Launder BE, Reece GI, Rodi W (1975) Progress in the development of reynolds-stress turbulent closure. J Fluid Mech 68:537–566
16. Launder BE, Spalding DB (1974) The numerical computation of turbulent flows. Comput Method Appl Mech Eng 3:269–289
17. Chou PY (1945) On the velocity correlations and the solution of the equation of turbulent fluctuations. Quart Appl Math 3:38
18. Jones WP, Launder BE (1972) The prediction of laminarization with a two-equation model of turbulence. Int J Heat and Mass Transf 15:301–314
19. Menter FR (1993) Zonal two-equation k-t turbulence models for aerodynamic flows. AIAA Technical Paper 93–2906
20. Rodi W, Scheurer G (1986) Scrutinizing the $k - \boldsymbol{g}$ model under adverse pressure gradient conditions. J Fluids Eng 108(1986):74–179
21. Menter FR (1992) Influence of freestream values on $k - \boldsymbol{T}$ turbulence model predictions. AIAA J 30(6):1992
22. Wicox D (2008) DCW industries, Inc., Private communications
23. Menter FR (1993) Zonal two equation $k - \boldsymbol{g}$ turbulence models for aerodynamic flows. AIAA Paper 93–2906
24. Menter FR (1994) Two-equation eddy-viscosity turbulence models for engineering applications. AIAA J 32:269–289
25. Menter FR, Kuntz M, Langtry R (2003) Ten years of experience with the SST turbulence model. In: Hanjalic K, Nagano Y, Tummers M (eds) Turbulence, heat and mass transfer, vol 4. Begell House Inc, pp 625–632
26. Menter FR (2008) CFX, Germany, Private communications

Chapter 10
Special Theory of Relativity

10.1 Introduction

Before publishing his General Theory of Relativity (GTR), Albert Einstein (1879–1950) published in 1905 a paper about the Special Theory of Relativity (STR) entitled "On the Electrodynamics of Moving Bodies" (for original German see [1][1]). This paper was about the inconsistency of Newtonian mechanics with Maxwell's equations of electromagnetism. Near the beginning of his career, Einstein thought that Newtonian mechanics was no longer enough to reconcile the laws of classical mechanics with the laws of the electromagnetic field. This led to the development of his special theory of relativity. This new theory, however, did not account for gravitational field. It took him more than ten years to complete his General Theory of Relativity that accounts for the gravitation and was published 1916, [2]. In 1920 Einstein published an article, [3], in which he explains his *thought experiments* (Gedankenexperimente) that has lead to the development of the mathematical framework for the Special and General Theory of Relativity. Both theories, STR and GTR, are subject of countless publications available through internet. Misner et al. give a comprehensive treatment of the subject in their book *Gravitation* [4]. The principal reference for the author of the current book is the original work by Einstein [1] through [3], which reflect Einstein's precise thinking and explaining the Special and particularly the General Theory formulated in his tensor equation. This equation is one of the most complex equation in theoretical physics. In this chapter we treat the STR followed by GTR in the next chapter.

Starting from an inertial frame of references that may move with zero or non-zero constant velocities (no acceleration), he formulated two postulates:

---

[1] At the end of this chapter the reader finds original sources in German as well as additional sources in English.

M. T. Schobeiri, *Tensor Analysis for Engineers and Physicists - With Application to Continuum Mechanics, Turbulence, and Einstein's Special and General Theory of Relativity*, https://doi.org/10.1007/978-3-030-35736-8_10

1. The laws of physics are invariant with respect to change of frame of reference.
2. The speed of light in a vacuum is the same for all observers, regardless of the motion of the light source.

In Newtonian Physics, the invariance of the laws of physics is based on Galilean transformation. In relativistic physics, however, the Galilean transformation cannot be used to verify the invariance of the physical laws with respect to change of frame of references. Furthermore, contrary to the Newtonian physics that is based on absolute time and absolute space that are defined separately, in STR and GTR space and time are interwoven into a single continuum known as spacetime. The crucial element for developing the STR and subsequently the GTR is the replacement of the Galilean transformations with the Lorentz transformations. The STR is applied to a spacial case where the curvature of spacetime due to gravity is negligible. In what follows, we first briefly present the Galilean transformation followed by Lorentz transformation. Using the Lorentz transformation, we will arrive at the equivalence of mass and energy.

10.2 Frames, Coordinate Systems, Lorenz, Transformation, Events

Galilean and Lorenz transformations take place between frame of references. A frame of reference is defined as a space occupied by objects with fixed mutual distances from each other. Frames may be stationary or moving. A laboratory with fixed walls is an example of a stationary frame. A moving train, a flying plane or a space ship are examples of moving frames. To identify the location of the objects within a reference frame, a coordinate system must be *embedded* in that frame. Thus, it is important to carefully distinguish between a frame of reference and a coordinate system. The coordinate system may be of any sort, Cartesian, cylindrical, spherical or generally curvilinear coordinate system.

The Galilean transformation is a transformation between inertial frames that move at a constant velocity relative to each other. In Galilean transformation the velocity of the moving reference frame is small compared to the speed of light. In Galilean transformation, the speed of light does not appear, while in Lorentz transformation [5], it plays a central role, whenever a transformation takes place between two frames of references. Thus, Lorenz transformation is a further development of Galilean transformation on which the Newtonian mechanics is based. The Newtonian mechanics is valid as long as the velocity of a moving material point at any time and in any place is negligibly small compared to the speed of light. As we will see in this chapter, for this particular case, the Lorenz transformation coincides with the Galilean.

10.2.1 Definition of an Event

Before starting with the Lorentz transformation and its consequences, we describe a term that appears frequently in Lorentz transformation namely the *event*. An event is a point in *spacetime* that occurs in a specific place at a specific time. Examples are found in daily life: Birth of a child that takes place in a particular hospital at a particular time is an event. A concert that takes place in a particular place of a particular city at a particular time is an event. These and many other examples can be defined in a spacetime coordinate that uniquely defines the location of the event and the time of its occurrence. Marking the birth of a child as the Event A with certain coordinates that uniquely identifies the location of that hospital and the time of birth, we can specify the event in a spacetime coordinate as:

$$E_A = f(t, x_1, x_2, x_3) \text{ or short } E_A = f(t, \boldsymbol{X}) \tag{10.1}$$

with t that specifies the moment at which the event occurs and the vector $\boldsymbol{X} = (x_1, x_2, x_3)$ as the spatial coordinates to describe the location of the event in Eq. (10.1). The event may take place in one direction for example in $x_1 \equiv x$ with x_2 and x_3 not being involved in the event.

10.2.2 Lorentz Transformation

This section deals with the transformation derived by Hendrik A. Lorentz (1853–1928), a Dutch theoretical physicist.

Considering two inertial frames, each of which has an observer on board, the Lorentz transformation connects the space and time coordinates measured by these observers. Assumed, an observer A sitting in a stationary inertial frame F_S that is assigned the position vector $x_1 \equiv x_S$ is passed by a moving inertial frame F_M that is occupied by the observer B and is moving with the velocity $V_1 = U_M$ in x_1-direction. At the time, when B arrives at the same location of A, both observers synchronize their clocks. At this time both observers are assigned the position $x_S = x_M = 0$ and time $t_S = t_M = 0$. After a certain time interval, the observer A sends a light signal (*event 1*), which hits a target (*event* 2). At this point both observers measure the time that elapsed between $t_S = t_M = 0$ and the time that the light signal has hit the target. Comparing their measured time, they found a difference. Considering the fact that both observers have synchronized their clocks, the reason for this difference lies in the fact that the Galilean transformation does not adequately reflect the transformation and another relation must be found to replace it. To find a replacement, we consider first the Galilean transformation. The position of the stationary frame is:

$$x_S = x_M + U_M t_S \tag{10.2}$$

with U_M as the *constant* velocity of the moving inertial frame F_M. From Eq. (10.2), it follows that

$$x_M = x_S - U_M t_M. \tag{10.3}$$

Note that the Galilei transformation is based on the assumption of an absolute time, which is the basis of the Newtonian physics. This implies that the time measured by an observer in a stationary frame is the same measured by an observer in a moving frame, provided that both observers have synchronized their clocks. This assumption leads to $t_M = t_S = t$. Using the transformation (10.3), the velocity as well as the acceleration in both frames can be calculated as:

$$W = \frac{dx_M}{dt} = \frac{d(x_S - U_M t)}{dt} = V_S - U_M$$

$$A = \frac{dW}{dt} = \frac{d(V_S - U_M)}{dt} = \frac{dV_S}{dt} = A_S \tag{10.4}$$

with W and A as the relative velocity and the relative acceleration of the observer within the moving frame. Equation (10.4) implies that with $A = A_S$ in Galilean transformation, the acceleration is invariant with respect to change of frames. Equations (10.2)–(10.4) are valid as long as the velocity of the moving frame is negligibly small compared with the speed of light $U_M << C$. In case that the order of magnitude comparison does not allow neglecting U_M, the times measured by both observers have to be modified. This time modification will result in different distance that must also be modified. To modify both, time and distance, Lorentz made the following Ansatz:

$$x_S = \gamma(x_M + U_M t_M) \tag{10.5}$$

and

$$x_M = \gamma(x_S - U_M t_S). \tag{10.6}$$

Because the speed of light is constant for all frame of references regardless of movement of the frame, it must hold that:

$$x_S = Ct_S,\ x_M = Ct_M. \tag{10.7}$$

Multiplying Eq. (10.5) with Eq. (10.6) under consideration of Eq. (10.7) we obtain:

$$x_S x_M = \gamma^2(x_S x_M + x_S U_M t_M - x_M U_M t_S - U_M^2 t_S t_M) \tag{10.8}$$

and considering Eq. (10.7) we obtain:

$$C^2 t_S t_M = \gamma^2(C^2 t_s t_M + C U_M t_S t_M - U_M^2 t_S t_M). \tag{10.9}$$

Canceling $t_S t_M$ from both sides of Eq. (10.9), we obtain the Lorentz correction factor as:

$$\gamma = \frac{1}{\sqrt{1 - \frac{U_M^2}{C^2}}} \tag{10.10}$$

with γ as the Lorentz factor and $C = 299,792$ km/s as the speed of light in vacuum. With this correction, relationships are found to adequately describe the distance:

$$x_M = \frac{x_S - U_M t_S}{\sqrt{1 - \frac{U_M^2}{C^2}}} \tag{10.11}$$

and by virtue of Eq. (10.7), the time measured by the moving observer:

$$t_M = \frac{t_S - \frac{U_M x_S}{C^2}}{\sqrt{1 - \frac{U_m^2}{C^2}}}. \tag{10.12}$$

In case that $U_M << C$ we have $\gamma = 1$, $t_M = t_S$ and the Lorenz transformation is reduced to Galilean transformation.

10.2.3 Consequences of Lorenz Transformation

Considering Eqs. (10.11) and (10.12), we establish the Lorentz transformation for two events. Event 1 has the stationary coordinates $x_S \equiv x_1$, $t_S \equiv t_1$, and $x_M \equiv x'_1$, $t_M \equiv t'_1$ and $U_M \equiv U$. Similarly, the event 2 has the coordinates $x_S \equiv x_2$, $t_S \equiv t_2$, and $x_M \equiv x'_2$, $t_M \equiv t'_2$. The indices *1* and *2* refer to event 1 and 2 respectively. Now we define the following differences:

$$\begin{aligned} \Delta x' &= x'_2 - x'_1,\ \Delta x = x_2 - x_1 \\ \Delta t' &= t'_2 - t'_1,\ \Delta t = t_2 - t_1. \end{aligned} \tag{10.13}$$

Implementing Eqs. (10.10) and (10.11) into Eq. (10.13) and considering Eq. (10.7), we obtain:

$$\begin{aligned} \Delta x' &= \gamma(\Delta x - U\Delta t) \qquad & \Delta x &= \gamma(\Delta x' + U\Delta t') \\ \Delta t' &= \gamma\left(\Delta t - \frac{U\Delta x}{c^2}\right) \qquad & \Delta t &= \gamma\left(\Delta t' + \frac{U\Delta x'}{c^2}\right). \end{aligned} \tag{10.14}$$

Equations (10.13) and (10.14) allow calculating the velocity of particles that move in frame 1, frame 2 and their velocities relative to each other.

$$V = \frac{\Delta x}{\Delta t} = \frac{x_2 - x_1}{\Delta t}$$

$$W = \frac{\Delta x'}{\Delta t'} = \frac{x'_2 - x'_1}{\Delta t'} = \frac{V - U}{1 - UV/C^2} \tag{10.15}$$

with V as the velocity measured by stationary observer, W the velocity measured by the observer in the moving frame that moves with the velocity U. Following the second Eq. of (10.15) we obtain:

$$V = \frac{W + U}{1 + UW/C^2}. \tag{10.16}$$

For $UW << C^2$, $UW/C^2 \approx 0$ we find the relationship between $\boldsymbol{U}$, $\boldsymbol{V}$, and $\boldsymbol{W}$ that is identical with the Galilean transformation:

$$\boldsymbol{V} = \boldsymbol{W} + \boldsymbol{U} \tag{10.17}$$

with V as the absolute velocity, W the relative velocity and U the frame velocity. For a rotating frame U is replaced by $U = R \times \omega$ with ω as the angular velocity of rotating frame. Equation (10.17) does not properly reflect the physics, whenever the magnitudes of the velocities U and W are not negligible compared with the magnitude of the speed of light.

A simple example should demonstrate the difference between the Lorenz and Galilean Transformation: Suppose the moving frame is a spaceship that travels with the velocity of $U = 0.8C$ and shoots a missile that travels with $W = 0.7C$. The Galilean transformation, Eq. (10.17), delivers a velocity of $V = 1.5C$. Lorenz transformation Eq. (10.16), however as shown below, delivers a result that is less than the speed of light.

$$V = \frac{0.8C + 0.7C}{1 + 0.8 \times 0.7} = \frac{1.5C}{1.56} < C. \tag{10.18}$$

10.3 Relativistic Length: Length Contraction

Given a stationary frame S and a moving frame S' that travels with the velocity U. In the moving frame S' we place a rod with the rest length L_0. Both end of the rod are measured simultaneously. This length is the *eigenlength* (also called proper length). An observer in frame S wants to measure the length of the rod placed in the moving frame S'. To analyze the experimental data the, we invoke Eq. (10.14)

$$\Delta x' = \gamma(\Delta x - U\Delta t) \tag{10.19}$$

in Eq. (10.19) $\Delta x' = L_0$ and $\Delta t = t_2 - t_1 = 0$ the length observed by the observer in frame S will be $\Delta x = L$ and Eq. (10.19) results in

$$L_0 = \gamma L = \frac{L}{\sqrt{1 - \frac{U^2}{C^2}}}. \tag{10.20}$$

According to Eq. (10.20), the length of the rod measured by the observer in stationary frame S is:

$$L = \frac{L_0}{\gamma} = \sqrt{1 - \frac{U^2}{C^2}} L_0. \tag{10.21}$$

Compared to the rest length L_0 the new length L appears shorter. This is the *length contraction*.

10.3.1 Relativistic Time: Time Dilation

Another consequence of Lorentz transformation is the time dilation. A clock placed in a moving frame measures the time elapsed between two events at the same location is $T_0 = t_2' - t_1'$. An observer in the stationary frame measures $T = t_2 - t_1$. These two time intervals are related by $\Delta t = t_2 - t_1$ in Eq. (10.14):

$$T = t_2 - t_1 = \frac{t_2' - t_1' + \frac{U \Delta x'}{C^2}}{\sqrt{1 - \frac{U^2}{C^2}}}. \tag{10.22}$$

Since the time interval $T_0 = t_2' - t_1'$ was measured at the same location, $\Delta x' = x_2' - x_1' = 0$, Eq. (10.22) is reduced to:

$$T = \frac{T_0}{\sqrt{1 - \frac{U^2}{C^2}}} = \gamma T_0. \tag{10.23}$$

Equation (10.23) means the time measured by an observer in stationary frame is running slower. As an example, the velocity of the moving frame assumed to be $U = 0.5C$, the time measured by the stationary observer is $T = 1.1547 T_0$.

10.3.2 Relativistic Mass: Mass Increase

Mass increase is another consequence of the Lorentz transformation. Given an object with the *rest mass* m_0, the Lorenz transformation provides a relationship between the velocity of the moving frame, the rest mass m_0, the speed of light and the *effective mass* m:

$$m = \frac{m_0}{\sqrt{1 - \frac{U^2}{C^2}}} = \gamma\, m_0. \tag{10.24}$$

Equation (10.24) means that if the velocity U approaches the speed of light, the effective mass approaches infinity. This limit in relativistic effective mass makes the speed of light c the speed limit. For a velocity of the moving frame of $U = 0.6\,C$, the effective mass becomes $m = 1.25\, m_0$. Having the infinity as the limit of the effective mass, the question that arises is why the photons are able to travel with the speed of light? The answer is the photons have zero rest mass $m_0 = 0$.

10.4 Einstein's Equivalence of Energy and Mass

Consider a light clock, Fig. 10.1 (left) placed in a stationary frame. In the same frame two mirrors are positioned at an arbitrary distance h from each other. Turning the light on at time $t = 0$ signifies the event E1. The light beam travels with speed of light C and after a time interval $\Delta t = t_0 - 0 = t_0$ reaches the above mirror. This signifies the event E2. Thus we are dealing with two events. Any subsequent back and forth reflections of the light between the two mirrors just constitute the repetition of E1 and E2 with the same time interval t_0.

Consider now the top mirror moving with the velocity U in positive x-direction as shown in Fig. 10.1. At the time $t = 0$ the light turns on and signifies the event E1. To reach the top mirror, a time interval from the triggering point of E1 until the light beams arrive at the top mirror, a time interval $\Delta t = t - 0 = t$ has elapsed. Once the light beam traveling with speed of light arrived at the top moving mirror, the event E2 is completed. During this time, however, the top mirror moving with the velocity U has reached a position in x-direction which is equal Ut_U. The spatial distance between the two events is calculated from:

$$C^2 t_U^2 = C^2 t_0^2 + U^2 t_U^2. \tag{10.25}$$

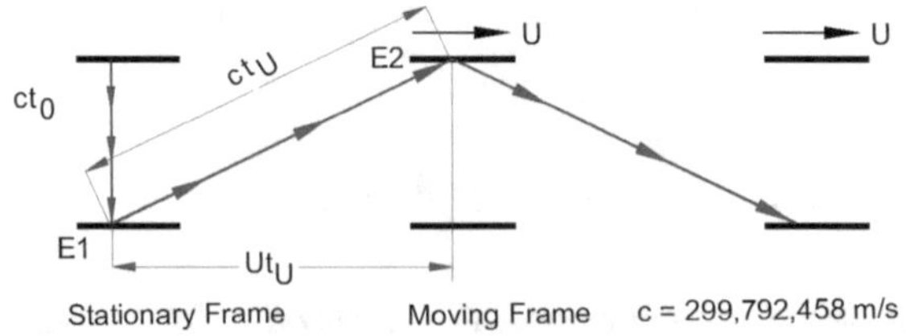

Fig. 10.1 Left two mirrors in a stationary frame, right two mirrors in a moving frame

From Eq. (10.25), the time t_U is calculated as:

$$t_U = \frac{t_0}{\sqrt{1 - \frac{U^2}{C^2}}} = \gamma\, t_0$$

$$\gamma = \frac{1}{\left(\sqrt{1 - \frac{U^2}{C^2}}\right)}. \tag{10.26}$$

Equation (10.26) states that if a moving object wants to move from point A to point B, the larger its velocity is, the longer it takes to arrive at point B. With other words, the moving object moves slower. With Eq. (10.26), Einstein arrived at the same result as Eq. (10.10) by Lorenz. Similar to time and distance the mass of a particle also will undergo a Lorenz transformation as specified below:

$$m_U = \frac{m_0}{\sqrt{1 - \frac{U^2}{C^2}}}$$

$$m_U = m_0 \left(1 - \frac{U^2}{C^2}\right)^{-1/2}. \tag{10.27}$$

In Equation (10.27) m_U is the mass of the moving particle and m_0 is its rest mass. Equation (10.27) has far-reaching consequences. It explicitly states that if a particle moves at higher velocity, its mass is getting heavier. For U approaching C, the denominator in (10.27) approaches zero and the mass of the moving particle approaches infinity.

We go back to Eq. (10.27) and expand the expression in the parentheses and neglect terms of secondary relevance:

$$m_U = m_0 \left(1 - \frac{U^2}{C^2}\right)^{-1/2} = m_0 \left(1 + \frac{1}{2}\frac{U^2}{C^2}\right). \tag{10.28}$$

Multiplying Eq. (10.28) with C^2 results in:

$$m_U C^2 = m_0 C^2 + \frac{1}{2} m_0 U^2. \tag{10.29}$$

The second term on the right hand side of Eq. (10.29) is the kinetic energy of the particle with the rest mass m_0 that moves with the velocity U. If the velocity $U << C$, then its kinetic energy compared to the first term on the right hand side and the term on the left hand side is extremely small. We may also set $U = 0$ and have the correct energy balance. As a result Eq. (10.29) can be written as

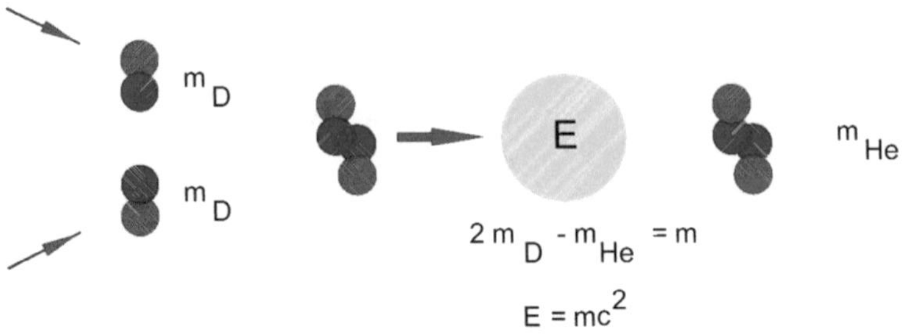

Fig. 10.2 Fusion of two Deuterium nuclei resulting in a He-nucleus. The mass difference is converted into thermo-nuclear energy

$$E = m_U C^2 = m_0 C^2. \tag{10.30}$$

Equation (10.30) is the result of special theory of relativity and states that the mass can be converted into energy and vice versa.

Thermo-nuclear Fusion: An example that confirms the physical validity of Eq. (10.30) is the thermo-nuclear fusion process that takes place on our sun and on billions other stars. The reason that stars can keep emitting thermal energy for millions of years is because of their continuous nuclear fusion as Fig. 10.2 shows. Two heavy hydrogen (Deuterium) nuclei, each containing one proton and one neutron come close enough to form a Helium nucleus. The difference in mass between the products and reactants is manifested as the release of large amounts of energy. Measuring the mass of two hydrogen nuclei and the mass of the product Helium, one finds a mass deficit:

$$2m_D - m_{He} = m. \tag{10.31}$$

This mass deficit constitutes the energy released as the result of the thermo-nuclear process.

$$E = mC^2. \tag{10.32}$$

Fission Process: Is another example of energy production as a result of mass-energy equivalence, Fig. 10.3.

Given a critical mass of enriched Uranium is hit by a neutron. The process of nuclear fission is triggered that releases several elements such as Barium, Krypton and others. The sum of the mass of all products shows a deficit:

$$(m_n + m_U) - (2m_n + m_{Ba} + m_{Kr}) = m. \tag{10.33}$$

This mass difference is converted into the energy.

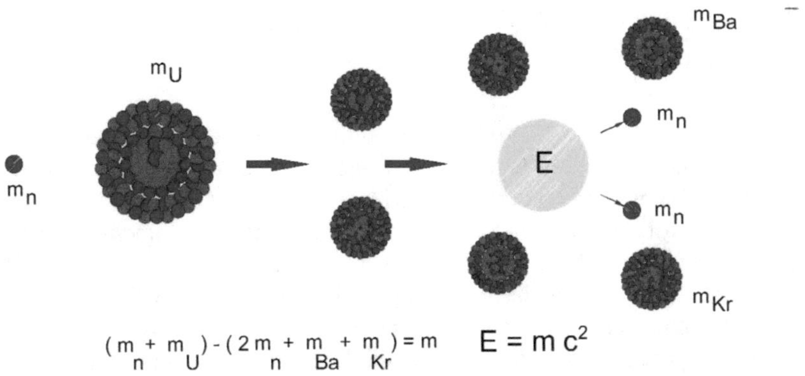

Fig. 10.3 Fission process: A mass of enriched Uranium is hit by a neutron that triggers a nuclear fission

10.5 Einstein's Four Vectors in Spacetime Coordinate

Before starting with the Einstein's theory of gravitation and his energy-stress tensor in four dimensional spacetime coordinate, we look into the vectors that make up the tensor. Each vector in the spacetime coordinate has four components following the Lorenz transformation. As seen in Chapter 2, a three-dimensional spatial vector and its magnitude is invariant with respect to a coordinate transformation. Equations (2.1) and (2.2) show their invariance with respect to coordinate transformation:

$$\boldsymbol{A} = e_i A_i = e'_j A'_j. \tag{10.34}$$

While the components change, with changing the coordinate system, the vector itself and its magnitude remains unchanged. Expressing Eq. (10.34) in components yields:

$$\boldsymbol{A} \cdot \boldsymbol{A} = A_i A_i = A'_i A'_i. \tag{10.35}$$

Renaming the scalar product in Eq. (10.35), we obtain:

$$\boldsymbol{A} \cdot \boldsymbol{A} = A_1^2 + A_2^2 + A_3^2 = A_1'^2 + A_2'^2 + A_3'^2. \tag{10.36}$$

Similarly the distance between two points is calculated as:

$$\Delta S^2 = \Delta X_1^2 + \Delta X_2^2 + \Delta X_3^2 = \Delta X_1'^2 + \Delta X_2'^2 + \Delta X_3'^2. \tag{10.37}$$

Equation (10.37) expresses the fact that the distance between two points in Euclidean space remains the same, regardless the rotation of the coordinate system. In this space, one is dealing with the geometric distances that can be measured.

10.5.1 Distance and Position Vectors in Spacetime Coordinate

In relativistic physics we are dealing with the spacetime frame, where any change of frame must follow the Lorentz transformation. This means that any point in the spacetime frame represents an event with a defined spatial and temporal location. For the sake of simplicity, in what follows, we assume a two dimensional spacetime coordinate with x and t, where two events A and B take place. The spacetime distance between these events are determined by the Lorentz transformation. To account for the time component, we introduce the product $c\Delta t$ with c as the speed of light and Δt the temporal distance from A to B. Similar to Eq. (10.37) one may be tempted to set:

$$c^2\Delta t^2 + \Delta x^2 = c^2\Delta t'^2 + \Delta x'^2. \tag{10.38}$$

Inserting for $\Delta t'$ and $\Delta x'$ the relations from Eq. (10.14) in conjunction with Eq. (10.26) we find that the ansatz in Eq. (10.38) fails. Trying a new ansatz:

$$c^2\Delta t^2 - \Delta x^2 = c^2\Delta t'^2 - \Delta x'^2 \equiv \Delta S^2 \tag{10.39}$$

and inserting for $\Delta t'$ and $\Delta x'$ the relations from Eq. (10.14) in conjunction with Eq. (10.26) as we did previously, we find the correct answer. This transformations preserve the magnitude of spacetime distance that we call spacetime metric. Invoking the other spatial components, Eq. (10.39) is re-written as:

$$(\Delta S)^2 = c^2\Delta t^2 - \Delta x^2 - \Delta y^2 - \Delta z^2. \tag{10.40}$$

Going back to the vector representation in spacetime coordinate, we obtain:

$$S^2 = c^2t^2 - \boldsymbol{X}\cdot\boldsymbol{X} \tag{10.41}$$

with $\boldsymbol{X} = e_i x_i$ in Euclidean system. To generalize (10.41) in spacetime coordinate we introduce the metric tensor

$$(g_{\mu\nu}) = \begin{pmatrix} 1 & 0 & 0 & 0 \\ 0 & -1 & 0 & 0 \\ 0 & 0 & -1 & 0 \\ 0 & 0 & 0 & -1 \end{pmatrix} \tag{10.42}$$

with the Greek indices $\mu, \nu = 0, 1, 2, 3$ for four dimensional spacetime coordinate as opposed to Latin indices $i, j = 1, 2, 3$ for Euclidean space. The Euclidean metric that corresponds to Eq. (10.42) is δ_{ij}. With Eqs. (10.41) and (10.42) we have:

$$S^2 = g_{\mu\nu}x_\mu x_\nu \tag{10.43}$$

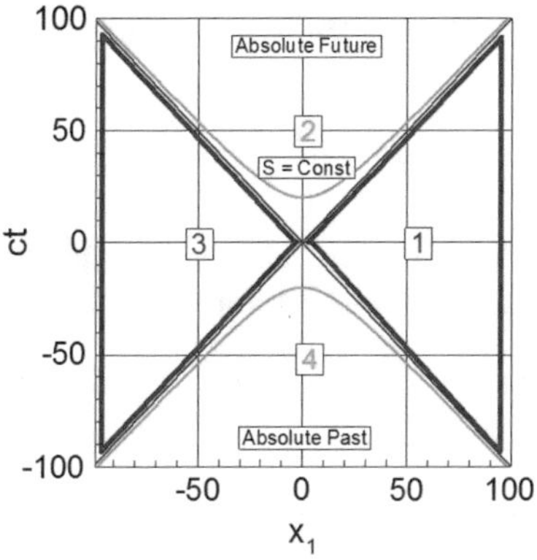

Fig. 10.4 The four quadrant in Minkowski's spacetime diagrams

with $\mu, \nu = 0, 1, 2, 3$. The four spacetime position vector is:

$$x_\mu = \begin{pmatrix} ct \\ x_1 \\ x_2 \\ x_3 \end{pmatrix} = \begin{pmatrix} x_0 \\ x_1 \\ x_2 \\ x_3 \end{pmatrix}. \tag{10.44}$$

In Equation (10.44) we have set $ct = x_0$ so we can sum over the index μ from 0 to 3. With Eqs. (10.41)–(10.44) we can now define the remaining vectors in the Einstein energy stress tensor.

For a two-dimensional spacetime, Eq. (10.41) is reduced to $S^2 = x_0^2 - x_1^2$ and interval S is plotted in Fig. 10.4.

Figure 10.4 illustrates schematically four quadrants in a two-dimensional spacetime diagram along with a spacetime interval S. The diagram was developed by H. Minkowski [6] in 1908. The diagram contains four quadrants. Quadrants 1 and 3, signify two zones within which the magnitude of the moving object has exceeded the speed of light of light. Since the speed of light is constant given by the two lines at 45 and 135 degree, no events can occur in these two quadrants. Events occurred in quadrant 2 and 4 signify what happens in the future and happened in the past.

10.5.2 Four Velocity and Momentum Vector in Spacetime Coordinate

In this section we use the original nomenclature by Einstein [2]. Consider an invariant infinitesimal length ds in spacetime coordinate, in conjunction with Eq. (10.41) we have:

$$dS = \sqrt{c^2 dt^2 - dx^2} = dt\sqrt{c^2 - dx^2} = c\,dt\sqrt{1 - u^2/c^2}. \tag{10.45}$$

At this point we introduce the *eigenzeit* translated in English as *proper time*:

$$d\tau = dt\sqrt{1 - \frac{u^2}{c^2}}. \tag{10.46}$$

Following Einstein [2], in the context of general relativity with the curvature of spacetime, we use the contravariant position vector x^μ, from which we obtain the velocity components:

$$V^\mu = \frac{dx^\mu}{d\tau}. \tag{10.47}$$

The product of the contravariant velocity components and the mass delivers the contravariant momentum components with τ as the proper time. With Eq. (10.47), we define the four momentum vector as:

$$\boldsymbol{P} = m\left(\frac{dx^0}{d\tau}, \frac{d\boldsymbol{X}}{d\tau}\right) = m\left(\frac{dx^0}{d\tau}, \frac{dx^1}{d\tau}, \frac{dx^2}{d\tau}, \frac{dx^3}{d\tau}\right) = m\left(\frac{cdt}{d\tau}, \frac{dx^1}{d\tau}, \frac{dx^2}{d\tau}, \frac{dx^3}{d\tau}\right) \tag{10.48}$$

the four momentum vector in spacetime coordinate. The components in the third parenthesis of Eq. (10.48) can be rearranged as follows:

$$p^\mu = mV^\mu = \left(\frac{E}{c}, p^i\right) = (\gamma mc, \gamma mV^i). \tag{10.49}$$

The expression mV^μ in Eq. (10.49) is the matrix of the energy and the three momentum components mV^i in Euclidean system. The totality of the four P-vectors build the components of a second order tensor $T^{\mu\nu}$ with the matrix:

$$(T^{\mu\nu}) = \begin{pmatrix} T^{00} & T^{01} & T^{02} & T^{03} \\ T^{10} & T^{11} & T^{12} & T^{13} \\ T^{20} & T^{21} & T^{22} & T^{23} \\ T^{30} & T^{31} & T^{32} & T^{33} \end{pmatrix} \tag{10.50}$$

with $\mu, \nu = 0, 1, 2, 3$. The physical quantities in each element $T^{\mu\nu}$ are per unit of volume. The convention for the superscripts μ, ν is as follows: The first superscript μ refers to the unit area, Fig. 10.5, where the physical quantity goes through. The second superscript ν refers to the direction normal to the above mentioned area. Considering Eqs. (10.48)–(10.50) the components of the tensor $T^{\mu\nu}$ can be redefined as $T^{\mu\nu} = \rho_0 V^\mu V^\nu$ with ρ_0 as the proper density and $\rho = \gamma^2\rho_0$. Considering $\mu = \nu = 0$, we have:

$$T^{00} = \rho_0\, V^0\, V^0 = \rho_0\left(\frac{d(ct)}{d\tau}\right)^2 = \gamma^2\rho_0 c^2 = \rho c^2. \tag{10.51}$$

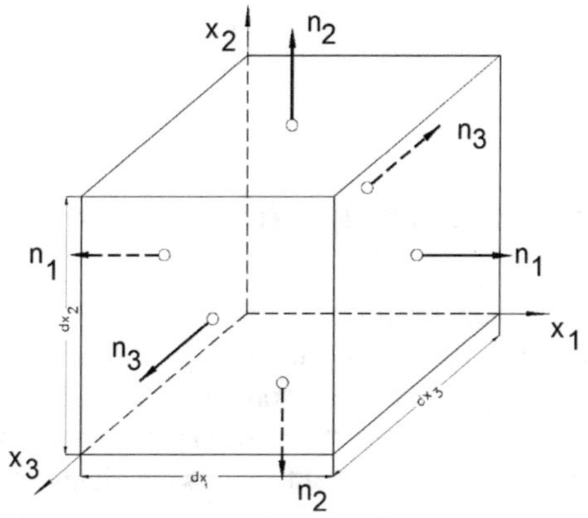

Fig. 10.5 Energy flux into and out of a unit volume. The unit vectors n_i refer to the inlet and outlet surfaces for energy flow

This is the relativistic energy density of the field occupied by the matter. Equation (10.51) is the energy per unit volume as explained above. The other components in first row are:

$$\begin{aligned} T^{01} &= \rho V^0 V^1 = \rho c V^1 \\ T^{02} &= \rho V^0 V^2 = \rho c V^2 \\ T^{03} &= \rho V^0 V^3 = \rho c V^3. \end{aligned} \tag{10.52}$$

The above components correspond to the momentum density in x^μ-direction. The components in first column correspond to the energy flow though a surface normal to a unit area per unit of time:

$$\begin{aligned} T^{10} &= \rho V^1 V^0 = \rho c V^1 \\ T^{20} &= \rho V^2 V^0 = \rho c V^2 \\ T^{30} &= \rho V^3 V^0 = \rho c V^3. \end{aligned} \tag{10.53}$$

Replacing all elements of Eq. (10.50) by Eqs. (10.52) and (10.53) and considering mV^μ from Eq. (10.49) we have:

$$(T^{\mu\nu}) = \rho \begin{pmatrix} c^2 & cV^1 & cV^2 & cV^3 \\ cV^1 & V^1 V^1 & V^1 V^2 & V^1 V^3 \\ cV^2 & V^2 V^1 & V^2 V^2 & V^2 V^3 \\ cV^3 & V^3 V^1 & V^3 V^2 & V^3 V^3 \end{pmatrix}. \tag{10.54}$$

Equations (10.52) and (10.53) contain the elements of the first row and the first column of the energy stress tensor $T^{\mu\nu}$ shown in Eq. (10.54). As seen, the tensor (10.54) is a symmetric second order tensor. This is the energy momentum tensor, which constitutes the right hand side of the Einstein tensor equation that will be presented in the following chapter.

10.6 Divergence of the Energy Stress Tensor

Since the tensor (10.54) includes mass, energy and momentum that are conserved quantities, the question arises first of all whether its divergence disappears. To present the proof, we take the divergence of the second order tensor $\boldsymbol{T}$ namely $T^{\mu\nu}_{,\nu}$. We start with the frame invariant $\boldsymbol{T}$ in a flat space where all Christoffel symbols vanish, $\Gamma^{n}_{ij} = 0$ and all covariant and contravariant vectors are identical and upper and lower indices are used indifferently. Under this assumption, Eq. (6.85) is reduced to:

$$\nabla \cdot \boldsymbol{T} = \boldsymbol{g}_{\mu}(T^{\nu\mu}_{,\nu}) \text{ with } \mu, \nu = 0, 1, 2, 3. \tag{10.55}$$

10.6.1 Divergence of Energy Component

We first investigate the divergence of the energy component. Since the energy is a scalar quantity and the divergence of a scalar quantity can be construct only, if we consider the energy flows into the surfaces of a volume normal to the direction of its movement and flows out from the opposite surfaces as shown in Fig. 10.5.

By doing so we implicitly allocate to the energy flow in and out of a unit volume an energy flux vector. This approach is analog to the continuity equation of classical fluid mechanics. To this end, we investigate the first component of $T^{\mu\nu}_{,\nu}$ and set $\mu = 0$ as the following steps shows:

$$T^{0\nu}_{,\nu} = T^{00}_{,0} + T^{01}_{,1} + T^{02}_{,2} + T^{03}_{,3}$$

$$T^{0\nu}_{,\nu} = \frac{\partial \rho}{\partial t} + \frac{\partial(\rho V^1)}{\partial x_1} + \frac{\partial(\rho V^2)}{\partial x_2} + \frac{\partial(\rho V^3)}{\partial x_3}. \tag{10.56}$$

The second equation in (10.56) is exactly the continuity equation of fluid mechanics [7], as shown below:

$$\begin{aligned}
&(a)\ \frac{\partial \rho}{\partial t} + \nabla \cdot (\rho \boldsymbol{V}) = 0\\
&(b)\ \frac{\partial \rho}{\partial t} + \frac{\partial(\rho V^i)}{\partial x_i} = 0\\
&(c)\ \frac{\partial \rho}{\partial t} + \frac{\partial(\rho V^1)}{\partial x_1} + \frac{\partial(\rho V^2)}{\partial x_2} + \frac{\partial(\rho V^3)}{\partial x_3} = 0.
\end{aligned} \tag{10.57}$$

Equation (10.57) shows the continuity equation in (a) coordinate invariant form,(b) its Cartesian index notation and (c) its expansion. This means that for the component $\mu = 0$, the conservation law is, from a physical point of view, fully satisfied.

10.6.2 Divergence of Momentum Component

Before investigating the momentum component on its divergence behavior in space-time coordinate, we have to proof first that it is divergence free in a simple flat space. For this purpose we resort to the integral analysis of classical fluid mechanics Eq. (5.11) in [6]:

$$\frac{D}{Dt}(m\boldsymbol{V}) = \boldsymbol{F_S} + \boldsymbol{F_G}. \tag{10.58}$$

The left hand side of Eq. (10.58) is the substantial change of the momentum whereas the right hand side includes the surface and gravitational forces. The conservation of momentum requires $\boldsymbol{F_S} + \boldsymbol{F_G} = 0$. The surface force $\boldsymbol{F_S}$ includes normal as well as shear stress forces, the latter translates into entropy change and is non-zero in real fluid that excludes the conservation of momentum (for details we refer to Eq. (4.56) in [7]). The condition for the momentum being conserved is that the right hand side of Eq. (10.58) has to disappear. An order of magnitude comparison may result in neglecting these two terms compared to the magnitude of $|m\boldsymbol{V}|$. Assuming that $\boldsymbol{F_S} + \boldsymbol{F_G} \approx 0$, it follows that:

$$\frac{D}{Dt}(m\boldsymbol{V}) = \int_{\nu(t)} \left(\frac{\partial(\rho \boldsymbol{V})}{\partial t} + \nabla \cdot (\rho \boldsymbol{V}\boldsymbol{V}) \right) d\nu = 0. \tag{10.59}$$

The integral (10.59) can vanish only if the integrand is zero. This means:

$$\frac{\partial(\rho \boldsymbol{V})}{\partial t} + \nabla \cdot (\rho \boldsymbol{V}\boldsymbol{V}) = 0. \tag{10.60}$$

The second term my be modified as detailed below.

$$\begin{aligned}
&(a)\ \frac{\partial(\rho \boldsymbol{V})}{\partial t} + \nabla \cdot (\rho \boldsymbol{V}\boldsymbol{V}) = 0\\
&(b)\ \frac{\partial(\rho \boldsymbol{V})}{\partial t} + \nabla \cdot (\rho g_\mu V^\mu \boldsymbol{V}) = 0\\
&(c)\ \frac{\partial(\rho \boldsymbol{V})}{\partial t} + \frac{\partial}{\partial x_\mu}(\rho V^\mu \boldsymbol{V}) = 0\\
&(d)\ \frac{\partial}{\partial t}(\rho \boldsymbol{V}) + \frac{\partial}{\partial x_1}(\rho V^1 \boldsymbol{V}) + \frac{\partial}{\partial x_2}(\rho V^2 \boldsymbol{V}) + \frac{\partial}{\partial x_3}(\rho V^3 \boldsymbol{V}) = 0
\end{aligned} \tag{10.61}$$

with $(\rho \boldsymbol{V} \boldsymbol{V})$ as a second order tensor, (a) as the invariant form, (b) modified version of the first vector, (c) index notation of first vector and (c) the full expanded version. After the divergence free proof through the last equation of (10.61) is presented for a flat space, its results can be transferred to spacetime coordinate. This however requires that we use the covariant derivative:

$$\nabla_\nu T^{\mu\nu} = T^{\mu\nu}_{,\nu} = \nabla_\nu (\rho_0 \, V^\mu V^\nu). \tag{10.62}$$

The tensor $\boldsymbol{T}$ constitutes the energy stress tensor of the right hand side equation of Einstein that we will discuss in detail in the following Chapter. Expanding Eq. (10.62) for spacetime coordinate, we find:

$$\nabla_\nu \boldsymbol{T}^{\mu\nu} = \boldsymbol{T}^{\mu\nu}_{,\nu} + \Gamma^\mu_{\lambda\nu} \boldsymbol{T}^{\lambda\nu} + \Gamma^\nu_{\lambda\nu} \boldsymbol{T}^{\mu\nu} = 0. \tag{10.63}$$

References

1. Einstein Albert (1905) Zur Elektrodynamik bewegter Körper. Annalen der Physik 322(10):891–921
2. Einstein A (1916) Die Grundlage der allgemeinen Relativitaetstheorie. Annalen der Physik. Vierte Folge Band 49
3. Einstein Albert (1920) Über die spezielle und allgemeine Relativitaetstheorie (gemeinverstaendlich), Fuenfte edn. Druck und Verlag von Friedr. Vieweg & Sohn in Braunschweig
4. Misner CW, Thorne KS, Wheeler J (1970) Gravitation. W. H. Freeman Company
5. Lorentz HA(1904) Electromagnetic phenomena in a system moving with any velocity smaller than that of light. In: Proceedings of the royal Netherlands academy of arts and sciences, vol 6
6. Minkowski H (1907–1908) Raum un Zeit. phisicalische Zeitschrift 10:75–88
7. Schobeiri M (2014) Engineering applied fluid mechanics. Graduate textbook. McGraw Hill. ISBN 978-0-07-180004-4

Chapter 11
Tensors in General Theory of Relativity

After completing the Special Theory of Relativity (STR) in 1905 [1], Albert Einstein developed the General Theory of Relativity (GTR) in 1915 [2]. The General Theory deals with the gravitation in a four-dimensional space-time frame. It has gone through several modifications and enhancements as published in [3] to arrive at the final form, which is the Einstein Field Equation as we know today. It is probably the most complex equation in theoretical physics. It reflects the theoretical framework in which Einstein structured his *thought experiments* (Gedankenexperimente). The reader can find a comprehensive explanation of GTR in [4] and in the Bibliography at the end of this chapter.

While the STR exclusively deals with relativistic kinematics in inertial frames, the GTR deals with the accelerated frames and gravity. As we saw in the previous chapter, the STR has led to famous equation of the equivalence of mass and energy. General theory of relativity generalizes STR and the Newton's law of universal gravitation, providing a unified description of gravity as a geometric property of space and time, or spacetime. The central point of his GTR is the curvature of spacetime and its direct relation to the mass, energy and momentum, where matter is present. The relation is specified by the Einstein field equations which is a system of partial differential equations. The left hand side of the field equation contains the Riemann-Ricci curvature tensor, the Ricci curvature scalar and the less important cosmological constant. The right hand side of the field equation contains a second order tensor with the spacetime components of mass, energy, momentum and stress. Because of the imminent importance of GTR to modern physics and the complex nature of the Einstein tensor, this chapter is considered an introduction into the GTR with special focus on understanding the mathematical background that has led to the Einstein tensor.

M. T. Schobeiri, *Tensor Analysis for Engineers and Physicists - With Application to Continuum Mechanics, Turbulence, and Einstein's Special and General Theory of Relativity*, https://doi.org/10.1007/978-3-030-35736-8_11

11.1 Operator Commutator

In the following sections we present features that are necessary for deriving the Einstein tensor. As the first step, we introduce the operator *Commutator* that is followed by the *Parallel Transport* leading to Riemann's tensor, which is a fourth order tensor.

In engineering and physics we have a number of operators, some of them we already have introduced in previous chapters. A new operator called *commutator* is used in conjunction with the parallel transport discussed in this chapter. The use of the commutator reduces the amount of work necessary to arrive at the Riemann tensor. The commutator is defined as:

$$[X, Y] = XY - YX. \tag{11.1}$$

The commutator (11.1) acting on an argument A results in:

$$[X, Y](A) = X(YA) - Y(XA). \tag{11.2}$$

The operator $[X, Y]$ acts on any argument that directly follows it. Starting with the first product on the right hand side of (11.1), namely XY, its second element Y multiplies with any argument such as A that might follow it. The first element X will act on the product consisting of Y and the argument A. Using the first product XY, an example should clarify this. Given the following operator:

$$[X, Y]V(x) = \left[\frac{\partial}{\partial x}, f(x)\right] V(x) \text{ show that: } [X, Y]V(x) = \frac{\partial f(x)}{\partial x} V(x) \tag{11.3}$$

with $X = \partial/\partial x$ as a differential operator and $Y = f(x)$ as an arbitrary tensor valued function. Assuming the argument is a vector $V(x)$, we first build a product consisting of the element Y and the argument which is $(f(x)V(x))$ and then let $X = \partial/\partial x$ acts on this product:

$$X(f(x)V(x)) = \frac{\partial}{\partial x}(f(x)V(x)). \tag{11.4}$$

Differentiating the product in the parentheses of Equation (11.4) using the chain rule gives:

$$X(YA) = \frac{\partial}{\partial x}(f(x)V(x)) = \frac{\partial f(x)}{\partial x} V(x) + f(x)\frac{\partial V(x)}{\partial x}. \tag{11.5}$$

Now we take the second product YX and apply the same rule as in Eq. (11.5):

$$YXA = f(x)\left(\frac{\partial}{\partial x}\right) V(x) = f(x)\frac{\partial V(x)}{\partial x}. \tag{11.6}$$

Finally, we build the difference of Equations (11.5)–(11.6) and obtain:

$$\left[\frac{\partial}{\partial x}, f(x)\right] V(x) - \left[f(x), \frac{\partial}{\partial x}\right] V(x) = \frac{\partial f(x)}{\partial x} V(x). \tag{11.7}$$

Which is identical with:

$$[X, Y]V(x) = [XY, YX]V(x) = \left[\frac{\partial}{\partial x}, f(x)\right] V(x) = \frac{\partial f(x)}{\partial x} V(x). \tag{11.8}$$

In the following we apply the commutator to a concrete case. As an example we set $X = \partial/\partial t$, $Y = e^t$ and the argument $A = \sin \omega t$. Before rearranging terms, we have:

$$[X, Y](A) = X(YA) - Y(XA) = \frac{\partial}{\partial t}(e^t \sin \omega t) - e^t \left(\frac{\partial \sin \omega t}{\partial t}\right). \tag{11.9}$$

After taking the differentiation of the first and the second term on the write hand side we arrive at the final result:

$$[X, Y](A) = (e^t \sin \omega t) + (\omega e^t \cos \omega t) - (\omega e^t \cos \omega t) = e^t \sin \omega t. \tag{11.10}$$

In the following sections we will apply the commutator to several operations.

11.2 Parallel Transport

The parallel transport is an effective way to determine whether a space is flat or curved. Take a sheet of paper representing a flat space and draw a closed curve and start at an arbitrary point on the curve. Draw a vector in any direction and a reference line segment that starts from the intersection of the vector and the line segment. Note the angle δ between this vector and the reference line segment. Starting from point A, move the vector along the closed curve by keeping the angle δ. This operation is called parallel transport. Moving counter clockwise along the entire closed curve, one arrives at the end point which is identical with the starting point at the same angle δ (Fig. 11.1).

Performing the parallel transport along a closed curve of a curved space for example a segment of a sphere, Fig. 11.2, the start and end points will have different angles. The change $\alpha = \lambda - \delta$ is an indication that the space is curved. In Fig. 11.2 we start from point A on the reference tangent plane and draw a vector from A tangent to the great circle arc AB. From A we continue to move tangent to the arc until we arrive at B. From here on we move continuously to point C, where the vectors lie in the corresponding tangent planes. Once arrived at C we move back to point A. As seen from Fig. 11.2, there is a substantial angle between the first and the last vector. In contrast performing the parallel transport along a triangle, results a difference of $\alpha = 0$.

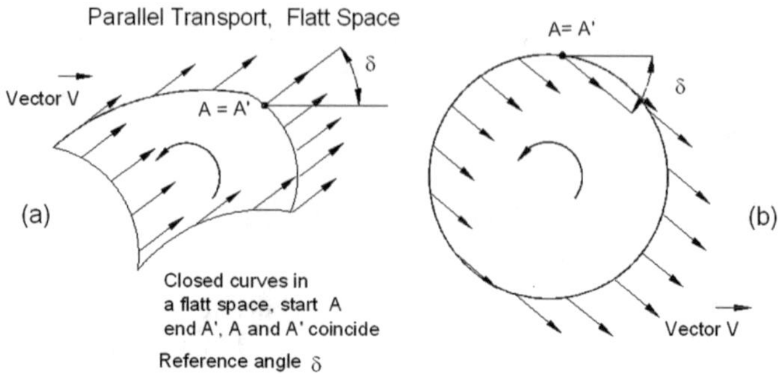

Fig. 11.1 Parallel transport along closed curves **a**, **b** in a flat space. This is shown in this Figure **a** and **b**

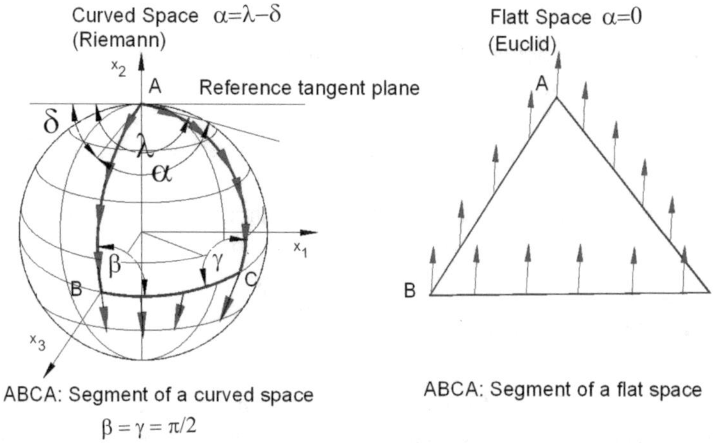

Fig. 11.2 Parallel transport along a segment of a sphere described by arcs AB, BC, and CA of great circles

11.2.1 Parallel Transport, Riemann Tensor

Given a curved space characterized by a surface that is covered by the grid x_μ and x_ν, shown in Fig. 11.3. We consider an infinitesimally small parallelogram $ABCDA$ with the sides dx_μ and dx_ν. We parallel transport from A to B to C to D and to A. To distinguish between the begin and the end point, we may label the end point A'. Along dx_μ in x_μ-direction we take the differences $(\boldsymbol{V_C} - \boldsymbol{V_D})$ and $(\boldsymbol{V_B} - \boldsymbol{V_A})$. We do the same along dx_ν in x_ν-direction, which is $(\boldsymbol{V_C} - \boldsymbol{V_B})$ and $(\boldsymbol{V_D} - \boldsymbol{V'_A})$. We now construct the difference of differences:

$$[(\boldsymbol{V_C} - \boldsymbol{V_D}) - (\boldsymbol{V_B} - \boldsymbol{V_A})] - [(\boldsymbol{V_C} - \boldsymbol{V_B}) - (\boldsymbol{V_D} - \boldsymbol{V'_A})] = \boldsymbol{V_A} - \boldsymbol{V'_A} = \delta V. \tag{11.11}$$

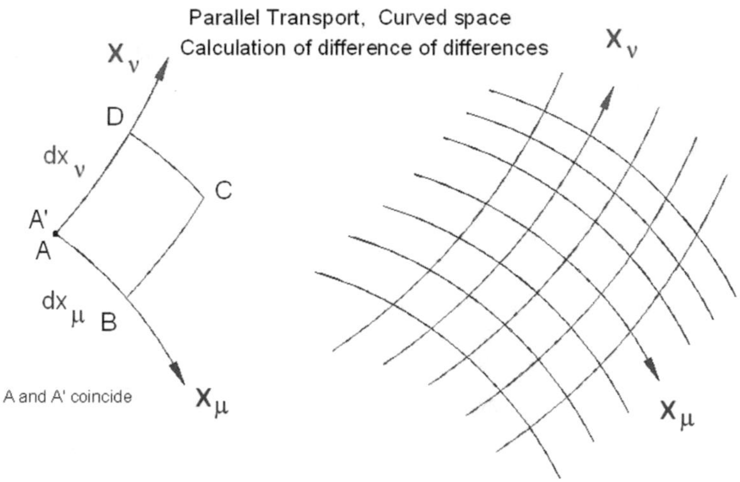

Fig. 11.3 Parallel transport along an infinitesimally small parallelogram $ABCDA'$ for calculating the changes in velocity components of a given vector. Note: the magnitude of the vector does not change

The first bracket in Eq. (11.11) is the difference of differences in x_μ-direction and the second bracket is the difference of differences in x_ν-direction. The individual differences are expressed in terms of covariant derivatives:

$$[(\boldsymbol{V_C} - \boldsymbol{V_D}) - (\boldsymbol{V_B} - \boldsymbol{V_A})] = dx_\mu dx_\nu \nabla_\mu \nabla_\nu \boldsymbol{V}. \tag{11.12}$$

The expression in each of the parentheses within the bracket of Eq. (11.12) exhibits a small difference. The difference of these small differences is expressed in terms of a second derivative. Similarly, the second bracket in Eq. (11.11) is the difference of differences in x_ν-direction. The individual differences are expressed in terms of a second covariant derivative:

$$(\boldsymbol{V_C} - \boldsymbol{V_B}) - (\boldsymbol{V_D} - \boldsymbol{V_A}) = dx_\nu dx_\mu \nabla_\nu \nabla_\mu \boldsymbol{V}. \tag{11.13}$$

Subtracting Eq. (11.13) from Eq. (11.12) we obtain the difference in velocity vector after we performed the parallel transport:

$$\delta \boldsymbol{V} = dx_\mu dx_\nu (\nabla_\nu \nabla_\mu - \nabla_\mu \nabla_\nu) \boldsymbol{V}. \tag{11.14}$$

The expression in the parentheses of Equation (11.14) has the same formation as a commutator described in previous section.

$$\delta \boldsymbol{V} = dx_\mu dx_\nu [\nabla_\nu, \nabla_\mu] \boldsymbol{V}. \tag{11.15}$$

Now we expand the covariant operators of Equation (11.14) and, for the sake of simplicity, at this point we use the Christoffel symbols with the *dominant* indices, in this case, μ and ν only. Once we have completed the following simple algebraic operations where several terms will disappear, we will complete the Christoffelsymbols with the appropriate indices.

$$\boldsymbol{\delta V} = dx_\mu dx_\nu \left[\left(\frac{\partial}{\partial x_\nu} + \Gamma_\nu\right)\left(\frac{\partial}{\partial x_\mu} + \Gamma_\mu\right) - \left(\frac{\partial}{\partial x_\mu} + \Gamma_\mu\right)\left(\frac{\partial}{\partial x_\nu} + \Gamma_\nu\right)\right]\boldsymbol{V}. \tag{11.16}$$

Now we multiply the parentheses and obtain:

$$\delta \boldsymbol{V} = \left[\left(\frac{\partial}{\partial x_\nu}\frac{\partial}{\partial x_\mu} - \frac{\partial}{\partial x_\mu}\frac{\partial}{\partial x_\nu}\right) - \left(\frac{\partial}{\partial x_\mu}\Gamma_\nu - \Gamma_\nu\frac{\partial}{\partial x_\mu}\right) + \left(\frac{\partial}{\partial x_\nu}\Gamma_\mu - \Gamma_\mu\frac{\partial}{\partial x_\nu}\right) + \Gamma_\nu\Gamma_\mu - \Gamma_\mu\Gamma_\nu\right]\boldsymbol{V}. \tag{11.17}$$

The expressions in the first parentheses of Equation (11.17) are equal to zero. The second and third parentheses are commutators:

$$\left(\Gamma_\nu\frac{\partial}{\partial x_\mu} - \frac{\partial}{\partial x_\mu}\Gamma_\nu\right) = -[\partial_\mu, \Gamma_\nu] = -\frac{\partial \Gamma_\nu}{\partial x_\mu}$$

$$\left(\frac{\partial}{\partial x_\nu}\Gamma_\mu - \Gamma_\mu\frac{\partial}{\partial x_\nu}\right) = +[\partial_\nu, \Gamma_\mu] = +\frac{\partial \Gamma_\mu}{\partial x_\nu} \tag{11.18}$$

with $\partial_\mu = \frac{\partial}{x_\mu}$ and $\partial_\nu = \frac{\partial}{x_\nu}$. Inserting the correct indices for Christoffel symbols in Eq. (11.17) and considering Eq. (11.18), we obtain the angle difference that a velocity vector experiences when it is subjected to a parallel transport:

$$dx_\nu dx_\mu\left(\frac{\partial}{\partial x_\nu}\Gamma^\alpha_{\mu\beta} - \frac{\partial}{\partial x_\mu}\Gamma^\alpha_{\nu\beta} + \Gamma^\alpha_{\nu\delta}\Gamma^\delta_{\mu\beta} - \Gamma^\alpha_{\mu\delta}\Gamma^\delta_{\nu\beta}\right) = \delta V^\alpha. \tag{11.19}$$

The expression within the parentheses is the Riemann tensor.

$$R^\alpha_{\nu\mu\beta} = \left(\frac{\partial}{\partial x_\nu}\Gamma^\alpha_{\mu\beta} - \frac{\partial}{\partial x_\mu}\Gamma^\alpha_{\nu\beta} + \Gamma^\alpha_{\nu\delta}\Gamma^\delta_{\mu\beta} - \Gamma^\alpha_{\mu\delta}\Gamma^\delta_{\nu\beta}\right). \tag{11.20}$$

Equation (11.20) is a forth order mixed tensor that includes the first and the second derivatives of the metric tensor. While the first and second expression in the parenthesis deal with the second derivative of the metric tensor, the third and fourth expressions include the squares of the metric tensor.

11.2.2 Properties of the Riemann Tensor

To discuss the properties of the Riemann tensor, it is more appropriate to convert the mixed tensor into a covariant one. Thus, the upper index of the Riemann tensor can be lowered, so that we obtain the covariant version of the Riemann tensor:

$$R_{\alpha\beta\mu\nu} \equiv g_{\alpha\lambda} R^{\lambda}_{\beta\mu\nu}. \tag{11.21}$$

Considering the covariant version (11.21), it is anti-symmetric in the first two and last two indices, and symmetric under interchanging those pairs:

$$\begin{aligned} R_{\alpha\beta\mu\nu} &= -R_{\alpha\beta\nu\mu} \\ R_{\alpha\beta\mu\nu} &= -R_{\beta\alpha\mu\nu} \\ R_{\mu\nu\alpha\beta} &= R_{\alpha\beta\mu\nu}. \end{aligned} \tag{11.22}$$

To obtain non-zero results, when contracting the Riemann tensor, one has to be careful not to choose cases, where a symmetric tensor is multiplied with an antisymmetric one. In addition to the properties (11.22), the Riemann tensor satisfies the *Bianchi identity*

$$R_{\alpha\beta\mu\nu} + R_{\alpha\nu\beta\mu} + R_{\alpha\mu\nu\beta} = 0. \tag{11.23}$$

11.3 Construction of Einstein Space-Time Geometry

The Einstein Field equation has gone through several modifications before it was presented in its final form. It consists of two part: (1) the left hand side that constitutes the spacetime geometry. It is based on the fourth order Riemann curvature tensor that has in a four dimensional spacetime coordinate $4^4 = 256$ components. (2) the right hand side contains the energy, momentum and stress tensor $T_{\mu\nu}$ is a second order tensor that we will discuss later in this chapter. Thus, the order of the Riemann tensor is not compatible with the order of energy stress tensor. The compatibility condition requires that the tensors on both sides of Einstein equation must have the same order. This means that the order of the Riemann tensor has to be contracted. As a result of this contraction we obtain the Ricci tensor which is a second order tensor.

11.3.1 Ricci Tensor and Curvature Scalar

To obtain the Ricci tensor we multiply the metric tensor with the Riemann tensor. Since the metric tensor is symmetric, if it is multiplied with any antisymmetric tensor the result will be zero. The following examples show the results of the contraction of an antisymmetric and a symmetric Riemann tensor:

$$g^{\mu\nu} R^{\alpha}_{\mu\nu\beta} = 0$$
$$g^{\alpha\beta} R^{\alpha}_{\mu\nu\beta} = 0$$
$$g^{\alpha\nu} R^{\alpha}_{\mu\nu\beta} = R^{\alpha}_{\mu\alpha\beta} = R_{\mu\beta}. \tag{11.24}$$

As seen from the last equation of (11.24), the non-zero tensor was obtained as the result of contracting two symmetric tensors. The resulting second order tensor called the Ricci-Riemann tensor. The expansion of the Ricci tensor is:

$$R_{\mu\beta} = \left(\frac{\partial}{\partial x_\nu} \Gamma^{\nu}_{\beta\mu} - \frac{\partial}{\partial x_\beta} \Gamma^{\nu}_{\nu\mu} + \Gamma^{\nu}_{\nu\lambda} \Gamma^{\lambda}_{\beta\mu} - \Gamma^{\nu}_{\beta\lambda} \Gamma^{\lambda}_{\nu\mu} \right). \tag{11.25}$$

In addition to satisfying the order compatibility condition, the Einstein equation has also to satisfy the condition to be divergence free. This is true for the energy, momentum and stress tensor $\nabla \cdot \boldsymbol{T} = 0$ but not for the Ricci tensor. Taking the divergence of the Ricci tensor, the result is:

$$\nabla^{\mu} R_{\mu\nu} = \nabla^{\mu} \left(-\frac{1}{2} g_{\mu\nu} R \right). \tag{11.26}$$

The expression on the right hand side of Equation (11.26) involves the *curvature scalar* R which is the contraction of the Ricci tensor

$$g^{\beta\mu} R_{\beta\mu} = R. \tag{11.27}$$

With Eqs. (11.26) and (11.27) and we arrive at:

$$\nabla^{\mu} \left(R_{\mu\nu} - \frac{1}{2} g_{\mu\nu} R \right) = 0. \tag{11.28}$$

With adding the product of the metric tensor and the curvature scalar, the divergence of the spacetime tensor becomes zero, thus the spacetime geometry is now complete. Ricci tensor and curvature scalar are of fundamental importance to GTR.

11.4 Einstein Tensor, Field Equation of General Relativity

To construct the right hand side of the Einstein Field Equation, we resort to the Newton second law of motion expressed in terms of Navier–Stokes arrangement (5.56). The right hand side of (5.56) is just one component of the Einstein energy, momentum and stress tensor which is a second order tensor $T_{\mu\nu}$ with 16 components. These components are packed into the following matrix presented in Eq. (10.54):

$$T_{\mu\nu} = \begin{pmatrix} \gamma^2\rho c^2 & \gamma^2\rho V^1 c & \gamma^2\rho V^2 c & \gamma^2\rho V^3 c \\ \gamma^2\rho c V^1 & \gamma^2\rho V^1 V^1 & \gamma^2\rho V^2 V^1 & \gamma^2\rho V^3 V^1 \\ \gamma^2\rho c V^2 & \gamma^2\rho V^1 V^2 & \gamma^2\rho V^2 V^2 & \gamma^2\rho V^3 V^2 \\ \gamma^2\rho c V^3 & \gamma^2\rho V^1 V^3 & \gamma^2\rho V^2 V^3 & \gamma^2\rho V^3 V^3 \end{pmatrix}. \tag{11.29}$$

With the *second order stress-energy-momentum tensor*, Eq. (11.29), we have the right hand side of the Einstein Field Equation:

$$\frac{8\pi G T_{\mu\nu}}{c^4} \tag{11.30}$$

with G as the gravitational constant and c the speed of light. The introduction of speed of light was done to maintain the dimensional integrity of the equation (see following Sect. 11.5). For the completion of the field equation, Einstein required that the tensor of the *spacetime* geometry on the left hand side be also a second order tensor that represents a curved space geometry. A possible candidate is the Ricci tensor (11.25), thus Einstein made the first ansatz:

$$R_{\mu\nu} = \frac{8\pi G T_{\mu\nu}}{c^4}. \tag{11.31}$$

As indicated above, the stress-energy-momentum tensor $T_{\mu\nu}$ contains conserved quantities. This requires that divergence of $T_{\mu\nu}$ must vanish. In fact, as shown in Chap. 10, with $\nabla_\lambda T_{\mu\nu} = 0$ the tensor $T_{\mu\nu}$ meets this requirement. However, the Ricci tensor the on left hand side does not meet the divergence requirement. This deficiency could be removed by adding the curvature scalar (11.27) to the Ricci tensor resulting in:

$$R_{\mu\nu} - \frac{1}{2} g_{\mu\nu} R, \qquad \nabla_\lambda (R_{\mu\nu} - \frac{1}{2} g_{mn} R) = 0. \tag{11.32}$$

The multiplication of the curvature scalar R with the metric tensor $g_{\mu\nu}$ cause the scalar to become a second order tensor with $\nabla_\lambda g_{\mu\nu} = 0$. After adding this expression, Einstein added another term, the *cosmological constant* $\mathbf{\Lambda} = 1.01055 \times 10^{-52} m^2$ that he converted into a second order tensor by multiplying it with the metric tensor leading to $g_{\mu\nu}\mathbf{\Lambda}$. Thus Einstein arrived at:

$$R_{\mu\nu} - \frac{1}{2} g_{\mu\nu} R + g_{\mu\nu}\mathbf{\Lambda} = \frac{8\pi G T_{\mu\nu}}{c^4}. \tag{11.33}$$

This is the Einstein Field Equation completed in 1915. It took Einstein ten years to develop this equation.

11.5 Newton' Gravitation as the Special Case of the GTR

The Field Equation is the general equation that describes the gravity. For a weak gravitational field and low velocity it must conform to the Newton's second law. In the following we look at the second law from a relativistic point of view. The second law reads:

$$mA = F, \qquad A = \frac{F}{m} \tag{11.34}$$

with m as the mass of the particle, $\boldsymbol{A}$ as the particle acceleration vector and $\boldsymbol{F}$ the external force vector acting on the particle. We assume that the force vector can be thought of as the gradient of a gravitational potential field:

$$\boldsymbol{F} = -\boldsymbol{n}\frac{GMm}{r^2}, \; |F| = \frac{GMm}{r^2} \tag{11.35}$$

with M as the large mass located in the center of a sphere with radius r, m the small mass that can be set $m = 1$, $\boldsymbol{n}$ unit vector that points away from the spherical surface and G the Newton gravitational constant. For a constant radius we build the surface integral:

$$\int_S \boldsymbol{F} \cdot dS = \int_S \frac{GM}{r^2} ds = \frac{4\pi GMr^2}{r^2} = 4\pi GM. \tag{11.36}$$

The surface integral (11.36) is converted into the volume integral:

$$\int_S \boldsymbol{F} \cdot dS = \int_\nu (\nabla \cdot \boldsymbol{F}) d\nu = \int_\nu \nabla \cdot \nabla\phi \, d\nu = 4\pi G \int_\nu \rho d\nu. \tag{11.37}$$

With ϕ as the gravitational potential and $\boldsymbol{F} = \nabla\phi$ to be determined. Taking the last two integrals in Eq. (11.37), we find:

$$\int_\nu (\nabla \cdot \nabla\phi - 4\pi G\rho) d\nu = 0. \tag{11.38}$$

Since the entire integral is zero, the integrand must be zero:

$$\nabla^2\phi = \Delta\phi = 4\pi G\rho \tag{11.39}$$

with $\Delta = \nabla \cdot \nabla = \nabla^2$ as the Laplace operator in Euclid space. This is the Poisson equation for gravitational potential. At this juncture, we need to know, how this potential is related to the spacetime geometry. To this end, we resort to geodesic equation (8.18) for spacetime coordinate, where s is replaced by the proper time

$$\frac{d^2x^\lambda}{d\tau^2} + \Gamma^\lambda_{\mu\nu} \frac{dx^\mu}{d\tau} \frac{dx^\nu}{d\tau} = 0. \tag{11.40}$$

We assume that

- Gravitational field is very weak
- Gravitational field varies slowly (no gravitational wave)
- $V << c$

with these assumptions, the metric coefficient $g_{\mu\nu}$ with $g_{\mu\nu}g^{\nu\lambda} = \delta^{\lambda}_{\mu}$ is reduced to Minkowski metric $\eta_{\mu\nu}$ that has a non-zero diagonal of $(-1, 1, 1, 1)$. Thus, the metric $g_{\mu\nu}$ can be written as:

$$\begin{aligned} g^{\mu\nu} &= \eta^{\mu\nu} - \varepsilon^{\mu\nu} \\ g_{\mu\nu} &= \eta_{\mu\nu} + \varepsilon_{\mu\nu} \end{aligned} \tag{11.41}$$

with $\varepsilon_{\mu\nu}$, $\varepsilon^{\mu\nu}$ as small perturbation of $(-1, 1, 1, 1)$. Implying the above assumptions and considering that all spatial velocity components compared to the time component are negligibly small:

$$\frac{\partial g_{\mu\nu}}{\partial x^0} = \frac{\partial \eta^{\mu\nu}}{\partial x^0} = \frac{\partial \eta^{\mu\nu}}{\partial x^0} = 0, \ \frac{dx^i}{d\tau} << 1, \ \frac{dx^0}{d\tau} = \frac{d(ct)}{d\tau} = c\frac{dt}{d\tau} \approx 1 \tag{11.42}$$

and Eq. (11.40) becomes:

$$\frac{d^2x^{\lambda}}{d\tau^2} + \Gamma^{\lambda}_{00}\frac{dx^0}{d\tau}\frac{dx^0}{d\tau} = 0$$

$$\frac{d^2x^{\lambda}}{d\tau^2} + \Gamma^{\lambda}_{00}c^2\left(\frac{dt}{d\tau}\right)^2 = 0. \tag{11.43}$$

The Christoffel symbol in Eq. (11.43) is according to Eq. (6.87)

$$\Gamma^{\lambda}_{00} = \frac{1}{2}g^{\lambda\mu}\left(\frac{\partial g_{0\mu}}{\partial x^0} + \frac{\partial g_{0\mu}}{\partial x^0} - \frac{\partial g_{00}}{\partial x^{\mu}}\right). \tag{11.44}$$

The first two terms in Eq. (11.44) are negligibly small reducing (11.44) to:

$$\Gamma^{\lambda}_{00} = -\frac{1}{2}g^{\lambda\mu}\frac{\partial g_{00}}{\partial x^{\mu}}. \tag{11.45}$$

Going back to Eq. (11.43) and setting $\frac{dt}{d\tau} \approx 1$ we get

$$\frac{d^2x^{\lambda}}{d\tau^2} + c^2\Gamma^{\lambda}_{00} = 0. \tag{11.46}$$

Inserting Eq. (11.45) into Eq. (11.46) we have

$$\frac{d^2x^\lambda}{d\tau^2} = -\frac{1}{2}g^{\lambda\mu}\frac{\partial g_{00}}{\partial x^\mu}c^2. \tag{11.47}$$

Without loss of generality, we can set $\lambda = 1$ and run μ from 0 to 3:

$$\frac{d^2x^\lambda}{d\tau^2} = -\frac{c^2}{2}\left[g^{10}\frac{\partial g_{00}}{\partial x^0} + g^{11}\frac{\partial g_{00}}{\partial x^1} + g^{12}\frac{\partial g_{00}}{\partial x^2} + g^{13}\frac{\partial g_{00}}{\partial x^3}\right]. \tag{11.48}$$

Of the four terms in Eq. (11.48) only the second term is non-zero because in Eq. (11.41) $\eta^{\lambda\mu} = 0\, for\ \lambda \neq \mu$. As a result we find:

$$\frac{d^2x^\lambda}{d\tau^2} = -\left(\frac{1}{2}\frac{\partial g_{00}}{\partial x^1}\right)c^2. \tag{11.49}$$

Going back to Newton equation (11.34) and setting $m = 1$ and replacing the force by the gravitational potential, we have

$$\frac{d^2\boldsymbol{X}}{dt^2} = \boldsymbol{F} = -\nabla\phi$$

$$\frac{d^2x_i}{dt^2} = F_i = -\frac{\partial\phi}{\partial x_i}. \tag{11.50}$$

We now set $i = 1,\ x_1 = x,\ \tau = 1$, we equate

$$\frac{d^2x}{dt^2} = -\left(\frac{1}{2}\frac{\partial g_{00}}{\partial x}\right)c^2$$

$$\frac{d^2x}{dt^2} = -\frac{\partial\phi}{\partial x}. \tag{11.51}$$

Equating the content of Equation (11.51), it follows that:

$$\left(\frac{1}{2}\frac{\partial g_{00}}{\partial x}\right)c^2 = \frac{\partial\phi}{\partial x}. \tag{11.52}$$

The integration delivers

$$g_{00} = \frac{2}{c^2}\phi + \text{Const.},\ \phi = \frac{1}{2}g_{00}c^2 + \text{const.} \tag{11.53}$$

Replacing in Poisson Equation (11.39) the gravitational potential Equation (11.53), we obtain

$$\nabla^2\left(\frac{1}{2}g_{00}c^2 + \text{const.}\right) = 4\pi G\rho, \qquad \nabla^2 g_{00} = \frac{8\pi G\rho}{c_2}. \tag{11.54}$$

Expressing the density in terms of $T_{00} = \rho c^2$ from which the density is calculated as $\rho = T_{00}/c^2$, we find:

$$\begin{aligned} \nabla^2 g_{00} &= \frac{8\pi G T_{00}}{c^4} \text{ or} \\ \nabla^2 \phi &= 4\pi G\rho. \end{aligned} \tag{11.55}$$

The second equation in (11.55) relates the Einstein Field equation to Newton's second law of motion through Eq. (11.39). It shows that Newton's second law of motion is a special case of the Einstein Field equation.

11.6 An Appendix from Chap. 6

11.6.1 Covariant Derivative of Second Order Tensors

Although the tensor operations in this chapter are explained in Sect. 6.6.1, they have gone through some re-arrangements to account for the spacial nomenclature inherent to the GTR. This appendix is intended to help the reader to easily relate the material in Sect. 6.6.1 to GTR. A second order tensor $g^i g^j T_{ii}$ its covariant derivative is obtained using the same procedure described in Sect. 6.6.1:

$$\nabla_k T_{ij} = \frac{\partial T_{ij}}{\partial \xi_k} - T_{mj}\Gamma^m_{ki} - T_{im}\Gamma^m_{kj}. \tag{11.56}$$

As we see in Chap. 9, it is necessary to prove that the covariant derivative of the metric tensor is $\nabla_k g_{ij} = 0$. For this purpose first we replace in Eq. (11.56) the second order tensor T_{ij} by g_{ij}:

$$\nabla_k g_{ij} = \frac{\partial g_{ij}}{\partial \xi_k} - g_{mj}\Gamma^m_{ki} - g_{im}\Gamma^m_{kj}. \tag{11.57}$$

Then, we convert the Christoffel symbol of second kind in Eq. (11.57) into the one of first kind by multiplying the symbols in (11.57) with g_{lk}

$$\Gamma^l_{ji} g_{lk} = \Gamma_{jik}. \tag{11.58}$$

Thus, Eq. (11.57) becomes:

$$\nabla_k g_{ij} = \frac{\partial g_{ij}}{\partial \xi_k} - \Gamma_{jik} - \Gamma_{ijk}. \tag{11.59}$$

The first term on the right hand side of Equation (11.59) is:

$$\frac{\partial g_{ij}}{\partial \xi_k} = \Gamma_{jik} + \Gamma_{ijk}. \quad (11.60)$$

Inserting Eq. (11.60) into Eq. (11.59) we have

$$\nabla_k g_{ij} = 0. \quad (11.61)$$

The contraction of Riemann's tensor that leads to Ricci tensor and its subsequent contraction to Ricci curvature scalar form the left hand side of the Einstein Field Equation. The development of the right hand side of the Einstein tensor equation concludes the chapter.

References

1. Einstein A (1905) Zur Elektrodynamik bewegter Körper. Annalen der Physik (ser. 4), 17:891–921
2. Einstein A (1915) Feldgleichungen der Gravitation, Preussische Akademie der Wissenschaften, Sitzungsberichte, 1915 (part 2), 844–847
3. Einstein A, Grossmann M (1913) Entwurf einer veralgemeinerten Relativitaetstheorie und einer Theorie der Gravitation, Teubne Verlag Leipzig und Berlin
4. Misner CW, Thorne KS, Wheeler JA, Gravitation. W. H. Freeman, San Francisco
5. Norton J (1984) How Einstein found his field equations. Hist Stud Phys Sci 14:253–316
6. Howard D, Stachel J (eds) (1989) Einstein and the history of general relativity. Birkhäuser, Boston
7. Eisenstaedt J, Kox A (eds) (1992) Studies in the history of general relativity. Birkhäuser, Boston
8. Earman J, Janssen M, Norton JD (eds) (1993) The attraction of gravitation. New studies in the history of general relativity. Birkhäuser, Boston
9. Einstein A (1988) Mein Weltbild. Frankfurt/M./Berlin, Ullstein Materialien
10. Einstein A (1915) Zur allgemeinenRelativitätstheorie (Nachtrag), Sitzungsberichte der Preussischen Akademie der Wissenschaften. 2. Halbband, pp 799–801
11. Einstein A (1915) Zur allgemeinen Relativitätstheorie, Sitzungsberichte der Preussischen Akademie der Wissenschaften. 2. Halbband, pp 778-786
12. Einstein A, Grossmann M (1914) Entwurf einer verallgemeinerten Relativitätstheorie und einer Theorie der Gravitation, Leipzig, Teubner (1913), wiederabgedruckt in Zeitschrift für Mathematik und Physik 62:225–259
13. Klein M et al (eds) (1993) The collected papers of Albert Einstein, vol 5. Princeton University Press, Princeton
14. Einstein A (1912) Lichtgeschwindigkeit und Statik des Gravitationsfeldes. Ann Phys 38:355–369
15. Einstein A (1912) Zur Theorie des statischen Gravitationsfeldes. Ann Phys 38:443–458
16. Klein M et al (eds) (1995) The collected papers of Albert Einstein, vol 4. Princeton University Press, Princeton
17. Castagnetti G, Damerow P, Heinrich W, Renn J, Sauer T (1994) Wissenschaft zwischen Grundlagenkrise und Politik: Einstein in Berlin, Arbeitsbericht der Arbeitsstelle Albert Einstein 1991-1993, Max Planck Institut für Bildungsforschung

Index

M. T. Schobeiri, *Tensor Analysis for Engineers and Physicists - With Application to Continuum Mechanics, Turbulence, and Einstein's Special and General Theory of Relativity*, https://doi.org/10.1007/978-3-030-35736-8

W

Printed in the United States
by Baker & Taylor Publisher Services